AF411404

Natural Antioxidants: Innovative Extraction and Application in Foods

Natural Antioxidants: Innovative Extraction and Application in Foods

Editors

Monica Rosa Loizzo
Ana Sanches Silva

MDPI • Basel • Beijing • Wuhan • Barcelona • Belgrade • Manchester • Tokyo • Cluj • Tianjin

Editors
Monica Rosa Loizzo
University of Calabria
Italy

Ana Sanches Silva
National Institute of Agrarian and
Veterinary Research
Portugal

Editorial Office
MDPI
St. Alban-Anlage 66
4052 Basel, Switzerland

This is a reprint of articles from the Special Issue published online in the open access journal *Foods* (ISSN 2304-8158) (available at: https://www.mdpi.com/journal/foods/special_issues/Natural_Antioxidants_Extraction_Application_Foods).

For citation purposes, cite each article independently as indicated on the article page online and as indicated below:

LastName, A.A.; LastName, B.B.; LastName, C.C. Article Title. *Journal Name* **Year**, *Volume Number*, Page Range.

ISBN 978-3-0365-2556-3 (Hbk)
ISBN 978-3-0365-2557-0 (PDF)

Contents

About the Editors

Monica Rosa Loizzo graduated cum laude in pharmacy at the University of Calabria (Italy) in 1999. She obtained a PhD with her thesis "Methodology for the Development of New Molecules of Pharmacological Interest" in 2005 at the University of Calabria, and also obtained a master's in clinical pathology from the same institution. From 2002 to 2004 she spent time as a researcher at the 'Centre for Bioactivity Screening of Natural Products' at King's College London (UK). In 2008, she spent a period of time in the 'Area of Food Industries and Technologies' laboratory of Politecnica delle Marche University (Ancona, Italy). From 18.12.2008 to 29.12.2017 she was a researcher in food science technology at the Department of Pharmacy, Health Science, and Nutrition of the University of Calabria. In 2017 she become an associate professor at the University of Calabria.

Prof. Loizzo is the author of 204 research papers, 17 book chapters, and 10 editorials. Her h index is 45 and she has over 6805 citations.

Monica Rosa Loizzo developed specific abilities for main separative techniques using different detector systems to determine the chemical composition of different food matrices. She developed specific abilities for the main spectrophotometric methods used to evaluate the health properties via different in vitro methods (e.g., antioxidant, hypoglycemic, hypolipidemic, etc.) of phytochemicals rich-food. Moreover, her research interests include the influence of technological processes on the overall quality of food, with particular reference to compounds capable of prolong the shelf-life (e.g., antioxidant, antimicrobial, anti-browning) of food items She has also studied the chemical composition, sensory evaluation, and health properties of extra virgin olive oil, wines, liquor, fish products and sausages. She is referee of several national and international institutions and was a member of the Management Committee of COST Action "Eurocaroten" until 2020.

Editorial board Member of Antioxidants (2019-today); FOODS (2019-today); Italian Journal Food Science (2019-today); The Open Nutraceuticals Journal (2008-today); International Journal of Nutrition and Dietetics (2012-today); American Journal of Experimental Agriculture (2012-today); Foods (2014-today); Annals of Nutrition and Food Science (2017-today); Food Science and Nutrition Research (2018-today).

Ana Sanches Silva obtained a degree in pharmaceutical sciences at the Pharmacy Faculty of the University of Coimbra (FFUC), Portugal, and received her Ph.D. with honors in pharmacy from the University of Santiago de Compostela (USC), Spain. In addition, she was awarded with two awards for best Ph.D. thesis. She is a member of the executive board of the Animal Science Studies Center and was invited to be professor at the FFUC. Ana has a remarkable track record, namely as co-author of papers in peer-reviewed journals with high impact factors, numerous book chapters, and as a co-editor of scientific books in the field of food science. Ana's research has focused on antioxidants, namely natural antioxidants, and their potential for use in active food packaging. In addition, she has a special interest in the development and validation of analytical methodologies (especially related to mass spectrometry) to determine food and food packaging components and contaminants.

foods

Editorial

Natural Antioxidants: Innovative Extraction and Application in Foods

Monica Rosa Loizzo [1,*] and Ana Sanches Silva [2,3]

[1]   Department of Pharmacy, Health and Nutritional Sciences, University of Calabria, 87036 Arcavacata di Rende, Italy
[2]   National Institute for Agricultural and Veterinary Research (INIAV), I.P., Rua dos Lágidos, Lugar da Madalena, Vairão, 4485-655 Vila do Conde, Portugal; anateress@gmail.com or ana.silva@iniav.pt
[3]   Center for Study in Animal Science (CECA), ICETA, University of Oporto, 4051-401 Oporto, Portugal
*   Correspondence: monica_rosa.loizzo@unical.it

Citation: Loizzo, M.R.; Silva, A.S. Natural Antioxidants: Innovative Extraction and Application in Foods. *Foods* **2021**, *10*, 937. https://doi.org/10.3390/foods10050937

Received: 19 April 2021
Accepted: 22 April 2021
Published: 25 April 2021

Publisher's Note: MDPI stays neutral with regard to jurisdictional claims in published maps and institutional affiliations.

Research has devoted great attention to the study of the biological properties of plants, animal products, microorganisms, marine species, and fungi, among others, often driven by the need to discover new medicines. Many times, in order to enhance biological activities, extracts are prepared. One of the most well-studied biological activities is antioxidant capacity related to anticancer and antiaging properties, improvement of immune function, and protection against cardiovascular diseases and neurological disorders. Moreover, in foods, antioxidants allow for delayed oxidation onset and enhancing food shelf life.

Changes in lifestyle patterns and world population growth demand safe, nutritious, flavourful, colourful, affordable, and convenient food, and high-quality standards have increased the use of food additives, especially antioxidants. The effects of some food additives on human health are controversial, and synthetic food additives are often associated with potential public health risk. Therefore, there is a tendency to substitute synthetic food additives with natural compounds.

We have organized a Special Issue titled "Natural Antioxidants: Innovative Extraction and Application in Foods" in the Foods (ISSN 2304-8158; CODEN: FOODBV; https://www.mdpi.com/journal/foods, accessed on 24 April 2021). This thematic issue focused on the application of innovative extraction techniques for the recovery of natural antioxidants from foods and their possible application in food industries. This Special Issue, now converted into a book, includes 11 chapters, which are important contributions to this topic made by distinguished experts in this area. Ten of these chapters are research papers, and one is a review paper.

Chapter 1 is titled "The Effect of Blanching on Phytochemical Content and Bioactivity of *Hypochaeris* and *Hyoseris* Species (Asteraceae), Vegetables Traditionally Used in Southern Italy" [1]. This chapter regards the effect of blanching on the bioactivity and phytochemical content of *Hypochaeris* and *Hyoseris* species, traditionally used in Southern Italy. The results of this study indicated that these wild plants are a good source of bioactive compounds; however, their antioxidant capacity decreased after blanching. In fact, blanching water presented higher antioxidant capacity than the blanched samples. Therefore, the reuse of blanching water is recommended in food preparation because it is a good source of bioactives, and its consumption can increase the uptake of micronutrients.

Chapter 2 ("Effect of Microwave Pretreatment of Seeds on the Quality and Antioxidant Capacity of Pomegranate Seed Oil") [2] is a very interesting study on the consequences of microwave pretreatment on pomegranate seeds and on the antioxidant capacity of pomegranate seed oil.

A considerable number of quality attributes were evaluated in three different pomegranate cultivars, including yellowness index, refractive index, oil yield, p-anisidine value, total oxidation value, conjugated dienes, total phenolic content, peroxide value, total carotenoid content, phytosterol composition, fatty acid composition, and antioxidant capacity through ferric reducing antioxidant power (FRAP) and 2,2-diphenyl-1-picrylhydrazyl (DPPH) radical scavenging capacity. Most of the parameters were increased after microwave pretreatment; however, punicic acid and beta-sitosterol were decreased.

Pomegranate seed oil may be enhanced if microwave pretreated seeds are used, although oil quality varies with cultivar.

Chapter 3 ("Chemical Composition and Antioxidant Activity of Thyme, Hemp, and Coriander Extracts: A Comparison Study of Maceration, Soxhlet, UAE, and RSLDE Techniques") [3] is a valuable chapter comparing different extraction techniques for the obtainment of thyme, hemp, and coriander extracts. The selected techniques were maceration, Soxhlet, ultrasound-assisted extraction, and rapid solid–liquid dynamic extraction (RSLDE). Several parameters were measured in order to compare the extraction techniques. ABTS·+, FRAP, and DPPH· assays were used to evaluate the antioxidant capacity. The total phenolic content by Folin–Ciocalteu method was also evaluated. The results revealed that all the evaluated techniques are valid extraction methods to extract bioactives and preserve their activity. However, extracts obtained by RSLDE showed to have slightly higher antioxidant capacity, and the technique is easy to use besides allowing the standardization of the extraction process.

Chapter 4 regards the paper "Organic Selenium as Antioxidant Additive in Mitigating Acrylamide in Coffee Beans Roasted via Conventional and Superheated Steam" [4]. In this chapter, the effect of coffee beans pretreated with selenium in the formation of acrylamide was evaluated in coffee beans roasted using two different methods (conventional versus superheated steam). The results showed that the antioxidant capacity of the organic selenium suppressed the formation of acrylamide during coffee roasting by 73%. Superheated steam roasting increased antioxidant activity and significantly reduced acrylamide (up to 32%), which was only noticed in the untreated coffee beans.

Chapter 5 is titled "Antioxidant Compounds for the Inhibition of Enzymatic Browning by Polyphenol Oxidases in the Fruiting Body Extract of the Edible Mushroom *Hericium erinaceus*" [5]. This noteworthy chapter focused on the identification of the cause of the dark brown pigmentation via the enzymatic reaction of the polyphenol oxidase (PPO) family with oxidation activity, and the reduction of the occurrence of this pigmentation. The mushroom contained relatively high amounts of natural antioxidant compounds for the inhibition of tyrosinase and the scavenging of free radicals. These antioxidants could diminish the browning reaction via PPO inhibitory mechanisms in the fruiting body of the *H. erinaceus* mushroom.

These results of this chapter allow for understanding the metabolites and PPO enzymes responsible for the enzymatic browning reaction of *H. erinaceus*.

Chapter 6 comprises the paper "Impact of Stability of Enriched Oil with Phenolic Extract from Olive Mill Wastewaters" [6]. This notable chapter evaluated the effect of phenolic extract addition in the oxidative deterioration of sunflower oil. XAD-7-HP resin was used to recover phenolic compounds from olive mill wastewaters. The extract was evaluated in terms of single phenol concentration by ultra-high-performance liquid chromatography. The highest amount was found for hydroxytyrosol. The oxidation state of fortified sunflower oil was evaluated during 90 days for different physicochemical parameters (refractive index, peroxide value, and oxidative resistance to degradation) and antioxidant assays (DPPH, ABTS, and ORAC). The study revealed that there was an increase of 50% in the oxidative stability of fortified oil compared with control. This indicates that olive mill wastewaters can be valorized through an efficient extraction method.

Chapter 7 concerns the paper "A Novel and Simpler Alkaline Hydrolysis Methodology for Extraction of Ferulic Acid from Brewer's Spent Grain and Its (Partial) Purification through Adsorption in a Synthetic Resin" [7].

This chapter interestingly developed a simple method to extract ferulic acid (FA) from brewer's spent grain (BSG), produced by brewing companies. The method includes an autoclave step to perform the alkaline hydrolysis, which allows for simplifying the postextraction process and increasing the ferulic acid yield. Finally, the extracted ferulic acid carries out a partial purification in a synthetic resin.

Chapter 8 is titled "Radical Scavenging and Antimicrobial Properties of Polyphenol Rich Waste Wood Extracts" [8]. This chapter evaluated the radical scavenging and antimicrobial capacities of wood waste extracts from black locust (*Robinia pseudoacacia* L.), mulberry (*Morus alba* L.), myrobalan plum (*Prunus cerasifera* Ehrh.), wild cherry (*Prunus avium* L.), and different species of oaks (*Quercus petraea* (Matt.) Liebl., *Q. robur* L., and *Q. cerris* L.) in order to conclude about their potential use in the food and pharmaceutical industries. Phenolic compounds were separated by using high-performance thin-layer chromatography (HPTLC), while radical scavenging activity was determined using DPPH-HPTLC. DPPH-HPTLC identified gallic, ferulic, and/or caffeic acids as the compounds with the highest contribution to antioxidant capacity. Regarding antimicrobial capacity, mulberry extract showed the lowest minimum inhibitory concentration (MIC) and minimum bactericidal concentration (MBC) values against methicillin-resistant *Staphylococcus aureus*. The growth rate of *Listeria monocytogenes* was significantly inhibited by extracts of myrobalan plum, wild cherry, and mulberry. *Candida albicans* showed poor sensitivity to the action of all extracts, with the exception of the wild cherry extract. *Escherichia coli* was less sensitive to the tested extracts.

The study concluded that due to their antimicrobial activities, cherry and mulberry wood extracts can be useful in preserving short-shelf-life foods.

Chapter 9 comprises the paper "Extract from Broccoli Byproducts to Increase Fresh Filled Pasta Shelf Life" [9]. This is a very interesting chapter that aims to find an alternative to chemical/conventional preservation strategies for fresh filled pasta. The idea was to evaluate the suitability of an extract from broccoli by-products for this purpose. The study monitored microbiological and sensory qualities besides phenolic compound content before and after in vitro digestion of pasta samples.

Results revealed that the shelf life of the natural extract increased by 18 days in comparison with control. The addition of the by-products' extract to pasta increased phenolic content after in vitro digestion. Consequently, it was concluded that broccoli by-products could be valorized for obtaining extracts able to enhance shelf life and improve the nutritional content of fresh filled pasta.

Chapter 10 is titled "Bioactive Compounds from Norway Spruce Bark: Comparison among Sustainable Extraction Techniques for Potential Food Applications" [10]. This is a great chapter comparing different techniques to extract antioxidants from Norway spruce bark (*Picea abies* (L.) Karst), a wood industry waste. Supercritical fluid extraction (SFE), pressurized liquid extraction (PLE), and ultrasound-assisted extraction (UAE) were compared, and results showed that PLE, using ethanol as solvent, was the most effective method for extracting total flavonoid compounds, with the highest antioxidant capacity according to ABTS˙+. On the other hand, UAE extract contained the maximum phenolic concentration and the highest antioxidant capacity by FRAP. UAE revealed the greatest efficiency in the extraction of *trans*-resveratrol with ethanol 70% (*v/v*), therefore suggesting its potential to be used to record antioxidants to be further applied in the food and pharmaceutical industries.

Chapter 11 comprises the review titled "A New Insight on Cardoon: Exploring New Uses besides Cheese Making with a View to Zero Waste" [11]. This chapter is the only review of this Special Issue/book. It regards cardoon (*Cynara cardunculus* L.), which is a plant native to the Mediterranean area whose flowers are used in cheese making, as vegetal rennet. The aim of the review was to address the properties of cardoon leaves, considered a by-product of this crop, and discuss their potential uses. The findings indicated that cardoon leaves are recognized for their potential health benefits, (e.g., diuretic, hepatoprotective, choleretic, hypocholesterolemic, anticarcinogenic, and antibacterial properties), and they can have new potential uses. In particular, they can be used for the preparation of extracts to be incorporated into active food packaging. In sum, the chapter concluded that the new uses of cardoon leaves will contribute to zero waste of this crop.

The chapters of this book address the unequivocal importance of natural antioxidants. There is a plethora of matrices that can be used to obtain natural antioxidants, including different parts of plants (e.g., leaves, bark, seeds), food by-products, and fungi. In conclusion, the choice of extraction technique is critical in order to improve the biological properties, especially the antioxidant and antimicrobial capacities, of the extracts and is strictly related to their potential application in the food industry. Nowadays, the food industry is looking for environmentally friendly extraction procedures as well as extraction procedures that allow for upscaling from lab to industry. Finally, the choice of pretreatment and processing methods can also have a great influence on the antioxidant capacity of the extracts.

Funding: This research received no external funding.

Conflicts of Interest: The authors declare no conflict of interest.

Abbreviations

ABTS$^{\cdot}$+	2:20-azino-bis(3-ethylbenzothiazoline-6-sulphonic acid
DPPH$^{\cdot}$	2,2-diphenyl-1-picrylhydrazyl
FRAP	ferric reducing antioxidant power
HPTLC	high-performance thin-layer chromatography
MBC	minimum bactericidal concentration
MIC	minimum inhibitory concentration
PLE	pressurized liquid extraction
PPO	polyphenol oxidase
SFE	supercritical fluid extraction
UAE	ultrasound-assisted extraction

References

1. Sicari, V.; Loizzo, M.R.; Silva, A.S.; Romeo, R.; Spampinato, G.; Tundis, R.; Leporini, M.; Musarella, C.M. The effect of blanching on phytochemical content and bioactivity of *hypochaeris* and *hyoseris* species (Asteraceae), vegetables traditionally used in Southern Italy. *Foods* **2021**, *10*, 32. [CrossRef]
2. Kaseke, T.; Opara, U.L.; Fawole, O.A. Effect of microwave pretreatment of seeds on the quality and antioxidant capacity of pomegranate seed oil. *Foods* **2020**, *9*, 1287. [CrossRef] [PubMed]
3. Palmieri, S.; Pellegrini, M.; Ricci, A.; Compagnone, D.; Lo Sterzo, C. Chemical composition and antioxidant activity of thyme, hemp and coriander extracts: A comparison study of maceration, Soxhlet, UAE and RSLDE techniques. *Foods* **2020**, *9*, 1221. [CrossRef] [PubMed]
4. Alafeef, A.K.; Ariffin, F.; Zulkurnain, M. Organic selenium as antioxidant additive in mitigating acrylamide in coffee beans roasted via conventional and superheated steam. *Foods* **2020**, *9*, 1197. [CrossRef]
5. Kim, S. Antioxidant compounds for the inhibition of enzymatic browning by polyphenol oxidases in the fruiting body extract of the edible mushroom *Hericium erinaceus*. *Foods* **2020**, *9*, 951. [CrossRef]
6. Romeo, R.; De Bruno, A.; Imeneo, V.; Piscopo, A.; Poiana, M. Impact of Stability of Enriched Oil with Phenolic Extract from Olive Mill Wastewaters. *Foods* **2020**, *9*, 856. [CrossRef]
7. Ideia, P.; Sousa-ferreira, I.; Castilho, P.C. A Novel and simpler alkaline hydrolysis brewer's spent grain and its (partial) purification. *Foods* **2020**, *9*, 600. [CrossRef]
8. Smailagić, A.; Ristivojević, P.; Dimkić, I.; Pavlović, T.; Zagorac, D.D.; Veljović, S.; Akšić, M.F.; Meland, M.; Natić, M. Radical scavenging and antimicrobial properties of polyphenol rich waste wood extracts. *Foods* **2020**, *9*, 319. [CrossRef]
9. Angiolillo, L.; Spinelli, S.; Conte, A.; Alessandro, M.; Nobile, D. Extract from Broccoli Byproducts to Increase Fresh Filled Pasta Shelf Life. *Foods* **2019**, *8*, 621. [CrossRef]
10. Spinelli, S.; Costa, C.; Conte, A.; La Porta, N.; Padalino, L.; Del Nobile, M.A. Bioactive compounds from Norway spruce bark: Comparison among sustainable extraction techniques for potential food applications. *Foods* **2019**, *8*, 524. [CrossRef]
11. Barbosa, C.H.; Andrade, M.A.; Vilarinho, F.; Castanheira, I.; Fernando, A.L.; Loizzo, M.R.; Silva, A.S. A new insight on cardoon: Exploring new uses besides cheese making with a view to zero waste. *Foods* **2020**, *9*, 564. [CrossRef]

Article

The Effect of Blanching on Phytochemical Content and Bioactivity of *Hypochaeris* and *Hyoseris* Species (Asteraceae), Vegetables Traditionally Used in Southern Italy

Vincenzo Sicari [1], Monica R. Loizzo [2,*], Ana Sanches Silva [3,4], Rosa Romeo [1], Giovanni Spampinato [1], Rosa Tundis [2], Mariarosaria Leporini [2] and Carmelo M. Musarella [1]

[1] Department of Agraria, "Mediterranea" University of Reggio Calabria, Cittadella Universitaria, Località Feo di Vito, 89122 Reggio Calabria (RC), Italy; vincenzo.sicari@unirc.it (V.S.); rosa.romeo@unirc.it (R.R.); gspampinato@unirc.it (G.S.); carmelo.musarella@unirc.it (C.M.M.)
[2] Department of Pharmacy, Health Science and Nutrition, University of Calabria, Via P. Bucci, Edificio Polifunzionale, 87036 Rende (CS), Italy; rosa.tundis@unical.it (R.T.); mariarosarialeporini@tiscali.it (M.L.)
[3] National Institute for Agricultural and Veterinary Research (INIAV), I.P., Vairão, 4485-655 Vila do Conde, Portugal; anateress@gmail.com
[4] Center for Study in Animal Science (CECA), ICETA, University of Oporto, 4051-401 Oporto, Portugal
* Correspondence: monica_rosa.loizzo@unical.it; Tel.: +39-984-493246

Abstract: The impact of blanching on the phytochemical content and bioactivity of *Hypochaeris laevigata* (HL), *Hypochaeris radicata* (HR), *Hyoseris radiata* (HRA), and *Hyoseris lucida* subsp. *taurina* (HT) leaves was studied and compared to fresh plant materials and residual blanching water. For this purpose, total phenols, flavonoids, carotenoids, and chlorophyll contents were quantified. The antioxidant effect was investigated by using different in vitro tests (β-carotene, ferric reducing ability power (FRAP), 2,2′-Azino-bis(3-ethylbenzothiazoline-6-sulfonic acid) (ABTS) and 2,2-diphenyl-1-picrylhydrazyl (DPPH), whereas the potential inhibitory activity of key enzymes linked to obesity was screened against lipase, α-amylase, and α-glucosidase. Generally, the phytochemical content followed the trend: fresh > blanching water > blanched samples. The same trend was observed in the antioxidant activity independently of the applied test as well as in the inhibition of lipase and carbohydrates-hydrolysing enzymes. In particular, fresh *Hypochaeris laevigata* (HL1) showed the lowest inhibitory concentration 50% (IC_{50}) values of 31.3 and 42.7 µg/mL, against α-glucosidase and α-amylase, respectively, whereas fresh *Hyoseris radiata* (HRA1) showed the most promising hypolipidemic activity (IC_{50} value of 39.8 µg/mL). Collectively, these results support the health effect of these wild plants and demonstrated that blanching water should be reused in food preparation since it is a good source of bioactive compounds and its consumption should be recommended in order to increase the uptake of micronutrients.

Keywords: *Hyoseris radiata*; *Hyoseris taurina*; *Hypochaeris laevigata*; *Hypochaeris radicata*; phytochemicals; antioxidants; obesity; diabetes type 2

Received: 27 November 2020
Accepted: 22 December 2020
Published: 24 December 2020

1. Introduction

Hyoseris L. and *Hypochaeris* L. species are largely used in Italy not only to prepare salads, omelettes or boiled in soups, but also for medicinal use through the use of infusions using blanching water. *Hyoseris radiata* L. and *Hypochaeris radicata* L. are the most studied species and have several ethnobotanical uses in Italy [1–8]. *Hyoseris lucida* L. subsp. *taurina* (Pamp.) Peruzzi & Vangelisti and *Hypochaeris laevigata* (L.) Ces., Pass. & Gibelli have not been much studied, probably due to their relatively narrow geographical distribution [9,10]. Guarrera and Savo [11,12] reported the use of leaves from *H. radicata* and *H. radiata* as boiled vegetables or salad in Piedmont, Liguria, Marche, Latium, Sardinia, Sicily, and Calabria. In this last region, these species are preserved in olive oil to be consumed as vegetable side dish.

Metabolic syndrome is a cluster of conditions that occur together. These conditions include increased blood pressure, high blood sugar levels, excess body fat around the waist, and abnormal cholesterol or triglyceride levels. [13]. The World Health Organization (WHO) Global Health Observatory estimated through Prophet models that the prevalence of diabetes and obesity in 2030 is likely to increase by 10.1% [14].

Oxidative stress plays a key role in the development of Diabetes Mellitus Type 2 (DMT2) complications [15]. In fact, the metabolic irregularities of diabetes cause an overproduction of superoxide in endothelial cells of both large and small vessels, and, also in the myocardium. Moreover, Găman et al. [16] demonstrated that these irregularities impair tissue glucose uptake and reduces β cell insulin secretion.

The consumption of wild plants represents a key part of the Mediterranean diet, recently recognized by United Nations Educational, Scientific and Cultural Organization UNESCO (UNESCO) as an Intangible Cultural Heritage of Humanity (UNESCO). Although several plants of *Hyoseris* and *Hypochaeris* species have been consumed daily for centuries, there is still no in-depth study on their micronutrient content and health properties after preparation prior to consumption. In the food industry and at a domestic level, blanching is a pretreatment largely used to inactivate enzyme activity, which can affect micronutrient content and preserve vegetables [17]. However, if it is effective in reducing degradation during shelf-life, on the other hand it produces modifications in cell structure and composition with a consequent significant loss of micronutrients in the food matrix [18].

In this context, the evaluation of the impact of blanching on Calabrian traditional vegetables, namely *H. laevigata, H. radicata, H. radiata,* and *H. lucida* subsp. *taurina,* was investigated. For this purpose, basal leaves were screened for their Total Phenols Content (TPC), Total Flavonoids Content (TFC), chlorophylls and Total Carotenoids Content (TCC) before and after blanching process. β-Carotene bleaching, FRAP, ABTS, and DPPH tests were applied to test the antioxidant activity. The inhibitory effects of enzymes linked to obesity and DMT2 such as lipase, α-glucosidase, and α-amylase were assessed. In order to evaluate the possible loss of bioactive compounds after blanching, the residual blanching water was also investigated.

2. Materials and Methods

2.1. Chemicals and Reagents

Chemicals and reagents used in this research were purchased from Sigma-Aldrich Chemical Co. Ltd. (Milan, Italy) and VWR International (Milan, Italy). Acarbose from *Actinoplanes* spp. was obtained from Serva (Heidelberg, Germany) whereas β-carotene was from Extrasynthese (Genay-France).

2.2. Plant Materials

The collection (Table 1) of the basal rosettes of *H. radicata* and *H. radiata* was carried out at 1157 m a.s.l. in the area of the Aspromonte Massif known as "Cucullaro" (38.171987° N–15.815650° E), a mountain resort in the municipality of Santo Stefano in Aspromonte. The collection of the basal rosettes of *H. lucida* subsp. *taurina* was carried out in the municipality of Scilla (40 m a.s.l., 38.253987° N–15.716493° E), while those of *H. laevigata* at the mountain resort of Trepitò, located at 952 m a.s.l. (38.284906° N–16.046640° E) in the municipality of Molochio. The collection of the basal rosettes was carried out before flower stem development, as traditionally these and similar plants are collected by local people in this way [8,19]. All the collection sites are located in the province of Reggio Calabria (South Italy)

Table 1. Sampling locations.

Sampling Locations	Latitude	Longitude	Altitude m a.s.l.
Cucullaro	38.166667°	15.850000°	1300
Scilla	38.250000°	15.733333°	73
Trepitò	38.316667°	16.150000°	880

For each of the 4 taxa examined, a minimum quantity of 1000 g of basal rosettes was collected and, afterwards, in the laboratory, a cleaning and separation of the edible leaves was performed manually to obtain a minimum quantity of 200 g of leaves. These were washed several times with distilled water in order to ensure the elimination of any type of residue. Finally, they were placed on absorbent paper and gently dabbed several times, so as to eliminate as much water as possible and thus avoid problems of contamination.

2.3. Blanching

To blanche, leaves were placed on a low heat in 1000 mL of hot water (90 °C) for 10 min. The ratio of sample to water was 1:5 (*w/w*). After blanching, the samples were cooled to room temperature under running tap water. Finally, the edible parts were drained and the blanching water collected.

2.4. Extraction Procedure

Extracts were prepared by mixing 10 g of each sample (fresh and blanched, both ground) with 50 mL of water. The mixture was shaken using a Ultraturrax T-25 (Ika Labortechnik, Janche & Kunkel, Milan, Italy). It was subsequently centrifuged by refrigerated centrifuge Nùve NF 1200R (Saracalar Kümeevleri, Ankara, Turkey), 10 min at 5000 rpm. Then, the supernatant was filtered through a 0.45 mm Millipore filter (GMF Whatman, Carlo Erba, Milan, Italy) before analysis.

2.5. Determination of pH, Titratable Acidity and Total Soluble Solid Content

The pH was determined by direct measurement in a digital potentiometer (Crison Instruments S.A., Milan, Italy). The Total soluble solids (TSS) were measured at 20 °C using a digital Atago Model PR-101 α refractometer (Atago Co. Ltd., Milan, Italy), results were reported as Brix degrees (°Brix). The titratable acidity (TA) was determined using 0.1 N NaOH to pH 8.1. Results were expressed as percentage of monohydrate citric acid.

2.6. Total Phenol Content (TPC) and Total Flavonoid Content (TFC)

TPC was determined as described by Sicari et al. [20]. An aliquot of 350 μL of aqueous extract was mixed with Folin-Ciocalteu reagent (1 mL) and 20% Na_2CO_3 solution (10 mL). The absorbance was measured at 760 nm in a spectrophotometer (UV-VIS-Agilent 8453) and the results were expressed in mg of gallic acid equivalent (GAE)/100 g fresh weight (FW).

A spectrophotometric method was used to measure the total flavonoid content. TFC was determined using a method based on the formation of a flavonoid-aluminium complex [21]. The extract was mixed with 2% aluminium chloride solution. The samples were incubated at room temperature for 15 min and then measured against a blank at 510 nm. TFC was calculated based on a standard curve and expressed as mg quercetin equivalents (QE)/100 g FW.

2.7. Spectrophotometric Determination of Carotenoids

For the carotenoid determination, the spectrophotometric analysis was carried out after extraction [22]. The analyses were carried out in darkness to prevent carotenoid degradation and isomerisation. Before chemical extraction, leaves were homogenised in a blender and an aliquot of 5 g of the sample was weighed into a 50 mL amber coloured flask wrapped with aluminium foil. Then, 100 mL of the solvent mix (hexane/acetone/methanol

2:1:1 $v/v/v$) was added to the flask and sonicated continuously for 10 min (Misonix Ultrasonic Liquid Processor, Misonix, Inc. 1938, New Highway, Farmingdale, NY, USA).

The extraction was repeated until the sample became colorless. The combined extract was transferred to a separating funnel and 5 mL of distilled water was added to separate polar and nonpolar phases. The nonpolar hexane layer containing carotenoids was collected and concentrated in a rotary evaporator (Heidolph, Schwabach, Germany) until dry. The residue was dissolved in 10 mL of hexane. The total carotenoid content was determined by a spectrophotometric method using a UV-Vis spectrophotometer (Agilent 845, Technologies, Agilent, Milan, Italy). The absorbance was read at 450 nm. All analyses were performed in triplicate and the results were expressed as mean $\pm$ standard deviation (SD).

The results were expressed as g β-carotene/100 g fresh weight of sample.

2.8. Chlorophyll Determination

The chlorophyll content was determined spectrophotometrically. The leaves were placed into a mortar and then ground in the dark until the green colour disappeared.

Ground leaf material was extracted with ethanol, filled to 10 mL, centrifuged at 5000 rpm for 3 min and the absorbance of the supernatant solution in a 1 cm cell was read at 440, 649 and 665 nm. Chlorophyll was calculated using the equations previously reported by Lichtenthaler and Buschmann [23].

2.9. Radical Scavenging Activity Assays

In ABTS radical scavenging ability test, potassium persulphate solution and ABTS solution were mixed to obtain ABTS radical cation solution [21]. After 12 h, this solution was stabilized (absorbance of 0.70) at 734 nm employing a UV-Vis spectrophotometer (Jenway 6003, Carlo Erba, Milan, Italy). ABTS$^+$ solution was mixed with different concentrations of aqueous extracts (from 1 to 400 µg/mL) and the absorbance was read at 734 nm after 6 min.

Another test used to evaluate the radical scavenging activity of our samples is the 2,2-diphenyl-1-picrylhydrazyl (DPPH) assay [21]. The DPPH radical solution was mixed with aqueous extracts (at concentrations in the range from 1 to 1000 µg/mL) and after 30 min the absorbance was read at 517 nm.

2.10. Ferric Reducing Ability Power (FRAP) Assay

In the ferric reducing ability power (FRAP) assay, FRAP reagent, tripyridyltriazine (TPTZ), HCl, FeCl$_3$, acetate buffer (pH 3.6) were mixed and added to the sample (2.5 mg/mL) [24]. The absorption reaction was read at 595 nm after 30 min of incubation.

2.11. β-Carotene Bleaching Test

The ability of samples to inhibit lipid peroxidation was evaluated by using the β-carotene bleaching test [21]. The aqueous extracts (from 2.5 to 100 µg/mL) were mixed with linoleic, Tween 20 acid and β-carotene solution. The absorbance was read at t = 0, and after 30 and 60 min of incubation at 470 nm.

2.12. Pancreatic Lipase Inhibitory Activity

The hypolipidemic potential of aqueous extracts was studied by the inhibition of pancreatic lipase [25]. A mixture of lipase (1 mg/mL), 4-nitrophenyl octanoate (substrate) and samples (from 25 to 4000 µg/mL) was prepared. After 30 min of incubation at 37 °C the Tris-HCl buffer (pH 8.5) was added and the absorbance was measured at 405 nm.

2.13. Carbohydrate-Hydrolysing Enzyme Inhibitory Activity

The hypoglycaemic activity of samples was evaluated through inhibition of α-amylase and α-glucosidase enzymes [21]. For α-amylase assay, aqueous extracts (from 25 to 1000 µg/mL) and starch solution were added to the enzyme at room temperature for 5 min, and the absorbance at 540 nm was measured. For α-glucosidase inhibitory activity

ity test, samples extracts (from 25 to 1000 µg/mL) were mixed with enzyme solution, a maltose solution, peroxidase/glucose oxidase (PGO) system-colour reagent solution and *o*-dianisidine (DIAN) solution. This mixture was incubated for 30 min and the absorbance was measured.

2.14. Statistical Analysis

All the investigations were performed in triplicate and results were expressed as means of three different experiments ± standard deviation (S.D.). They were processed by analysis of variance (ANOVA). Differences among the samples were analyzed by Turkey's test using SPSS statistics software (version 17.0, SPSS Inc., Chicago, IL, USA). All the *p* values at < 0.05 were observed as significant.

Principal Component Analysis (PCA) was applied using SPSS software for Windows, version 17.0 (SPSS Inc., Chicago, IL, USA).

3. Results and Discussion

In this work the phytochemical content and bioactivity of traditionally consumed wild plants namely *H. laevigata*, *H. radicata*, *H. radiata*, and *H. lucida* subsp. *taurina* were investigated. The edible portions (leaves) were studied, both fresh and after blanching, to assess the impact of processing on these food matrices. Among bioactive phytochemicals, TPC, TFC, lycopene, β-carotene, and chlorophylls were quantified. Samples were studied for their antioxidant potential using different approaches and as inhibitors of enzymes linked to obesity and hyperglycaemia.

3.1. Phytochemical Content

Used in the experiment procedures had a significant effect on changes of total soluble solids (TSS), titratable acidity (TA), and pH. The blanching effect of leaves is shown in Table 2. Fresh *Hypochaeris radicata* (HR) and *H. laevigata* (HL) leaves showed a higher TSS contents than the values obtained in the leaves after blanching and in the blanching water. The same trend was observed in the titratable acidity values.

Table 2. Total soluble solid content (TSS), Titratable acidity (TA), and pH of *Hypochaeris* and *Hyoseris* extracts.

Sample	TSS	TA	pH
Hypochaeris leavigata (HL)			
HL1	0.6 ± 0.05 [a]	5.2 ± 2.0 [a]	6.0 ± 2.6 [b]
HL2	0.1 ± 0.01 [c]	0.8 ± 0.07 [c]	6.6 ± 2.7 [a]
HL3	0.4 ± 0.04 [b]	1.9 ± 1.3 [b]	6.2 ± 2.4 [b]
Sign	**	**	**
Hypochaeris radicata (HR)			
HR1	0.6 ± 0.07 [a]	4.5 ± 1.9 [a]	6.0 ± 2.6 [b]
HR2	0.2 ± 0.02 [b]	0.7 ± 0.06 [c]	6.4 ± 2.3 [a]
HR3	0.7 ± 0.08 [a]	2.1 ± 1.4 [b]	6.5 ± 2.5 [a]
Sign.	**	**	**
Hyoseris radiata (HRA)			
HRA1	0.4 ± 0.03 [a]	6.8 ± 2.9 [a]	5.8 ± 2.2 [b]
HRA2	0.3 ± 0.02 [b]	1.4 ± 1.1 [c]	6.4 ± 2.4 [a]
HRA3	0.4 ± 0.04 [a]	1.7 ± 1.2 [b]	6.2 ± 2.3 [a]
Sign.	**	**	**

Table 2. *Cont.*

Sample	TSS	TA	pH
Hyoseris lucida **subsp.** *taurina* **(HT)**			
HT1	0.9 ± 0.1 [a]	6.3 ± 2.7 [a]	5.9 ± 2.2 [b]
HT2	0.7 ± 0.09 [b]	2.3 ± 1.5 [c]	6.2 ± 2.4 [a]
HT3	0.5 ± 0.06 [a]	2.1 ± 1.2 [b]	6.4 ± 2.5 [a]
Sign.	**	**	**

Data are reported to mean ± Standard Deviation (SD) ($n = 3$). 1: Extract fresh plant materials; 2: Extract plant materials after blanching; 3: Blanching water extract. ** Significance at $p < 0.01$. Results followed by different letters in a same column are significantly different ($p < 0.05$) by Tukey's multiple. range test.

Carotenoid content in raw and cooked leaf samples are shown in Table 3. The reported values show a greater concentration of total carotenoids (expressed in β-Carotene equiv.) in the fresh leaves than in the blanched ones. The blanching water had a significantly higher ($p < 0.05$) carotenoid content than the blanched leaves, but lower than the fresh leaves. A similar trend was found in leaves of both *Hypochaeris* (*leavigata* and *radicata*) and *Hyoseris* (*radiata* and *lucida* subsp. *taurina*). Compared to the initial concentration in the fresh leaves (HL1, HR1, HRA1 and HT1), the lowest carotenoid concentration in blanching water was found in sample HT3, with a value of 205.2 g/100 g, whereas the highest value was found in sample HRA3 (298.1 g/100 g). The presence of carotenoids in blanching water was reported by Parmar et Rupasinghe [26], who observed that an infusion of wild berry stems in hot water had a carotenoid content in the range of 275–417 mg/L. They further showed a high carotenoid content in commercial green tea with average values of 410 and 1332 mg/L for extraction in methanol and hot water respectively. In a study carried out by Loranty et al. [27] on 25 infused fruit and herbal teas, only lutein was present in the infusions, other carotenoids not being found. The highest level of lutein (24.3 mg/200 mL) was found in a Tilias infusion (Edward Tea). Suzuki et Shioii [28] found chlorophyll and carotenoids in seven teas infused from *Camellia sinensis* leaves.

Table 3. Phytochemical content of *Hypochaeris* and *Hyoseris* extracts.

Sample	TCC (mg βC/100 g FW)	Chlorophylls (mg/kg FW)	TPC (mg GAE/100 g FW)	TFC (mg QE/100 g FW)
Hypochaeris leavigata **(HL)**				
HL1	351.8 ± 5.1 [a]	170.2 ± 3.2 [a]	946.5 ± 6.2 [a]	732.7 ± 6.0 [a]
HL2	215.9 ± 4.7 [c]	106.8 ± 3.1 [c]	378.1 ± 5.1 [c]	179.4 ± 3.2 [b]
HL3	298.1 ± 4.9 [b]	163.0 ± 3.2 [b]	623.0 ± 5.9 [b]	569.6 ± 5.7 [c]
Sign	**	**	**	**
Hypochaeris radicata **(HR)**				
HR1	449.5 ± 5.5 [a]	191.7 ± 3.6 [a]	1103.6 ± 6.5 [a]	769.2 ± 6.1 [a]
HR2	236.9 ± 4.5 [c]	118.4 ± 3.3 [c]	434.3 ± 5.4 [c]	322.5 ± 4.2 [c]
HR3	266.0 ± 4.6 [b]	143.2 ± 3.4 [b]	550.8 ± 5.6 [b]	375.0 ± 4.3 [b]
Sign	**	**	**	**
Hyoseris radiata **(HRA)**				
HRA1	386.7 ± 5.2 [a]	212.3 ± 4.0 [a]	975.3 ± 6.2 [a]	855.6 ± 5.9 [a]
HRA2	181.3 ± 3.5 [c]	89.3 ± 2.9 [c]	380.1 ± 4.2 [c]	277.1 ± 3.8 [c]
HRA3	240.6 ± 4.5 [b]	136.7 ± 3.2 [b]	550.7 ± 5.2 [b]	370.4 ± 4.0 [b]
Sign	**	**	**	**

Table 3. Phytochemical content of *Hypochaeris* and *Hyoseris* extracts.

Sample	TCC (mg βC/100 g FW)	Chlorophylls (mg/kg FW)	TPC (mg GAE/100 g FW)	TFC (mg QE/100 g FW)
	Hyoseris lucida **subsp.** *taurina* **(HT)**			
HT1	532.4 ± 5.8 [a]	187.4 ± 3.4 [a]	997.6 ± 6.3 [a]	796.8 ± 6.0 [a]
HT2	202.6 ± 3.5 [b]	106.4 ± 3.1 [c]	416.1 ± 5.3 [c]	436.2 ± 5.3 [b]
HT3	205.2 ± 3.6 [b]	109.4 ± 3.2 [b]	627.5 ± 5.8 [b]	447.3 ± 5.4 [b]
Sign.	**	**	**	**

Data are reported to mean ± Standard Deviation (SD) ($n = 3$). Data are expressed as mean ± S.D. ($n = 3$). 1: Extract fresh plant materials; 2: Extract plant materials after blanching; 3: Blanching water extract. ** Significance at $p < 0.01$. Results followed by different letters in a same column are significantly different ($p < 0.05$) by Tukey's multiple range test.

A drastic reduction in chlorophyll was also observed between fresh and blanched samples (from 170.2 to 212.3 for HL1 and HRA1, vs. from 89.3 to 118.4 for HRA2 and HR2, respectively). Blanching reduced the chlorophyll content independently of the species (Table 3).

The concentration of total polyphenols and flavonoids was shown in Table 3. Phenolic compounds are known to be responsible for the antioxidant activity of the food matrix. All fresh samples were characterized by a high TPC content. Leaves of *hypochaeris* (*leavigata e radicata*) and *hyoseris* (*radiata e lucida* subsp. *taurina*), showed a significantly higher ($p < 0.05$) total polyphenol content in their raw samples compared with their cooked samples [29,30]. The aqueous extract of fresh *H. radicata* showed the highest TPC value of 1103.6 mg GAE/100 g FW followed by *H. lucida* subsp. *taurina* (997.6 mg GAE/100 g FW) (Table 3). Fresh *H. radiata* leaves (HRA1) showed the highest TFC with a value of 855.6 mg QE/100 g FW. It is interesting to note that the blanching process significantly affects both TPC and TFC and that the water in which the leaves are cooked retained a great amount of these bioactive phytochemicals.

The total polyphenol content decreased significantly ($p < 0.05$) after blanching as reported by [31–34]. In addition, Gawlik-Dziki [35] demonstrated that boiling significantly reduced the polyphenol content of fresh broccoli. Similarly, Sikora et al. [36] reported a significant decrease in total polyphenol and antioxidant components in boiled broccoli. Abu-Ghannam and Jaiswal [37] described a reduction in the total phenolic content of up to 45% at lower blanching temperatures (80–90 °C) within 2 min of blanching and reduction of the total polyphenol content continued within 6 min at high blanching temperatures. Furthermore, as some authors have observed, the degree of leaf fragmentation may be a factor in a greater diffusion of the bioactive compounds from the leaves to the water.

Medina et al. [38] compared infusions obtained from samples at various degrees of fragmentation. Levels of oleuropein were found to be 103 mg/kg in an infusion from a whole sample used as control, and 466 mg/kg in another sample passed through a blender and subsequently an Ultra-Turrax. The greater the degree of leaf fragmentation, the greater the quantities of phenols in the infusions. The total flavonoid content dropped significantly ($p < 0.05$) after boiling. Interestingly, despite heating, the concentration of flavonoids in the blanching water remained high. This may be due to the fact that after boiling there was a greater availability of flavonoids, and a more efficient extraction from the softened cell walls [39].

Moreover, several studies have reported that the increase in bioactive compounds in boiled vegetables may be partly due to the breakdown of cell walls and subcellular structures by boiling, which allows the release of antioxidants [40]. Thus, it is probable that the structural matrix of the cell walls is the factor that determines the cell's ability to hold onto or breakdown phytochemical compounds.

3.2. Antioxidant Activity

The antioxidant activities of *H. laevigata* (HL), *H. radicata* (HR), *H. radiata* (HRA), and *H. lucida* subsp. *taurina* (HT) fresh and blanched leaves were assessed employing in vitro methods: β-carotene bleaching, FRAP, ABTS, and DPPH tests. The resulting blanching water was also screened. To our knowledge this is the first report to evaluate the antioxidant potential of *H. laevigata*, and *H. lucida* subsp. *taurina* leaves, both fresh and processed. Generally, the following antioxidant trend was observed fresh samples > blanching water > blanched samples (Table 4).

Table 4. Antioxidant activity of *Hypochaeris* and *Hyoseris* extracts.

Samples	β-Carotene Bleaching Test IC$_{50}$ (µg/mL)		FRAP µMFe (II)/g	ABTS IC$_{50}$ (µg/mL)	DPPH IC$_{50}$ (µg/mL)
	t = 30 min	t = 60 min			
Hypochaeris leavigata (HL)					
HL1	46.7 ± 3.0 [b]	48.5 ± 3.1 [b]	84.4 ± 3.8 [b]	4.7 ± 0.8 [c]	24.7 ± 2.2 [c]
HL2	39.6 ± 2.3 [c]	30.1 ± 1.9 [c]	39.1 ± 3.4 [c]	7.9 ± 1.2 [a]	43.6 ± 3.5 [a]
HL3	48.7 ± 3.1 [a]	59.0 ± 3.6 [a]	92.6 ± 4.0 [a]	5.9 ± 0.9 [b]	37.6 ± 3.1 [b]
Sign.	**	**	**	**	**
Hypochaeris radicata (HR)					
HR1	54.6 ± 3.3 [b]	58.5 ± 3.4 [b]	41.6 ± 3.2 [c]	7.8 ± 1.2 [c]	18.7 ± 1.5 [c]
HR2	48.1 ± 2.8 [c]	35.7 ± 1.4 [c]	53.8 ± 3.5 [a]	12.3 ± 1.5 [a]	41.6 ± 3.0 [a]
HR3	60.5 ± 3.8 [a]	63.9 ± 3.8 [a]	49.2 ± 3.5 [b]	13.7 ± 1.6 [b]	23.7 ± 1.9 [b]
Sign.	**	**	**	**	**
Hyoseris radiata (HRA)					
HRA1	51.7 ± 3.5 [a]	41.7 ± 2.4 [a]	73.5 ± 4.1 [a]	8.8 ± 0.9 [c]	37.6 ± 2.2 [b]
HRA2	18.7 ± 1.3 [c]	14.1 ± 0.8 [c]	38.9 ± 3.1 [c]	11.8 ± 1.1 [a]	51.5 ± 3.2 [a]
HRA3	26.4 ± 1.6 [b]	29.2 ± 1.1 [b]	41.1 ± 2.9 [b]	10.1 ± 1.0 [b]	29.7 ± 2.0 [c]
Sign.	**	**	**	**	**
Hyoseris lucida subsp. *taurina* (HT)					
HT1	84.7 ± 3.9 [a]	99.1 ± 4.2 [a]	54.7 ± 3.6 [a]	1.8 ± 0.3 [c]	33.6 ± 2.0 [c]
HT2	41.4 ± 2.1 [b]	23.8 ± 1.8 [c]	41.0 ± 3.1 [c]	4.1 ± 0.9 [a]	54.5 ± 3.2 [a]
HT3	39.1 ± 2.5 [c]	31.1 ± 2.1 [b]	53.2 ± 3.5 [b]	2.9 ± 0.5 [b]	42.6 ± 2.5 [b]
Sign.	**	**	**	**	**

Data are expressed as mean ± Standard Deviation (SD) (n = 3). 1: Extract fresh plant materials; 2: Extract plant materials after blanching; 3: Blanching water extract. Ferric Reducing Antioxidant Power (FRAP); Antioxidant Capacity Determined by Radical Cation (ABTS+); DPPH Radical Scavenging Activity Assay. [a]: [100 µg/mL]. Propyl gallate (IC$_{50}$ = 0.09 ± 0.04 µg/mL after t = 30 min and t= 60 min of incubation) was used as control positive in β-carotene bleaching test, BHT (IC$_{50}$ = 63.2 ± 2.3 µMFe (II)/g) in FRAP assay and ascorbic acid in ABTS and DPPH radical scavenging test (IC$_{50}$ = 5.0 ± 0.8 and 1.7 ± 0.1 µg/mL, respectively). One-way ANOVA followed by Tukey's multiple range test was applied for statistical analysis. Different letters in the same column are significantly different ** at $p < 0.01$.

A great variability of results was observed in the β-carotene bleaching test. In this assay, the presence of antioxidant compounds minimized the oxidation of β-carotene by hydro-peroxides, which were counteracted by bioactive compounds in the extract. In the present study, both *H. radicata* and *H. laevigata* fresh leaves exerted a greater activity than the other investigated species (IC$_{50}$ = of 46.7 and 54.6 µg/mL after 30 min of incubation respectively). Blanched samples were less active with a percentage of inhibition of 39.6% and 48.1% at maximum concentration tested (100 µg/mL). A low activity was observed also with HT1 sample (IC$_{50}$ = 84.7 and 99.1 µg/mL after 30 and 60 min of incubation respectively).

In FRAP assay, a great ferric reducing power higher than that found for BHT (Butylated Hydroxytoluene) was observed with both HL3 and HL1 samples with values of 92.6 and 84.4 μM Fe(II)/g, respectively (Table 4). A promising ability was observed also with HRA1 (73.5 μM Fe(II)/g).

A different result was observed using ABTS radical cation. In fact, in this test fresh *H. lucida* subsp. *taurina* leaves (HT1) showed a comparable radical scavenging activity to that reported for ascorbic acid (IC_{50} = 1.8 and 1.7 μg/mL, respectively). A promising ABTS radical scavenging potential was noted with the blanching water from the blanching of the leaves of the same species (HT3, IC_{50} = 2.9 μg/mL).

The DPPH free radical scavenging method is based on electron-transfer reaction that produces a violet solution. The DPPH radical is stable at 25 °C. Among the investigated species, aqueous extract of fresh leaves of HR and HL exerted a higher radical scavenging potential (IC_{50} = 18.7 and 24.7 μg/mL, respectively) (Table 4). In addition, interest activity was found for HR3 and HRA3 (IC_{50} = 23.7 and 29.7 μg/mL, respectively).

Previously, Senguttuvan et al. [41] studied the radical scavenging potential of an infusion from the dried leaves of *H. radicata* and found IC_{50} values of 595.23 μg/mL and 2143.1 μmol of Trolox equivalent (TE)/dried weight (DW). Dried leaves are also able to exert their antioxidant activities through other mechanisms, including the protection of lipid peroxidation and ferric reducing power.

Values of 97.99% at 250 mg/mL and 38.69% at 5 mg/mL were recorded by Senguttuvan et al. [41] for Indian *H. radicata* dried leaves in DPPH and ABTS test, respectively. Successively the same research group showed that the oral administration of *H. radicata* methanolic leaf and root extracts and isolated compounds proved to be significant, promising candidates able to quench free radicals. The antioxidant effect was more pronounced in the animals treated with root extract, probably due its high content in alkaloids, flavonoids, saponins and terpenoids that could act as antioxidant compounds [41]. Lower antioxidant activity was recorded for *H. radicata* collected in Latium [42] where IC_{50} average values of 2.02 and 2.33 mg/mL were found for fresh and boiled leaves, respectively in DPPH assay. Average values of 6.2 and 2.3 mmol/kg FW were found for fresh and boiled leaves in ABTS test, respectively.

Compared to the Calabria sample HRA1, a lower ferric reducing power was recorded for *H. radiata* fresh leaves collected in Liguria that showed FRAP value of 31.1 mM Fe (II) /Kg. More recently, Souilah et al. [43] investigated the antioxidant effect of *n*-butanol, dichloromethane and ethyl acetate fractions of the aerial parts of *Hypochaeris laevigata* var. *hipponensis*. The highest DPPH radical scavenging activity was exhibited by *n*-Butanol extract (IC_{50} = 8.12 μg/mL) followed by ethyl acetate extract (IC_{50} = 8.70 μg/mL). This last extract was also able to exert a potent ABTS radical cation inactivation (IC_{50} = 4.32 μg/mL). The following rank dichloromethane, *n*-butanol and ethyl acetate in protection of lipid peroxidation was found.

3.3. Inhibition of Enzymes linked to Obesity

The study of bioactive foods useful in the prevention and management of metabolic diseases including obesity and diabetes is a topic of great interest for researchers in the area of food science. To our knowledge, no previous studies have investigated the species *Hypochaeris* and *Hyoseris* for their ability to inhibit carbohydrate-hydrolyzing and lipase enzymes. All investigated samples exerted inhibitory activity on enzymes linked to obesity and DMT2 in a concentration dependent manner (Table 5).

Table 5. Lipase, α-amylase, and α-glucosidase inhibitory activity [IC_{50} (µg/mL)] of *Hypochaeris* and *Hyoseris* extracts.

Sample	Lipase	α-Amylase	α-Glucosidase
Hypochaeris leavigata **(HL)**			
HL1	58.2 ± 1.5 [c]	42.7 ± 1.3 [c]	31.3 ± 1.3 [c]
HL2	85.0 ± 1.8 [a]	94.9 ± 2.0 [a]	81.6 ± 1.8 [a]
HL3	66.4 ± 1.7 [b]	56.8 ± 1.5 [b]	62.3 ± 1.6 [b]
Sign.	**	**	**
Hypochaeris radicata **(HR)**			
HR1	52.4 ± 1.5 [c]	56.9 ± 1.4 [c]	37.6 ± 1.3 [c]
HR2	81.4 ± 1.7 [a]	112.9 ± 2.1 [a]	118.8 ± 2.3 [a]
HR3	62.8 ± 1.7 [b]	90.1 ± 2.0 [b]	43.9 ± 1.4 [b]
Sign.	**	**	**
Hyoseris radiata **(HRA)**			
HRA1	39.8 ± 1.4 [c]	79.4 ± 1.7 [c]	41.9 ± 1.4 [c]
HRA2	85.7 ± 1.8 [a]	99.1 ± 1.9 [a]	84.5 ± 1.8 [a]
HRA3	65.7 ± 1.7 [b]	83.4 ± 1.8 [b]	51.9 ± 1.5 [b]
Sign.	**	**	**
Hyoseris lucida **subsp.** *taurina* **(HT)**			
HT1	73.5 ± 1.7 [c]	74.0 ± 1.7 [c]	63.1 ± 1.6 [c]
HT2	89.6 ± 1.9 [a]	94.2 ± 1.9 [a]	94.9 ± 1.9 [a]
HT3	78.8 ± 1.8 [b]	82.8 ± 1.6 [b]	74.4 ± 1.7 [b]
Sign.	**	**	**

Data are expressed to mean ± Standard Deviation (SD) ($n = 3$). 1: Extract fresh plant materials; 2: Extract plant materials after blanching; 3: Blanching water extract. Orlistat used as positive control in lipase test (IC_{50} = 37.4 ± 1.0). Acarbose used as positive control in α-amylase and α-glucosidase tests (IC_{50} = 50.1 ± 1.3 and 35.5 ± 0.9 respectively for α-amylase α-glucosidase). One-way ANOVA followed by Tukey's multiple range test was applied for statistical analysis. Different letters in the same column are significantly different ** at $p < 0.01$.

Fresh *H. laevigata* leaves (HL1) exerted a promising α-amylase inhibitory activity with IC_{50} value of 42.7 µg/mL which is lower compared to positive control acarbose (IC_{50} = 50.1 µg/mL). A notable activity was observed, also in the blanching water (HL5, IC_{50} = 56.8 µg/mL). The HL1 sample showed the greatest α-glucosidase inhibitory activity followed by *H. radicata* (HR1) with IC_{50} values of 31.3 and 37.6 µg/mL, respectively. Both results are comparable with those found for acarbose (IC_{50} = 35.5 µg/mL). Also, in this case, blanching significantly affected the bioactivity of the samples. Fresh *H. radiata* leaves (HRA1) showed the highest lipase inhibitory activity (IC_{50} = 39.8 µg/mL) whereas, values of 52.4 and 58.2 µg/mL were found for HR1 and HL1, respectively. Unlike what happened for the carbohydrate hydrolysing enzymes, the blanching water has only a minimal inhibitory activity (IC_{50} values ranged from 62.8 to 78.8 for HR3 and HT3, respectively) testifying that compounds able to inhibit lipase enzyme are retained in the matrix also after blanching.

Among our investigated samples, only *H. radicata* had been previously investigated for its hypoglycaemic activity. However, our data disagree with Ko et al. [44], who did not find α-glucosidase inhibitory activity at the maximum concentration tested of 1000 µg/mL. Probably, this lack of activity was due the relatively low content of TPC and TFC found by the authors.

3.4. Principal Component Analysis (PCA)

PCA was performed to identify accession groups and to determine the axes and the characters significantly contributing to the variation. In this procedure, the similarity matrix was used to generate eigenvalues and scores for the accessions. The first two principal components, which accounted for the highest variation, were then used to plot two-dimensional scatter plots [45].

PCA was applied to differentiate the four different taxa of *Hypochaeris* and *Hyoseris*. By choosing eigenvalues greater than one (>1), the dimensionality was reduced from 16 variables to two principal components (PC). PCA results revealed that the first two principal components explained total variance completely 78.4%. The loadings of first and second principal components (PC1 and PC2) accounted for 45.25 and 33.15% of the variance, respectively (Figure 1). The first component (PC1) is highly positively correlated with Chlorophylls, and FRAP. The second component (PC2) is positively correlated with TSS, β-carotene bleaching test t = 30 min, β-carotene bleaching test t = 60 min, while citric acid, β-Carotene, TPC, and TFC) are positively correlated with component 1 and component 2. pH, DPPH, α-glucosidase, and α-amylase show a negative correlation for PC1 and PC2.

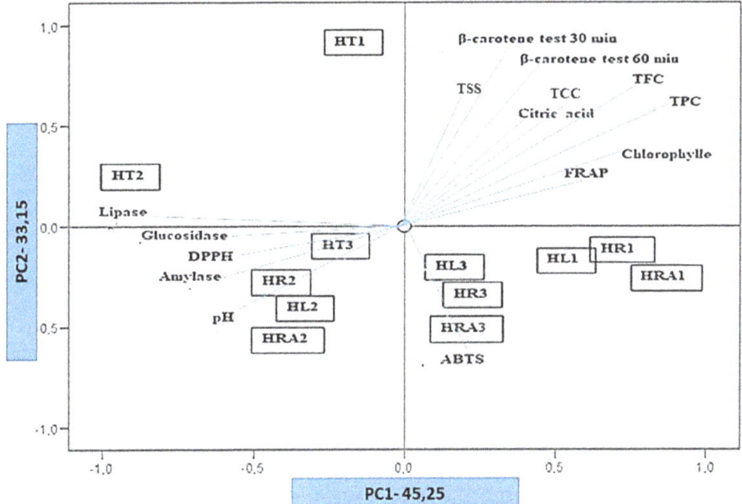

Figure 1. Factor loadings for principal components (PC) PC1 and PC2 and scatter plot of all. samples for principal component analysis.

The bi-dimensional PCA analysis clearly classifies the similarities or differences of the botanic species and the treatments performed. The score plot analysis clearly classifies the species HL1, HR1, HRA1, and HR3, HL3, HRA3 in the lower right region of the PCA score plot. This shows that fresh plants and blanching water maintain a higher bioactivity than blanched leaves.

4. Conclusions

The present study assessed for the first time the impact of the blanching process on the phytochemical content and bioactivity of the spontaneous plants namely *Hypochaeris laevigata*, *H. radicata*, *Hyoseris radiata*, and *H. lucida* subsp. *taurina*. Traditionally, these species are widely consumed in Central and South Italy, both fresh and after blanching. For this purpose, fresh and blanched samples as well as residual blanching water were studied. The blanching process determined a reduction in the content of all investigated phytochemical classes. At the same time, the analysis of the data showed that the blanching water retains most of the bioactive compounds and for this reason it is characterized by a good antioxidant and inhibitory activity against enzymes linked to obesity and related diseases such as diabetes type 2. For this reason, the consumption of the fresh spontaneous plant and the reutilization of residual blanching water should be promoted in order to ensure the right amounts of healthy micronutrients able to counteract oxidative stress and related diseases.

Author Contributions: Methodology, V.S. and M.R.L.; software, G.S.; validation, V.S., M.R.L., and R.T.; formal analysis, M.L. and R.R.; investigation, M.L. and R.R.; data curation, M.R.L., V.S., M.L and A.S.S.; writing—original draft preparation, M.R.L. C.M.M., and V.S.; writing—review and editing M.R.L., V.S., and R.T. project administration, V.S. and M.R.L. All authors have read and agreed to the published version of the manuscript.

Funding: This research received no external funding.

Conflicts of Interest: No potential conflict of interest was reported by the authors.

References

1. Gastaldo, P.; Barberis, G.; Fossati, F. Le piante della medicina tradizionale nei dintorni di Praglia (Appennino Ligure-Piemontese *Atti. Dell'accad. Ligur. Sci. Lett.* **1978**, *35*, 1–35.
2. Bellomaria, B.; Lattanzi, E. Le piante del territorio di Cupra Marittima (Marche) attualmente usate nella medicina popolare. *Arch Bot. Biogeogr. Ital.* **1982**, *58*, 1–19.
3. Guarrera, P.M. Usi tradizionali delle piante in alcune aree marchigiane. *Inf. Bot. Ital.* **1990**, *22*, 155–167.
4. Ghirardini, M.P.; Carli, M.; Del Vecchio, N.; Rovati, A.; Cova, O.; Valigi, F.; Agnetti, G.; Macconi, M.; Adamo, D.; Traina, M.; et a The importance of a taste. A comparative study on wild food plant consumption in twenty-one local communities in Italy. *Ethnobiol. Ethnomed.* **2007**, *3*, 22–33. [CrossRef] [PubMed]
5. Cornara, L.; La Rocca, A.; Marsili, S.; Mariotti, M.G. Traditional uses of plants in the Eastern Riviera (Liguria, Italy). *Ethnopharmacol.* **2009**, *125*, 16–30. [CrossRef] [PubMed]
6. Lentini, F.; Venza, F. Wild food plants of popular use in Sicily. *J. Ethnobiol. Ethnomed.* **2007**, *3*, 15. [CrossRef] [PubMed]
7. Giambanelli, E.; D'antuono, L.; Ferioli, F.; Frenich, A.; Romero-González, R. Sesquiterpene lactones and inositol 4 hydroxyphenylacetic acid derivatives in wild edible leafy vegetables from Central Italy. *J. Food Comp. Anal.* **2018**, *72* 1–6. [CrossRef]
8. Maruca, G.; Spampinato, G.; Turiano, D.; Laghetti, G.; Musarella, C.M. Ethnobotanical notes about medicinal and useful plants o the Reventino Massif tradition (Calabria region, Southern Italy). *Genet. Resour. Crop. Evol.* **2019**, *66*, 1027–1040. [CrossRef]
9. Brullo, S.; Minissale, P.; Siracusa, G.; Spampinato, G. Considerazioni fitogeografiche su *Hyoseris taurina* (Pamp.) Martinol (Asteraceae). *Giorn. Bot. Ital.* **1990**, *124*, 104.
10. Pignatti, S.; Guarino, R.; La Rosa, M. (Eds.) *Flora d'Italia*, 2nd ed.; Edagricole: Milano, Italy, 2018; Volume 3.
11. Guarrera, P.M.; Savo, V. Perceived health properties of wild and cultivated food plants in local and popular traditions of Italy A review. *J. Ethnopharmacol.* **2013**, *146*, 659–680. [CrossRef]
12. Guarrera, P.M.; Savo, V. Wild food plants used in traditional vegetable mixtures in Italy. *J. Ethnopharmacol.* **2016**, *5*, 202–234 [CrossRef]
13. Furukawa, S.; Fujita, T.; Shimabukuro, M.; Iwaki, M.; Yamada, Y.; Nakajima, Y.; Nakayama, O.; Makishima, M.; Matsuda, M. Shimomura, I. Increased oxidative stress in obesity and its impact on metabolic syndrome. *J. Clin. Investig.* **2004**, *114*, 1752–176 [CrossRef] [PubMed]
14. Ampofo, A.G.; Boateng, E.B. Beyond 2020: Modelling obesity and diabetes prevalence. *Diabetes Res. Clin. Pract.* **2020**, *167* [CrossRef] [PubMed]
15. Giacco, F.; Brownlee, M. Oxidative stress and diabetic complications. *Circ. Res.* **2010**, *107*, 1058–1070. [CrossRef]
16. Găman, M.; Epingeac, M.; Diaconu, C.; Găman, A. Oxidative stress levels are increased in type 2 diabetes mellitus and obesity *J. Hypertens.* **2019**, *37*, e265. [CrossRef]
17. Lucci, P.; Pacetti, D.; Loizzo, M.R.; Frega, N.G. Canning: Impact on food products quality attributes. In *Food Processing Technologies Impact on Product Attributes*; Jaiswal Amit, K., Ed.; Taylor & Francis Group; CRC Press: Boca Raton, FL, USA, 2016; Chapter 3 pp. 41–60, ISBN 9781315372365.
18. Lin, C.H.; Chang, C.Y. Textural change and antioxidant properties of broccoli under different cooking treatments. *Food Chem* **2005**, *90*, 9–15. [CrossRef]
19. Musarella, C.M.; Paglianiti, I.; Cano-Ortiz, A.; Spampinato, G. Indagine etnobotanica nel territorio del Poro e delle Preserre Calabresi (Vibo Valentia, S-Italia). *Atti. Soc. Tosc. Sci. Nat. Mem. Ser. B* **2019**, *126*, 13–28.
20. Sicari, V.; Loizzo, M.R.; Tundis, R.; Mincione, A. Pellicanò, *Portulaca oleracea* L. (Purslane) extracts display antioxidant and hypoglycaemic effects. *J. Appl. Bot. Food Qual.* **2018**, *91*, 39–46.
21. Loizzo, M.R.; Tundis, R.; Sut, S.; Dall'acqua, S.; Ilardi, V.; Leporini, M.; Falco, T.; Sicari, V.; Bruno, M. High-Performance Liquid Chromatography/Electrospray Ionization Tandem Mass Spectrometry (HPLC-ESI-MSn) analysis and bioactivity useful fo prevention of "diabesity" of *Allium commutatum* Guss. *Plant Foods Hum. Nutr.* **2020**, *75*, 124–130. [CrossRef]
22. Fish, W.W.; Perkins-Veazie, P.; Collins, J.K. A quantitative assay for lycopene that utilizes reduced volumes of organic solvents *J. Food Compos. Anal.* **2002**, *15*, 309–317. [CrossRef]

23. Lichtenthaler, H.K.; Buschmann, C. *Current Protocols in Food Analytical Chemistry*; John Wiley and Sons, Inc.: Hoboken, NJ, USA, 2001. Available online: http://onlinelibrary.wiley.com (accessed on 16 December 2020).
24. Loizzo, M.R.; Sicari, V.; Pellicanò, T.; Xiao, J.; Poiana, M.; Tundis, R. Comparative analysis of chemical composition, antioxidant and anti-proliferative activities of Italian *Vitis vinifera* by-products for a sustainable agro-industry. *Food Chem. Toxicol.* **2019**, *127*, 127–134. [CrossRef] [PubMed]
25. El-Shiekh, R.A.; Al-Mahdy, D.A.; Hifnawy, M.S.; Abdel-Sattar, E.A. *In-vitro* screening of selected traditional medicinal plants for their anti-obesity and anti-oxidant activities. *S. Afr. J. Bot.* **2019**, *123*, 43–50. [CrossRef]
26. Parmar, I.; Vasantha Rupasinghe, H.P. Antioxidant Capacity and Anti-diabetic Activity of Wild Berry Stem Infusions. *Eur. J. Med. Plant.* **2015**, *8*, 11–28. [CrossRef]
27. Loranty, A.; Rembiałkowska, E.; Rosa, E.A.S.; Bennet, R.S. Identification, quantification and availability of carotenoids and chlorophylls in fruit, herb and medicinal teas. *J. Food Comp. Anal.* **2010**, *23*, 432–441. [CrossRef]
28. Suzuki, Y.; Shioi, Y. Identification of chlorophylls and carotenoids in major teas by high-performance liquid chromatography with photodiode array detection. *J. Agric. Food Chem.* **2003**, *51*, 5307–5314. [CrossRef]
29. Kao, F.J.; Chiu, Y.S.; Chiang, W.D. Effect of water cooking on the antioxidant capacity of carotenoid-rich vegetables in Taiwan. *J. Food Drug Anal.* **2014**, *22*, 202–209. [CrossRef]
30. Prasanna, K.D.; Gunathilake, P.; Somathilaka Ranaweera, K.K.D.; Vasantha Rupasinghe, H.P. Effect of Different Cooking Methods on Polyphenols, Carotenoids and Antioxidant Activities of Selected Edible Leaves. *Antioxidants* **2018**, *7*, 117.
31. Amin, I.; Zamaliah, M.; Marjan, C.; Foong, W. Total antioxidant activity and phenolic content in selected vegetables. *Food Chem.* **2004**, *87*, 581–586.
32. Amin, I.; Lee, W.L. Effect of different blanching times on antioxidant properties in selected cruciferous vegetables. *J. Sci. Food Agric.* **2005**, *85*, 2314–2320. [CrossRef]
33. Bamidele, O.P.; Fasogbon, M.B.; Adebowale, O.J.; Adeyanju, A.A. Effect of Blanching Time on Total Phenolic, Antioxidant Activities and Mineral Content of Selected Green Leafy Vegetables. *Curr. J. Appl. Sci. Technol.* **2017**, *24*, 1–8. [CrossRef]
34. Hong-Wei, X.; Zhongli, P.; Li-Zhen, D.; El-Mashad, H.M.; Xu-Hai, Y.; Mujumdar, A.S.; Zhen-Jiang, G.; Qian, Z. Recent developments and trends in thermal blanching—A comprehensive review. *Inf. Process. Agric.* **2017**, *4*, 101–127.
35. Gawlik-Dziki, U. Effect of hydrothermal treatment on the antioxidant properties of broccoli (*Brassica oleracea* var. botrytis italica) florets. *Food Chem.* **2008**, *109*, 393–401. [CrossRef]
36. Sikora, E.; Cieslik, E.; Leszczynska, T.; Filipiak-Florkiewicz, A.; Pisulewski, P.M. The antioxidant activity of selected cruciferous vegetables subjected to aquathermal processing. *Food Chem.* **2008**, *107*, 55–59. [CrossRef]
37. Jaiswal, A.K. Blanching as a treatment process: Effect on polyphenols and antioxidant capacity of cabbage. In *Processing and Impact on Active Components in Food*, 5nd ed.; Abu-Ghannam, N., Amit, J., Eds.; Elsevier/Academic Press: London, UK, 2015; pp. 35–43.
38. Medina, E.; Romero, C.; García, P.; Brenes, M. Characterization of bioactive compounds in commercial olive leaf extracts, and olive leaves and their infusions. *Food Funct.* **2019**, *10*, 4716–4724. [CrossRef] [PubMed]
39. Wachtel-Galor, S.; Wong, K.W.; Benzie, I.F. The effect of cooking on Brassica vegetables. *Food Chem.* **2008**, *110*, 706–710. [CrossRef]
40. Yamaguchi, T.; Katsuda, M.; Oda, Y.; Terao, J.; Kanazawa, K.; Oshima, S.; Inakuma, T.; Ishiguro, Y.; Takamura, H.; Matoba, T. Influence of polyphenol and ascorbateoxidases during cooking process on the radical-scavenging activity of vegetables. *Food Sci. Technol. Res.* **2003**, *9*, 79–83. [CrossRef]
41. Senguttuvan, J.; Paulsamy, S.; Karthika, K. Phytochemical analysis and evaluation of leaf and root parts of the medicinal herb, *Hypochaeris radicata* L. for *in vitro* antioxidant activities. *Asian Pac. J. Trop. Biomed.* **2014**, *4*, S359–S367. [CrossRef]
42. Savo, V.; Salomone, F.; Mattoni, E.; Tofani, D.; Caneva, G. Traditional salads and soups with wild plants as a source of antioxidants: A comparative chemical analysis of five species growing in central Italy. *Evid. Based Compl. Alter. Med.* **2019**. [CrossRef]
43. Souilah, N.; Ullah, Z.; Bendif, H.; Medjroubi, K.; Hazmoune, T.; Hamel, T.; Ozturk, M., Nieto, G.; Akkal, S. Phenolic compounds from an algerian endemic species of *Hypochaeris laevigata* var. *hipponensis* and investigation of antioxidant activities. *Plants* **2020**, *9*, 514. [CrossRef]
44. Ko, Y.M.; Eom, T.K.; Song, S.K.; Jo, G.Y.; Kim, J.S. Tyrosinase and α-Glucosidase Inhibitory Activities and Antioxidant Effects of Extracts from Different Parts of *Hypochaeris radicata*. *Korean J. Med. Crop. Sci.* **2017**, *25*, 139–141. [CrossRef]
45. D'agostino, M.F.; Sanz, J.; Martínez-Castro, I.; Giuffrè, A.M.; Sicari, V.; Soria, A.C. Statistical analysis for improving data precision in the SPME GC–MS analysis of blackberry (*Rubus ulmifolius* Schott) volatiles. *Talanta* **2014**, *125*, 248–256. [CrossRef] [PubMed]

Article

Effect of Microwave Pretreatment of Seeds on the Quality and Antioxidant Capacity of Pomegranate Seed Oil

Tafadzwa Kaseke [1], Umezuruike Linus Opara [1,2,*] and Olaniyi Amos Fawole [3,*]

[1] Postharvest Technology Research Laboratory, Department of Food Science, Faculty of AgriSciences, Stellenbosch University, Stellenbosch 7602, South Africa; tafakaseqe@gmail.com

[2] Postharvest Technology Research Laboratory, Department of Horticultural Sciences, Faculty of AgriSciences, Stellenbosch University, Stellenbosch 7602, South Africa

[3] Postharvest Research Laboratory, Department of Botany and Plant Biotechnology, University of Johannesburg, Johannesburg 2006, South Africa

* Correspondence: opara@sun.ac.za (U.L.O.); olaniyif@uj.ac.za (O.A.F.)

Received: 19 August 2020; Accepted: 6 September 2020; Published: 14 September 2020

Abstract: Microwave pretreatment of oilseeds is a novel technique used to enhance oil nutraceutical properties. In this study, the effect of microwave pretreatment of seeds was investigated on pomegranate seed oil quality attributes including oil yield, yellowness index, refractive index, peroxide value, p-anisidine value, total oxidation value, conjugated dienes, total phenolic content, total carotenoids content, phytosterol composition, fatty acid composition, 2,2-diphenyl-1-picrylhydrazyl (DPPH) radical scavenging capacity, and ferric reducing antioxidant power (FRAP). The seeds of three different pomegranate cultivars ('Acco', 'Herskawitz', and 'Wonderful') were microwave heated at 261 W for 102 s. Pomegranate seeds microwave pretreatment enhanced oil yield, yellowness index, total carotenoids content, total phenolic content, FRAP and DPPH radical scavenging capacity, despite an increase in conjugated dienes, and peroxide value. Palmitic acid, oleic acid, linoleic acid, saturated, and monosaturated fatty acids were increased after pomegranate seeds microwave pretreatment, whilst the levels of punicic acid and β-sitosterol were reduced. Nevertheless, the refractive index, the ratio of unsaturated to saturated fatty acid of the extracted oil were not significantly ($p > 0.05$) affected by pomegranate seeds microwave pretreatment. Principal component analysis and agglomerative hierarchical clustering established that 'Acco' and 'Wonderful' oil extracts from microwave pretreated PS exhibited better oil yield, whilst 'Herskawitz' oil extracts showed higher total carotenoids content, total phenolic content, and antioxidant capacity.

Keywords: pomegranate seeds; oil; microwave pretreatment; total phenolic content; antioxidant capacity

---

1. Introduction

The demand for the use of natural products in preventing chronic and degenerative diseases has increased in recent decades, driven by increased consumer health awareness [1]. Attention has been given to functional foods that provide both nutritional functions and health benefits. Pomegranate fruit is rich in both nutritional and biological properties [2]. The fruit has been cultivated since ancient times throughout the Mediterranean region, mainly for its nutritional and pharmacological value [3]. Pomegranates have been used in the treatment of sore throats, coughs, ulcers, urinary infections, intestinal worms, digestive disorders, skin disorders, and arthritis for centuries [4]. Apart from being consumed as fresh fruits, pomegranates may be processed into various products such as juice, jam, wine, syrup, among other products. The fruit contains seeds that range between 37 and 143 g/kg of the total fruit weight depending on factors such as cultivar, growing region, growing conditions, and maturity stages [5,6]. Pomegranate seeds (PS) have oil that range between 12% to 20% (dry weight basis) and is a good source of bioactive compounds such as punicic acid, tocopherols, phenols, sterols, and carotenoids [3,7].

Epidemiological studies have revealed that pomegranate seed oil (PSO) has biological properties related to the prevention of microbial growth, lipoperoxidation, skin photoaging, cancer, diabetes, and obesity that are linked to the bioactive compounds [7,8]. In line with the biological activities, the oil can be used as a functional ingredient in nutraceutical, pharmaceutical, and functional foods preparations [9]. Despite the wealth of health benefits in PSO, the seeds are still regarded as waste after processing the fruits into juice and other products, thereby increasing the total pomegranate fruit postharvest losses. In some instances, the seeds are used as stock feed [10]. From an industrial and health perspective, valorization of the PS into oil presents a more valuable alternative utilization of the postharvest waste.

Among other conventional seed oil extraction techniques, such as cold pressing and supercritical carbon dioxide, extraction using organic solvents is the best technique with regards to oil extraction efficiency [11]. Needless to say, the use of organic solvents such as hexane has become unpopular due to its hazardous effects on humans and the environment [11,12]. Consequently, the avoidance or reduction in the use of hexane as an extraction solvent has become a requirement for the food industry [13]. Alternatively, short-chain alcohols such as ethanol are promising solvents in seed oil extraction. Ethanol is a less hazardous, bio-renewable, and cheaper organic solvent [14]. The high polarity of ethanol provides it with the ability to extract polar bioactive compounds, such as the phenolic compounds, and therefore oil extracted with ethanol has better biological activities [15]. The main drawback in using ethanol is that it produces low oil yield [16,17]. Therefore, treatment of the oilseeds before oil extraction is important for improving the oil extraction efficacy of ethanol.

The treatment of seeds with microwaves before oil extraction has received great interests due to obvious advantages including uniform energy delivery, high thermal conductivity to the interior of the material, energy saving, and precise process control [18]. The application of microwave radiation to seeds result in direct interaction of the electromagnetic waves with the polar oxygen group from the seeds moisture [19]. This results in rapid heating and evaporation of the moisture in the seeds, thereby creating an internal pressure that causes the rapture of seed matrices [20]. The seeds microstructure alterations facilitated by microwave heating increases the interaction of the extraction solvent with the intracellular materials and enhances the lipids and bioactive compounds mass transfer into the extraction solvent [21]. Zhang and Jin [22], Li et al. [23], Porto et al. [24], and Güneşer and Yilmaz [25] have reported an improvement in oil yield and bioactive compounds recovery after microwave pretreatment of the camellia oleifera, yellow horn, moringa, and orange seed, respectively.

In order to establish whether microwave pretreatment adds value or not to pomegranate seed oil, it is important to investigate different cultivars. Cultivar significantly influences the quality of seed oil from the perspective of genetic characteristics variation [26]. Therefore, the effect of microwave pretreatment on the quality of the extracted oil may also vary with cultivar. To optimize the economic benefits, cultivars with better oil quality after seeds microwave pretreatment are valuable to food processors. However, information about the application of microwave irradiation on seed from different cultivars to establish variation in the quality of the extracted oil is limited.

Therefore, this study aimed to investigate the effect of microwave pretreatment of seeds on the quality and antioxidant capacity of PS oil extracted from three pomegranate cultivars.

2. Materials and Methods

2.1. Experimental Material

Pomegranate fruits (cv. Wonderful, Herskawitz, Acco) free of quality defects were harvested at commercial maturity stage from a farm (33°48′0″ S, 19°53′0″ E) in Western Cape Province, South

Africa between February and April during the 2019 season. Pomegranate seeds (PS) extracted from the fruits were thoroughly cleaned before drying in an oven at 55 ± 2 °C for 24 h [27]. The dried seeds were stored at 4 ± 2 °C before use [28].

2.2. Sample Moisturizing

The moisture content of PS was measured in a moisture analyzer at 100 °C (DBS60-3, KERN, Balingen, Germany). The procedure of PS moisturizing was done following the method described by Rekas et al. [29]. PS (200 g) was sprayed with the pre-calculated amount of water, thoroughly mixed, sealed in zipped polyethylene bags, and equilibrated at 4 ± 2 °C for 48 h. This procedure was applied to moisturize the seeds to obtain a moisture content of 6% before microwave pretreatment. Water is a polar molecule and an important heat transfer medium during seeds microwave pretreatment and therefore PS moisturizing was vital [30]. A mass balance was used to calculate the amount of water to be added to the PS to obtain 6% moisture content. After 48 h of equilibration, the PS moisture content was checked in order to verify moisture homogeneity in the samples.

2.3. Microwave Pretreatment

2.3.1. Equipment Calibration

A 2450 MHz domestic microwave oven (Model: DMO 351, Defy Appliances, Cape Town, South Africa) with a nominal power of 900 W was used in the present study. The microwave power calibration was performed following the procedure described by Rekas et al. [29]. Briefly, 500 g of water was heated in a glass beaker (80 mm diameter) and the time of 10 ± 2 °C elevation of water temperature was measured. The procedure was done in triplicate. The microwave power absorbed by the water during the heating was calculated as:

$$W = m_w C_{pw} \frac{\Delta T}{\Delta t} \tag{1}$$

where W is the power absorbed by the water (W), m_w is the mass of water (kg), C_{pw} is the specific heat (J/C kg), ΔT is the difference in temperature (°C), and Δt is the time (s). The absorbed power by the water was 261 W for the applied 40% microwave power.

2.3.2. Pretreatment

Ground PS (30 g) of uniform particle size (<1 mm) were evenly spread in a glass petri dish (190 mm in diameter) inside the calibrated microwave. The seed powder was exposed to microwave irradiation at 2450 MHz and 261 W for 102 s. This condition was established in preliminary experiments using response surface methodology (RSM), which confirmed 261 W and 102 s as the optimum microwave conditions for higher oil yield and antioxidant activity (unpublished). The microwave treated PS powder was allowed to cool to ambient temperature and thoroughly mixed to ensure sample homogeneity. Each experiment was performed in triplicate.

2.4. Oil Extraction

An ultrasonic bath (Separation Scientific, Cape Town, South Africa) (700 W, 40 kHz and 25 L capacity) was used to extract the oil. The PS powder (20 g) was mixed with 100 mL ethanol in 500 mL plastic capped glass bottles. The samples were sonicated at 700 W, 40 ± 5 °C for 40 min before filtration through Whatman No. 1 filter paper and vacuum evaporation to recover the solvent (G3 Heidolph, Schwabach, Germany). Unmicrowaved PS powder was used as the control samples. Oil extractions were done twice on triplicated samples ($n = 3$). The yield of pomegranate seed oil (PSO) was calculated using Equation (2).

$$\text{PSO yield (\%)} = \frac{M_1}{M_2} \times 100 \tag{2}$$

where M_1 and M_2 are the mass of PSO and dry weight (dw) of the pomegranate seed powder, respectively. The extracted PSO samples were packed in brown bottles and stored at 4 ± 2 °C to minimize oxidation during analyses [31].

2.5. Pomegranate Seeds Microstructures Analysis

Scanning electron microscopy (SEM) studies assess changes in the PS morphology due to microwave treatment and were conducted using the field emission scanning electron microscope (FESEM) (Thermo Fisher Apreo, Hillsboro, OR, USA). The samples were mounted on aluminum stubs using a double-sided carbon tape before sputter-coating with a thin layer of gold (10 nm thick) using a gold sputter coater (EM ACE200, Leica, Wetzlar, Germany) to induce conductivity within the sample. A voltage of 2 kV was used to collect the images, which were recorded digitally.

2.6. Determination of PSO Quality Indices

2.6.1. Refractive and Yellowness Index

A calibrated Abbe 5 refractometer (Bellingham + Stanley, Kent, United Kingdom) was used to measure refractive index (RI) at ambient condition (25 °C). PSO colour properties including L* (lightness) and b* (yellowness) measured using a calibrated Chromameter CR-410 (Konica Minolta, INC, Tokyo, Japan) were used to calculate yellowness index (YI).

$$YI = \frac{142.86b^*}{L^*} \qquad (3)$$

2.6.2. Peroxide Value, Conjugated Dienes, p-Anisidine Value and Total Oxidation Value

PSO peroxide value (PV) was determined using the modified ferrous oxidation-xylenol orange (FOX) method [32]. Conjugated dienes (K232) and trienes (K270) were analyzed according to the standard [33]. The p-anisidine value (AV) was measured in accordance with [34]. Total oxidation (TOTOX) value was calculated from the PV and AV using the equation [35].

$$TOTOX = 2PV + AV \qquad (4)$$

2.7. Determination of Bioactive Compounds and Antioxidant Capacity

2.7.1. Total Carotenoids Content and Total Phenolic Content

Total carotenoids content (TCC) was measured following the method described by Ranjith et al. [36]. Briefly, PSO (0.2 g) was dissolved in hexane (5 mL) and 0.5 mL of 0.5% (*w/v*) sodium chloride (NaCl) was added. The mixture was vortexed and centrifuged (Centrifuge 5810R, Eppendorf, Germany) at 4000 rpm for 5 min. The absorbance of the supernatant was measured at 460 nm using a UV spectrophotometer (Helios Omega, Thermo Scientific, Waltham, MA, USA). The results were reported as mg β-carotene/100 g of PSO. Total phenolic content (TPC) was determined using the Folin–Ciocalteu method [37]. The reaction mixture contained 200 μL of PSO methanol extracts, 250 μL of the Folin–Ciocalteau reagent and 750 μL of 2% (*w/v*) sodium carbonate, and 3 mL of distilled water. The reaction mixtures were incubated in the dark for 40 min after which their absorbances were measured at 760 nm using a UV spectrophotometer (Helios Omega, Thermo Scientific, Waltham, MA, USA), and the final results were expressed as milligram gallic acid equivalent per g PSO (mg GAE/g PSO).

2.7.2. Phytosterol Composition

The phytosterol composition was determined following the method described by Fernandes et al. [6] with some modifications. PSO (100 mg) samples weighed in 15 mL glass vials were mixed with 2.5 mL of saponification reagent (94 mL of absolute ethanol, 6 mL of 33% (*w/v*) potassium hydroxide, 500 μL of 20% (*w/v*) ascorbic acid). A hundred microliters of 5α-Cholestane (1000 mg/L) in chloroform (internal standard) was added and the mixture vortexed before saponification in an oven at 60 °C for 1 h. After saponification, the samples were cooled in ice for 10 min, followed by the addition of 5 mL of distilled of water and 2 mL of chloroform. The mixture was vortexed before centrifugation at 3000 rpm for 4 min. The chloroform extracts (500 μL) were concentrated with a gentle stream of nitrogen to ± 200 μL. To 100 μL of the concentrated chloroform extracts, pyridine (100 μL), and N,O-Bis (trimethylsilyl) trifluoroacetamide (30 μL) were added, and the mixture was vortexed before derivatization at 100 °C for 1 h in an oven. The derivatized sterol fractions were analyzed using gas chromatography connected to mass spectrometry (GC-MS) (Thermo Scientific Co. Ltd., Milan, Italy). The samples were injected 100 °C and held for 2 min before they were heated to 250 °C at the speed of 7 °C/min. The temperature was maintained for 2 min. A split ratio of 5:1, and an injection volume of 1.0 μL were used. The flow rate of helium, the carrier gas was maintained at 1 mL/min. The detector was operated under electron impact mode at ionization energy of 70 eV, scanning between 40 and 650 m/z. For peak identification, a standard containing a mixture of sterols (β-sitosterol, stigmasterol and ergosterol) was used. Phytosterol compounds identification was done by comparing the retention times. The results were reported as mg/100 g of PSO.

2.7.3. Radical Scavenging Ability

PSO antiradical activity was evaluated using 2,2-Diphenyl-1-picryl hydrazyl (DPPH) assay [38]. Briefly, PSO methanol extracts (100 μL) were added to 2.5 mL of 0.0004% (*w/v*) freshly prepared DPPH in 80% (*v/v*) methanol. The mixture was vortexed before incubation in the dark for 60 min. The absorbance of the remaining DPPH was measured using a UV spectrophotometer (Helios Omega, Thermo Scientific, Waltham, MA, USA) at 517 nm. The absorbance of DPPH in 80% methanol was measured as the negative control. The final result was expressed as mmol Trolox/g of PSO.

2.7.4. Ferric Reducing Antioxidant Power

The ferric reducing antioxidant power (FRAP) of PSO methanol extracts was determined following the method described by Benzie and Strain [39]. Freshly prepared FRAP reagent consisting of 2.5 mL of 10 mM 2,4,6-Tri(2-pyridyl)-s-triazine (TPTZ) solution in 40 mM HCl, 2.5 mL of 20 mM $FeCl_3$ and 25 mL of 0.3 M acetate buffer, pH 3.6 was warmed at 37 °C for 10 min. In total, 40 microliters of PSO methanol extracts were mixed with 200 μL distilled water and 1.8 mL FRAP reagent. The samples were incubated at 37 °C for 30 min before the absorbances were measured at 593 nm using a UV spectrophotometer (Helios Omega, Thermo Scientific, Waltham, MA, USA). Trolox was used to prepare the standard curve (5-100 mM), and the final results were expressed as mmol Trolox/g of PSO.

2.8. Fatty Acid Composition

Gas chromatography–mass spectrometry (GC–MS) was used to determine the fatty acid composition of PSO following a procedure described in a previous study [40]. PSO (100 mg) was weighed into 15 mL glass vials after which 2.0 mL hexane, 50 μL heptadecanoic acid (1000 mg/L, internal standard), and 1.0 mL of 20% (*v/v*) H_2SO_4 in methanol were successively added. The samples were thoroughly vortexed before incubation at 80 °C for 1 h in an oven. To the cooled the samples, 3 mL of saturated NaCl was added, and the mixture vortexed and centrifuged at 3000 rpm for 3 min. The hexane extracts were transferred into vials for analysis with GC-MS (6890N, Agilent technologies network, Palo Alto, CA, USA) coupled to an Agilent technologies inert XL EI/CI Mass Selective Detector (MSD) (5975B, Agilent Technologies Inc., Palo Alto, CA, USA). Separation of the FAMEs was performed on a polar RT-2560 (100 m, 0.25 mm ID, 0.20 μm film thickness) (Restek, Bellefonte, Pennsylvania, USA) capillary column. Helium was used as the carrier gas at a flow rate of 0.017 mL/s. One microliter (1 μL) of the sample was injected in a split ratio of 10:1. The oven temperature was run as: 100 °C/min, 180 °C at 25 °C/min, and held for 3 min; 200 °C at 4 °C/min and held for 5 min; 280 °C at 8 °C/min, 310 °C at 10 °C/min, and held for 5 min. The PSO fatty acids profiles were identified using the NIST library.

Results were expressed as a percentage of the total and calculated by dividing the area peak of each fatty acid by the total area peaks of all the fatty acids.

2.9. Statistical Analysis

The results of all the studied variables are presented as mean ± SD (standard deviation). One-way analysis of variance (ANOVA) was performed to compare the means using Statistica software (Statistical v13, TIBC, Palo Alto, CA, USA) after which the means were separated using Duncan's multiple range test. Graphs were prepared using Microsoft Excel (Version: 16.0.13029.20344, Microsoft Cooperation, Washington, USA). The relationship between the PSO quality attributes and cultivars was determined by performing the principal component analysis (PCA) and agglomerative hierarchical clustering (AHC) and using Microsoft Excel software (XLSTAT 2019.4.1.63305, Addinsoft, New York, NY, USA).

3. Results and Discussion

3.1. Oil Yield and Seeds Microstructures

The results in Figure 1 shows that pretreating pomegranate seeds (PS) with microwaves significantly enhanced the oil yield between 10% and 14%. 'Acco' exhibited significantly higher oil yield (17.10%) (dw) than 'Wonderful' (15.77%) (dw) and 'Herskawitz' (13.10%) (dw) after PS microwave pretreatment, a phenomenon that can be explained by the differences in their genetic material [41]. Previously, Durdevic et al. [19] reported that microwave (100, 250 and 600 W for 2 and 6 min) pretreatment of PS could enhance oil yield between 23% and 32%. Compared to the current study, the difference in oil yield could be explained by variation in cultivar, seeds pretreatment conditions, oil extraction methods, and fruit growing region, among other factors. Depending on the concentration of cellulose and lignin, PS rheological properties such as hardness and toughness may vary with cultivar [42]. This could have affected the cultivars' response to microwave pretreatment and oil extraction. The scanning electron microscopy (SEM) images in Figure 2 confirmed that PS microwave pretreatment significantly deformed the cell walls. As shown in Figure 2b microwave pretreated PS were characterized by conspicuous perforations on the cell walls. Microwaves generate heat by interacting with polar substances and, therefore, water as a polar molecule is an essential heat transfer medium during seeds microwave pretreatment [30]. The heat energy causes a rapid increase in the seed temperature and vaporization of the water in the seeds creating an intracellular pressure that ruptures the oilseeds cell walls and membranes [20]. Figure 2c shows parenchymal cells from unmicrowaved PS with intact cell walls, which could have created a major resistance to solvent penetration into the seeds cells and could be the reason for the low oil yield observed from unmicrowaved PS [43]. On the other hand, Figure 2d shows extensively damaged PS parenchymal cells due to microwave treatment. Similar findings have been reported from microwave pretreatment of hazelnuts [44]. In addition to damaging the cell walls, microwave pretreatment could have deformed the lipoprotein membranes surrounding the individual lipid bodies [45]. These microstructural changes could have enhanced porosity of the PS cell walls and membranes that led to the improved efficiency of oil extraction with ethanol.

3.2. Refractive and Yellowness Index

Thermal treatment of the oilseeds could result in fatty acids conjugation and an increase in the oil refractive index [46]. Therefore, RI could be used as an indirect quality measure of oil. Neither cultivar nor PS microwave pretreatment significantly ($p > 0.05$) affected the RI of the oil extracts, despite the significant cell walls and membranes deformation (Table 1). The pomegranate seed oil RI values (1.5180–1.5181) in the current study were comparable to those reported by Costa et al. [47] (1.5091–1.5177) from cold pressed PSO further demonstrating that PS microwave pretreatment did not cause significant negative effect on the oil quality.

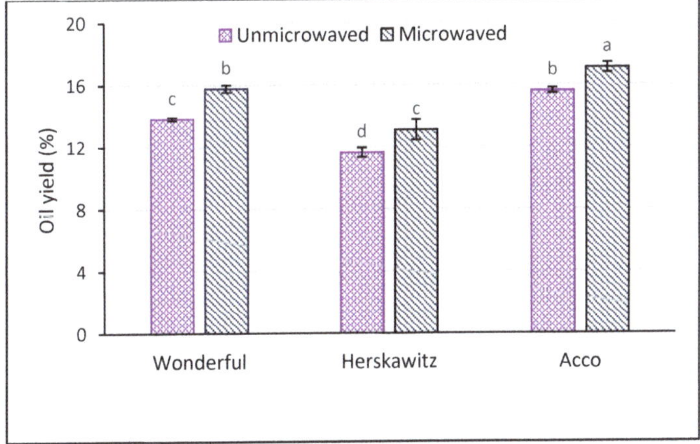

Figure 1. Oil yield from unmicrowaved and microwaved (261 W for 102 s) pomegranate seeds of three pomegranate cultivars. Within the same cultivar (unmicrowaved and microwave), columns followed by different letters are significantly different ($p < 0.05$) according to Duncan's multiple range test. Vertical bars indicate the standard deviation of the mean.

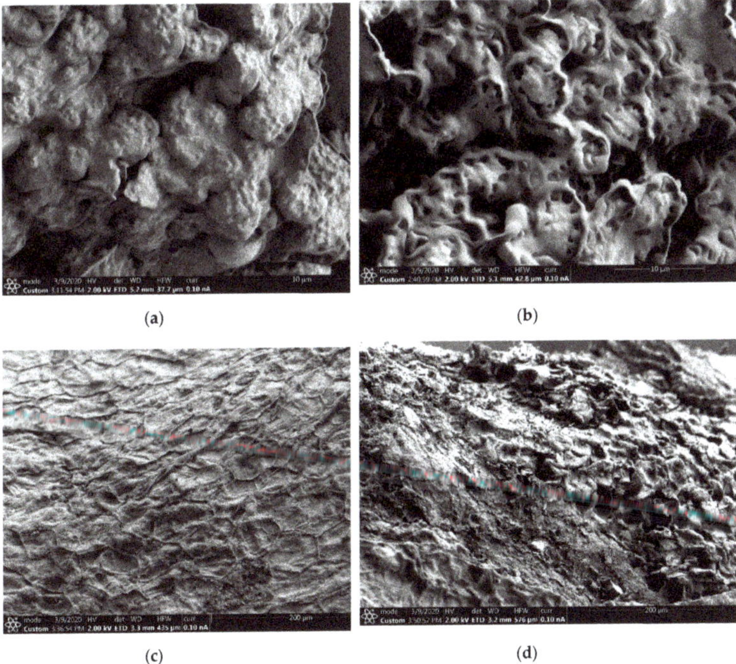

Figure 2. Representative scanning electron microscopy (SEM) micrographs show the effect of microwave pretreatment (261 W/102 s) on the pomegranate seeds microstructures. (**a**) Unmicrowaved pomegranate seeds, (**b**) microwaved pomegranate seeds, (**c**) parenchymal cells from unmicrowaved pomegranate seeds, and (**d**) parenchymal cells from microwaved pomegranate seed.

Table 1. Physicochemical characteristics of oil from unmicrowaved and microwave pretreated (261 W/102 s) pomegranate seeds of three pomegranate cultivars.

Cultivar	Treatment	RI	K232	K270	PV	AV	TOTOX
Wonderful	Unmicrowaved	1.5181 ± 0.00 [a]	0.22 ± 0.01 [b]	0.28 ± 0.005 [b]	0.04 ± 0.001 [d]	14.22 ± 0.58 [a]	14.30 ± 0.58 [a]
	Microwaved	1.5180 ± 0.00 [a]	0.28 ± 0.01 [a]	0.29 ± 0.009 [ab]	0.05 ± 0.005 [d]	12.50 ± 0.59 [a]	12.59 ± 0.59 [a]
Herskawitz	Unmicrowaved	1.5180 ± 0.00 [a]	0.30 ± 0.01 [a]	0.31 ± 0.008 [a]	0.17 ± 0.008 [e]	13.06 ± 0.34 [a]	13.40 ± 0.35 [a]
	Microwaved	1.5180 ± 0.00 [a]	0.19 ± 0.02 [b]	0.29 ± 0.016 [ab]	0.22 ± 0.011 [c]	12.90 ± 1.20 [a]	13.33 ± 1.19 [a]
Acco	Unmicrowaved	1.5180 ± 0.00 [a]	0.20 ± 0.01 [b]	0.31 ± 0.005 [a]	0.27 ± 0.005 [a]	2.00 ± 0.66 [c]	2.53 ± 0.65 [c]
	Microwaved	1.5180 ± 0.00 [a]	0.29 ± 0.01 [a]	0.31 ± 0.009 [ab]	0.35 ± 0.007 [b]	5.90 ± 1.15 [b]	6.60 ± 1.16 [b]

Values represent mean ± SD of triplicate determinations. Different superscript letters in the same column indicate statistical significance ($p < 0.05$) according to Duncan's multiple range test. RI= index (25 °C), PV = Peroxide value ($meqO_2$/kg PSO), $meqO_2$/kg = milli-equivalents of active oxygen per kg), AV = Anisidine value, TOTOX = Total oxidation value, RI = Refractive, K232 = Conjugated dienes, K270 = Conjugated triene.

Color is a valuable parameter that influences the consumer's preference and decision to purchase a food product. Yellowness index can be used to measure the influence of processing, including seeds microwave pretreatment on oil color [48]. The results in Figure 3 show that PS microwave pretreatment significantly improved the YI of 'Herskawitz' and 'Acco' oil extracts by 1.5 and 1.7 fold, respectively. The significant increase in the YI after PS microwave pretreatment could be ascribed to the improved extraction of the oil color pigments such as carotenoids as facilitated by the extensively damaged cell walls and membranes (Figure 2). In the study of Rekas et al. [49], YI significantly increased by 13% and 63% when dehulled rape seeds were microwaved at 800 W for 2 and 4 min, respectively. However, PS microwave pretreatment insignificantly ($p > 0.05$) changed the YI of 'Wonderful' oil extracts suggesting that the effect of pretreating PS with microwaves on oil color compounds differed among the cultivars.

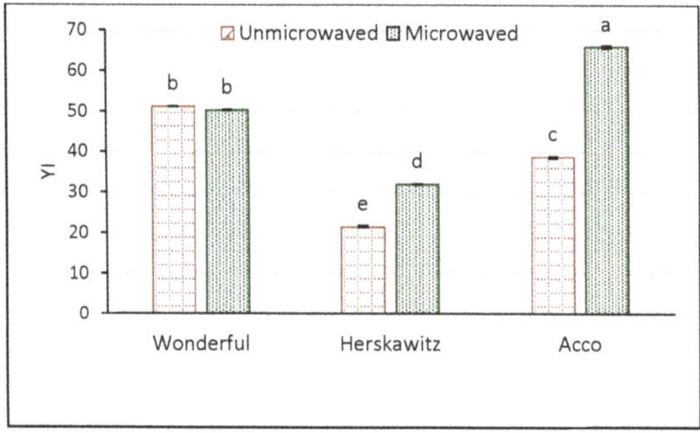

Figure 3. Yellowness index (YI) of pomegranate seed oil from unmicrowaved and microwave pretreated (261 W/102 s) seeds of three pomegranate cultivars. Within the same cultivar (unmicrowaved and microwaved), columns followed by different letters are significantly different ($p < 0.05$) according to Duncan's multiple range test. Vertical bars indicate the standard deviation of the mean.

3.3. Peroxide Value, Conjugated Dienes and Trienes, ρ-Anisidine Value and Total Oxidation Value

Peroxide value indicates the extent of fats and oils oxidation and therefore is one of the most widely used quality indicators in the food industry. As shown in Table 1, PS microwave pretreatment significantly increased the PV from 'Herskawitz' and 'Acco' oil extracts by 29% and 30%, respectively. The significant increase in PV could be explained by significant heat penetration into the seeds matrices during microwave pretreatment that could have induced lipid oxidation and hydroperoxides formation. Despite the significant increase in PV of 'Herskawitz' and 'Acco' oil extracts after seeds microwave pretreatment, the values (0.17–0.35 $meqO_2$/kg PSO) conformed to the Codex Alimentarius Commission standard for seed oil that permits a maximum of 15 $meqO_2$/kg in unrefined seed oils [50]. Microwave pretreatment of PS did not significantly oxidize 'Wonderful' oil extracts. Our PV results were lower than those reported by Basiri [51] (0.79 meqO$_2$/kg PSO) from petroleum ether extracted PSO, further demonstrating that microwave pretreatment of PS may not cause significant oil degradation.

During hydroperoxides formation, the non-conjugated double bonds of fatty acids may be converted to conjugated double bonds through isomerization [52]. Therefore, fatty acids conjugation can also be used as a quick indirect quality measure of oil. Conjugated dienes are part of fatty acid oxidation initial products. The results in Table 1, demonstrate that the level of K232 significantly increased in 'Wonderful' (27%) and 'Acco' (45%) oil extracts, whilst it significantly decreased in 'Herskawitz' (37%) oil extracts after PS microwave pretreatment. Although there was a significant increase in conjugated dienes in 'Wonderful' and 'Acco' oil extracts due to PS microwave pretreatment, the K232 values (0.19–0.30) were lower than the K232 values (4.15) reported by Amri et al. [53] from 'Tounsi' hexane PSO extracts, indicating that the oil from the present study was of higher quality.

A low PV is not the only marker for good oil quality because hydroperoxides are unstable and quickly decompose into secondary oxidation products. For this reason, the analysis of secondary products of seed oil oxidation is equally important. Unlike the K232 values, PS microwave pretreatment did not significantly influence the levels of K270 values in the oil from all the cultivars. p-Anisidine value measures the secondary products of fatty acids oxidation, such as aldehydes formed due to further hydroperoxides decomposition [52]. Despite the AV of oil extracted from 'Acco' significantly increasing by 3 fold after PS microwave pretreatment, the values (2.00–5.90) were lower than those from 'Wonderful' (14.22–12.50) and 'Herskawitz' (13.06–12.90) oil extracts that insignificantly changed after PS microwave pretreatment (Table 1). The observation that AV from 'Wonderful' and 'Herskawitz' oil extracts did not significantly change after PS microwave pretreatment suggests that there was minimum decomposition of hydroperoxides to form carbonyl compounds. In a previous study, Costa et al. [47] reported AV ranging from 13.8 to 18.6 from cold pressed PSO that were higher than the AV (5.90–14.22) results in the present study. High AV in freshly processed seed oil could indicate interference by other substances, leading to false positive overestimation of AV values.

Total oxidation value is a summation of the primary and secondary oxidation products and provides a better indication of fats and oils overall oxidative deterioration. As can be seen in Table 1, PS microwave pretreatment significantly increased the TOTOX value of 'Acco' oil extracts by 2.6 fold and this could be linked to either increased heat penetration into the seed matrices or increased lipolytic enzyme activity in the microwave damaged cells [54]. The result that the level of TOTOX value from 'Wonderful' and 'Herskawitz' oil extracts did not significantly change after PS microwave pretreatment suggests resistance to oxidation by the oil from the two cultivars, which could be attributed to enhanced total phenolic compounds [43].

3.4. Total Carotenoids Content, Total Phenolic Content, and Antioxidant Capacity

Epidemiological studies suggest that the consumption of carotenoid-rich foods such as seed oil is associated with the prevention of cancers, cardiovascular diseases, age-related cataracts, and immune system function improvement [55]. Therefore, the maximum extraction of these antioxidative compounds from plant materials such as seeds is important. The results in Table 2 indicate that PS microwave pretreatment significantly increased the total carotenoids content of 'Herskawitz' and 'Acco' oil extracts by 11% and 19%, respectively. Previously, Mazaheri et al. [56] also observed a significant improvement in carotenoids after microwave pretreatment of black cumin seeds. The extensive damage of the PS cell walls and membranes by microwave pretreatment could have increased the dissociation of carotenoids from the carotenoprotein complexes enhancing their mass transfer into the extraction solvent (Figure 2) [57]. On the other hand, the finding that TCC of oil extracted from 'Wonderful' did not significantly change after PS microwave pretreatment, whilst that of 'Herskawitz' and 'Acco' oil extracts significantly changed after seeds microwave pretreatment, indicating that the response of carotenoids compounds to PS microwave pretreatment was cultivar dependent.

Table 2. TCC, TPC and antioxidant capacity (DPPH, FRAP) of oil extracted from unmicrowaved and microwave pretreated (261 W/102 s) seeds of three pomegranate cultivars.

Cultivar	Treatment	TCC	TPC	DPPH	FRAP
Wonderful	Unmicrowaved	22.65 ± 0.96 [d]	1.67 ± 0.01 [c]	1.70 ± 0.05 [bc]	6.09 ± 1.44 [b]
	Microwaved	21.19 ± 1.81 [d]	2.09 ± 0.17 [b]	1.72 ± 0.02 [bc]	8.98 ± 0.41 [a]
Herskawitz	Unmicrowaved	30.27 ± 0.36 [b]	2.91 ± 0.11 [a]	1.66 ± 0.01 [c]	3.00 ± 0.17 [c]
	Microwaved	33.47 ± 0.43 [a]	3.12 ± 0.07 [a]	1.78 ± 0.01 [ab]	5.46 ± 0.90 [b]
Acco	Unmicrowaved	27.00 ± 0.96 [c]	2.05 ± 0.06 [c]	1.69 ± 0.03 [c]	1.95 ± 0.02 [c]
	Microwaved	32.08 ± 0.73 [ab]	2.39 ± 0.13 [b]	1.76 ± 0.02 [a]	1.80 ± 0.13 [c]

Values represent mean ± SD of triplicate determinations. Different superscript letters in the same column indicate statistical significance ($p < 0.05$) according to Duncan's multiple range test. TPC = Total phenolic content (mg GAE/g PSO, TCC = Total carotenoids content (mg β-carotene/100 g PSO), FRAP = Ferric reducing antioxidant power (mmol Trolox/g PSO), DPPH = 2,2-Diphenyl-1-picryl hydrazyl (mmol Trolox/g PSO), Trolox = 6-hydroxy-2,5,7,8-tetramethylchroman-2-carboxylic acid, PSO = Pomegranate seed oil, GAE = Gallic acid equivalence.

Phenolic compounds have been implicated in the anti-inflammatory and antioxidant properties of potential functional foods [26]. In this respect, polyphenol-rich foods intake may be associated with decreased risk of chronic diseases. PS microwave pretreatment significantly enhanced the total phenolic compounds of oil extracted from 'Wonderful' and 'Acco' by 25% and 17%, respectively, but did not significantly change the TPC of 'Herskawitz' oil extracts (Table 2). The results demonstrate that cultivar is an invaluable factor in PSO value addition. In addition, it has been reported that in plant materials phenolic compounds exist as either glycosylated, non-glycosylated, esterified, or free compounds, which could vary with cultivar and significantly influence their extraction [58]. The TPC (1.67–3.12 mg GAE/g PSO) results from the current study were higher than those reported by Pande and Akoh [59] (0.85–0.91 mg/g PSO) and Costa et al. [47] (0.00–0.17 mg/g PSO) from solvent extracted, and cold pressed PSO an indication that pomegranate cultivars from the current study could be valuable sources of phenolic compounds. Besides, factors such as cultivar, oil extraction technique, and fruit growing region could also be sources of variation in the TPC results among the studies.

Phytochemicals are complex; no single assay accurately reflects all antioxidants in a complex system such as seed oil. In this study, the antioxidant capacity of PSO was assessed using the DPPH and FRAP assays. PS microwave pretreatment significantly enhanced the DPPH radical scavenging capacity of 'Herskawitz' (7%) and 'Acco' (4%) oil extracts but did not significantly change the DPPH radical scavenging capacity of oil extracted from 'Wonderful'. FRAP significantly increased in the oil extracted from 'Wonderful' and 'Herskawitz' by 47 and 82%, respectively, after PS microwave pretreatment. In contrast to this finding, the reducing potential of 'Acco' oil extracts did not significantly change after PS microwave pretreatment. The finding that cultivar significantly influenced the antioxidant capacity of oil from microwave pretreated PS agrees with the results from Xi et al. [60], who reported significant variation in DPPH radical scavenging capacity and FRAP of seed oil from different lemon cultivars. The significant increase in the oil antioxidant capacity after PS microwave pretreatment, particularly from 'Herskawitz' could be related to the improved TCC and TPC. However, it is worth mentioning that PSO is also a good source of tocopherols, which have been reported to be associated with the oil antioxidant capacity in previous studies [61].

3.5. Phytosterol Composition

The ability of phytosterols to lower blood cholesterol may reduce the risk of coronary heart disease. Optimum extraction of these valuable compounds during seed oil processing is therefore essential to enhance the extracted oil health benefits. The effect of PS microwave pretreatment on phytosterol composition is presented in Figure 4. Three different phytosterols, including β-sitosterol (455.91–683.37 mg/100g PSO), stigmasterol (9.04–45.74 mg/100g PSO), and ergosterol (2.06–2.53 mg/100g PSO) were quantified in PSO from the studied cultivars. The levels of β-sitosterol and stigmasterol were consistent with the findings of Pande and Akoh [59] and Caligiani et al. [62] from PSO extracted using hexane and ethyl ether, respectively. In addition, the concentration of phytosterols was higher than those reported from other fruit seed oils such as apple, strawberry, and raspberry but comparable to those from sour cherry [63,64]. The level of β-sitosterol significantly decreased in 'Acco' and 'Wonderful' oil extracts by 26% and 29%, respectively, after PS microwave pretreatment, whilst it did not significantly change in 'Herskawitz' oil extracts, regardless of the PS cell walls and membranes extensive damage by microwave pretreatment (Figure 2). The finding that β-sitosterol significantly decreased in 'Wonderful' and 'Acco' oil extracts after PS microwave pretreatment suggests that the applied microwave pretreatment conditions thermally degraded this low-density lipoprotein (LDL) reducing phytosterol. Likewise, the levels of stigmasterol and ergosterol significantly decreased between 8 and 13% in 'Wonderful' and 'Herskawitz' oil extracts after treating the seeds with microwaves. Unlike in the present study, Azadmard-Damirchi et al. [21] and Fathi-Achachlouei et al. [43] reported significant improvement in phytosterols from microwave pretreated rape and milk thistle seed, respectively. The dissimilarity of our results with previous studies indicates that microwave pretreatment conditions are seed specific.

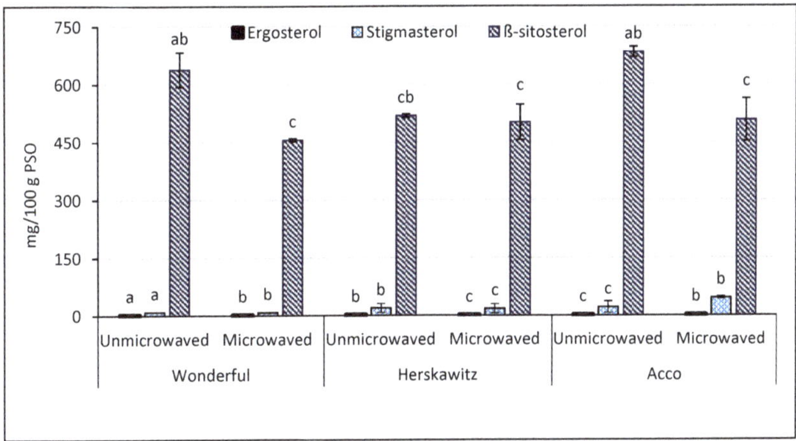

Figure 4. Phytosterol composition of oil extracted from unmicrowaved and microwave pretreated (261 W/102 s) seeds of three pomegranate cultivars. Within the same cultivar (unmicrowaved and microwaved), columns representing the same phytosterol and followed by different letters are significantly different ($p < 0.05$) according to Duncan's multiple range test. Vertical bars indicate the standard deviation of the mean.

On the other hand, microwave pretreatment of PS significantly improved the concentration of ergosterol and stigmasterol (9 and 111%, respectively) in 'Acco' oil extracts. Naturally, phytosterols exist as free compounds or conjugates in which they are either esterified to fatty acids or glycosylated with sugars [65]. The form in which they exist may therefore influence their dissociation and isolation from the seed matrix.

3.6. Fatty Acid Composition

The GC chromatogram shows that the primary fatty acids identified in PSO from the studied cultivars were palmitic acid, stearic acid, oleic acid, linoleic acid and punicic acid (Figure 5), which accounted for 5.64–7.74%, 2.34–3.08%, 7.43–9.62%, 11.59–16.54%, and 62.75–70.51%, respectively. The fatty acid composition was comparable to the findings of Tian et al. [66] and Aruna et al. [67]. However, the values of punicic acid were lower when compared with the findings of Khoddami et al. [5] and Fernandes et al. [4], which could be attributed to differences in fruit ripening index, seed oil extraction method, cultivar, and geographical location, among other factors. As can be seen in Table 3, PS microwave pretreatment significantly increased palmitic acid between 6% and 20%. Moreover, treatment of PS with microwaves significantly increased the stearic acid from 'Acco' oil extracts by 7%. Among the saturated fatty acids, stearic acid has unique properties, as it has been associated with a decrease in LDL cholesterol, cancer, and atherosclerosis risk [68]. The levels of stearic acid did not significantly change in 'Wonderful' and 'Herskawitz' oil extracts after PS microwave irradiation, which was comparable with the results reported by Durdevic et al. [69]. Oleic acid, the main monosaturated fatty acid in PSO significantly improved in 'Herskawitz' and 'Acco' oil extracts by 9% and 10%, respectively, after seeds microwave pretreatment. Improvement in oleic acid after seeds microwave pretreatment is desirable in oil oxidative stability as monosaturated fatty acids are less susceptible to oxidation. More so, PS microwave pretreatment significantly increased linoleic acid from 'Herskawitz' (37%) and 'Acco' (12%) oil extracts. The finding that PS microwave pretreatment enhanced the oil oleic acid and linoleic acid is essential to human health since oleic acid is associated with lowering low density lipoprotein (LDL) blood cholesterol, and linoleic acid has an important role in balancing fatty acid content proportions in body cells [20]. However, the concentration of punicic acid, the primary bioactive lipid with several biological properties significantly decreased in 'Herskawitz' and 'Acco' by 10% and 5%, respectively (Table 3). PSO health benefits potential is mostly attributed to punicic acid and therefore, its decrease after PS microwave pretreatment was not desirable. Arachidic acid significantly increased in 'Herskawitz' (13%) oil extracts, whilst it significantly decreased in 'Acco' (14%) oil extracts. Saturated fatty acids (SFA) significantly increased in 'Acco' (11%) and 'Herskawitz' (15%) oil extracts after PS microwave pretreatment. However, the polyunsaturated fatty acids (PUFA) and the ratio of unsaturated to saturated fatty acid (UFA: SFA) in 'Herskawitz' oil extracts decreased by 3% and 4%, respectively, after PS microwave pretreatment, indicating a loss in nutritional quality despite the significant increase in the antioxidant capacity (Tables 1 and 2). The decrease in the ratio of UFA: SFA could be attributed to the decline in punicic acid, the main polyunsaturated fatty acid in PSO. The amount of monosaturated fatty acids (MUFA) was insignificantly changed after microwave pretreatment of PS from all the cultivars. The levels of PUFA and UFA: SFA ratio in oil extracted from 'Acco' were not significantly affected by PS microwave pretreatment. Except for palmitic acid, PS microwave pretreatment had no significant effect ($p > 0.05$) on the fatty acid content of oil extracted from 'Wonderful'. Insignificant effect of seeds microwave pretreatment on the oil fatty acids content has also been reported in prior researches. For example, Wroniak et al. [20] and Guneser and Yilmaz [25] observed no significant change in the fatty acids content of oil extracted from orange and rape seeds, respectively, after microwave pretreatment.

3.7. Principal Component Analysis and Agglomerative Hierarchical Clustering Analysis

Principal component analysis (PCA) and agglomerative hierarchical clustering (AHC) were performed in order to provide an overview of the relationship between pomegranate cultivars, seeds microwave pretreatment, and the oil quality attributes. According to Kaiser's rule, only eigenvalues greater than 1 are considered significant descriptors of data variance [70]. The first two factors with the highest eigenvalues (F1 = 5.0, F2 = 3.1) accounted for 62.27% (F1: 38.10% and F2: 24.10, respectively) of the total variance in the original data and were considered more important (Figure 6). The first factor (F1), which was contributed by 'Acco' and 'Wonderful' oil extracts, was positively correlated with oil yield and PV, but negatively correlated with RI, AV, TOTOX, punicic acid, and FRAP. This points out that cultivars, which exhibited higher oil yield after seeds microwave pretreatment such as 'Acco', were associated with low FRAP. Although extensive damage of the PS cell walls and membranes by microwave pretreatment facilitated increased extraction of lipids, it could have exposed the oil to thermal degradation (Figure 2). The second factor (F2) that was contributed by 'Herskawitz' oil extracts from microwave pretreated seeds and 'Acco' oil extracts from unmicrowaved seeds was positively correlated with TCC, TPC, and DPPH radical scavenging capacity, but negatively correlated with β-sitosterol and YI. As shown in Table 2. microwave pretreatment of 'Herskawitz' PS improved the extraction of TCC and TPC that could have enhanced the oil DPPH radical scavenging capacity. The agglomerative hierarchical clustering (AHC) of PSO extracts from the different cultivars clustered the 'Acco' and 'Wonderful' oil extracts from microwaved seeds together that were correlated with higher oil yield and YI (Figure 7). These results concurred with the PCA analysis results. Like the PCA results, AHC separated 'Herskawitz' oil extracts from 'Acco' and 'Wonderful', illustrating that 'Herskawitz' was higher in oil quality attributes that were lower in 'Acco' and 'Wonderful', such as TCC, TPC, and DPPH radical scavenging capacity. Although microwave pretreatment of 'Acco' and 'Wonderful' PS may enhance oil yield, It was established that it also increases the oil oxidative degradation. On the other hand, microwave pretreatment of 'Herskawitz' PS may not produce oil yield results comparable to 'Wonderful' and 'Acco', but the oil has better bioactive compounds and antioxidant capacity.

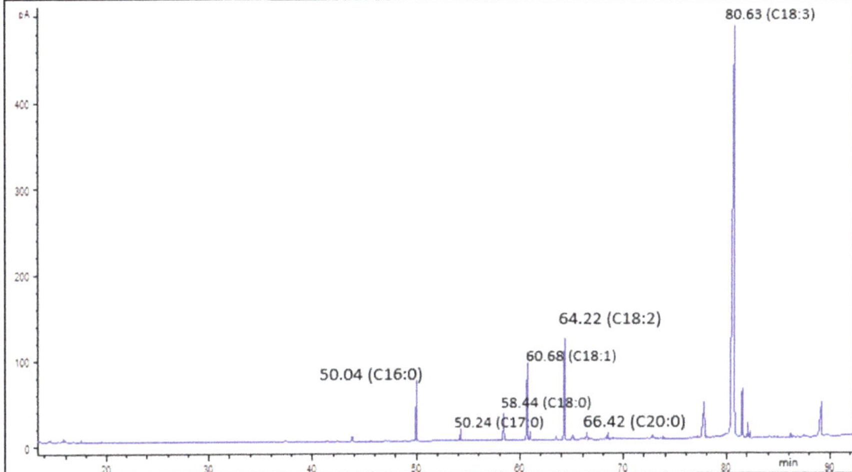

Figure 5. *A* typical gas chromatography-mass spectrometry (GC-MS) chromatograph of the major fatty acids identified in pomegranate seed oil and their retention times. C16:0 = palmitic acid, C17:0 = heptadecanoic acid (internal standard), C18:0 = stearic acid, C18:1 = oleic acid, C18:2 = linoleic acid, C18:3 = punicic acid, C20:0 = arachidic acid.

Table 3. Fatty acid composition (% relative area) of pomegranate seed oil from unmicrowaved and microwave pretreated (261 W/102 s) seeds of three pomegranate cultivars.

Fatty Acid	Cultivar/Treatment					
	Wonderful		Herskawitz		Acco	
	Unmicrowaved	Microwaved	Unmicrowaved	Microwaved	Unmicrowaved	Microwaved
Palmitic acid (C16:0)	5.64 ± 0.14 [c]	5.98 ± 0.17 [bc]	5.66 ± 0.35 [c]	6.82 ± 0.53 [ab]	6.72 ± 0.16 [b]	7.74 ± 0.27 [a]
Stearic acid (C18:0)	2.50 ± 0.08 [c]	2.49 ± 0.09 [bc]	2.34 ± 0.11 [ab]	2.35 ± 0.08 [ab]	2.87 ± 0.03 [b]	3.08 ± 0.02 [a]
Oleic acid (C18:1)	8.04 ± 0.47 [c]	8.59 ± 0.16 [bc]	7.43 ± 0.30 [c]	8.11 ± 0.40 [ab]	8.75 ± 0.12 [b]	9.62 ± 0.17 [a]
Linoleic acid (C18:2)	11.59 ± 0.23 [c]	11.62 ± 0.53 [bc]	12.09 ± 1.25 [c]	16.54 ± 1.53 [ab]	12.86 ± 0.42 [b]	14.35 ± 0.95 [a]
Punicic acid (C18:3)	68.95 ± 0.63 [c]	68.99 ± 0.71 [bc]	70.51 ± 1.96 [c]	63.55 ± 2.84 [ab]	66.30 ± 0.58 [b]	62.75 ± 1.84 [a]
Arachidic acid (C20:0)	0.45 ± 0.03 [c]	0.54 ± 0.01 [bc]	0.53 ± 0.02 [c]	0.60 ± 0.04 [ab]	0.88 ± 0.19 [b]	0.76 ± 0.02 [a]
SFA	8.59 ± 0.24 [cd]	9.01 ± 0.27 [cd]	8.53 ± 0.30 [d]	9.77 ± 0.63 [bc]	10.47 ± 0.36 [ab]	11.58 ± 0.30 [a]
MUFA	8.04 ± 0.47 [bc]	8.59 ± 0.16 [b]	7.43 ± 0.30 [c]	8.11 ± 0.40 [bc]	8.75 ± 0.12 [ab]	9.62 ± 0.17 [a]
PUFA	80.53 ± 0.43 [ab]	80.61 ± 0.19 [ab]	82.60 ± 0.70 [a]	80.09 ± 1.31 [b]	79.16 ± 0.17 [bc]	77.09 ± 0.90 [c]
UFA/SFA ratio	17.43 ± 0.25 [a]	17.55 ± 0.14 [a]	17.15 ± 0.15 [a]	16.40 ± 0.27 [b]	16.33 ± 0.14 [b]	16.29 ± 0.08 [b]

Values represent mean ± SD of triplicate determinations. Different superscript letters in the same row indicate statistical significance ($p < 0.05$) according to Duncan's multiple range test. SFA = Saturated fatty acid, MUFA = Monounsaturated fatty acid, PUFA = Polyunsaturated fatty acid, UFA = Unsaturated fatty acid.

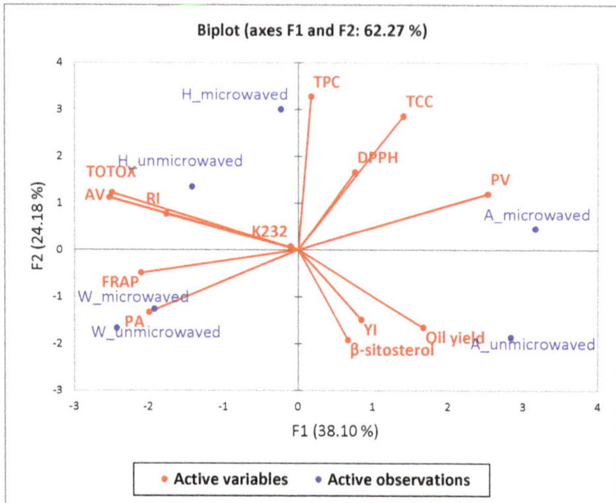

Figure 6. Principal component analysis data of pomegranate seed oil (PSO) quality attributes from unmicrowaved and microwaved (261 W/102 s) pomegranate seeds of three pomegranate cultivars. A = 'Acco', H = 'Herskawitz', W = 'Wonderful', AV = ρ-anisidine value, TOTOX = Total oxidation value, K232 = Conjugated dienes, RI = Refractive index, TPC = Total phenolic content, TCC = Total carotenoids content, PA = punicic acid, FRAP = Ferric reducing antioxidant power, DPPH = 2,2-diphenyl-1-picryl hydrazyl.

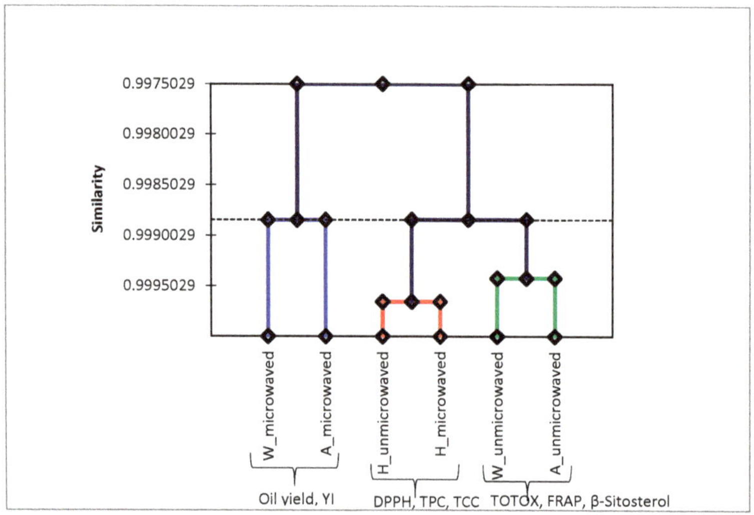

Figure 7. Agglomerative hierarchical clustering (AHC) of PSO extracts from unmicrowaved and microwaved (261 W/102 s) seeds. A = 'Acco', H = 'Herskawitz', W = 'Wonderful', TOTOX = Total oxidation value, TPC = Total phenolic content, TCC = Total carotenoids content, DPPH = 2,2-diphenyl-1-picryl hydrazyl, FRAP = Ferric reducing antioxidant power.

4. Conclusions

The present study established that PSO quality may be enhanced by seeds during microwave pretreatment, although oil quality varies with cultivar. Microwave pretreatment of PS improved oil yield, YI, TCC, TPC, DPPH radical scavenging capacity, and FRAP. This is a desirable development to the food industry given the increasing consumers' demand for natural and healthier foods. Moreover, the enhancement of bioactive compounds and antioxidant capacity after PS microwave pretreatment is valuable for the oil oxidative stability and storability. Despite an increase in K232 and PV, PS microwave pretreatment slightly decreased the oil TOTOX value.

Regarding fatty acid composition, PS microwave pretreatment increased palmitic acid, oleic acid, linoleic acid, SFA, and MUFA, but reduced the level of punicic acid. Pretreating PS with microwaves did not significantly affect the RI, PUFA, ratio of UFA: SFA, and phytosterol composition of the extracted oils. According to the PCA and AHC, 'Acco' and 'Wonderful' oil extracts from microwave pretreated PS exhibited better oil yield, whilst 'Herskawitz' oil extracts showed higher TCC, TPC, and DPPH radical scavenging valuable in functional foods formulation. In conclusion, 'Herskawitz' is a desirable cultivar for exploitation in nutraceutical and functional foods formulations.

Author Contributions: Conceptualization, O.A.F. and U.L.O.; Formal analysis, T.K.; Funding acquisition, O.A.F. and U.L.O.; Investigation, T.K.; Methodology, T.K.; Supervision, O.A.F. and U.L.O.; Validation, O.A.F. and U.L.O.; Visualization, T.K.; Writing-original draft, T.K.; Writing-review and editing, T.K.; O.A.F. and U.L.O. All authors have read and agreed to the published version of the manuscript.

Funding: This research was funded by the National Research Foundation of South Africa (grant number 64813) and APC was partly funded by Stellenbosch University.

Acknowledgments: This work is based on the research supported wholly or in part by the National Research Foundation of South Africa (Grant Numbers: 64813). The opinions, findings and conclusions or recommendations expressed are those of the author(s) alone, and the NRF accepts no liability what so ever in this regard.

Conflicts of Interest: The authors declare that they have no conflict of interest.

References

1. Jimenez-Monreal, A.M.; Garcia-Diz, L.; Martinez-Tome, M.; Mariscal, M.; Murcia, M.A. Influence of cooking methods on antioxidant activity of vegetables. *J. Food Sci.* **2009**, *74*, 97–103. [CrossRef] [PubMed]

2. Fawole, O.A.; Opara, U.L. Developmental changes in maturity indices of pomegranate fruit: A descriptive review. *Sci. Hortic.* **2013**, *159*, 152–161. [CrossRef]

3. Khoddami, A.; Bin, Y.; Man, C.; Roberts, T.H. Physicochemical properties and fatty acid profile of seed oils from pomegranate (*Punica granatum* L.) extracted by cold pressing. *Eur. J. Lipid Sci. Technol.* **2014**, *116*, 553–562. [CrossRef]

4. Ismail, T.; Sestili, P.; Akhtar, S. Pomegranate peel and fruit extracts: A review of potential anti-inflammatory and anti-infective effects. *J. Ethnopharmacol.* **2012**, *143*, 397–405. [CrossRef] [PubMed]

5. Eikani, M.H.; Golmohammad, F.; Saied, S. Extraction of pomegranate (*Punica granatum* L.) seed oil using superheated hexane. *Food Bioprod. Process.* **2012**, *90*, 32–36. [CrossRef]

6. Fernandes, L.; Pereira, J.A.; Lopez-Cortes, I.; Salazar, D.M.; Ramalhosa, E.; Casal, S. Fatty acid, vitamin E and sterols composition of seed oils from nine different pomegranate (*Punica granatum* L.) cultivars grown in Spain. *J. Food Compos. Anal.* **2015**, *39*, 13–22. [CrossRef]

7. Aruna, P.; Venkataramanamma, D.; Singh, A.K.; Singh, R.P. Health benefits of punicic acid: A review. *Compr. Rev. Food Sci. Food Saf.* **2016**, *15*, 16–27. [CrossRef]

8. Lansky, E.P.; Newman, R.A. Punica granatum (Pomegranate) and its potential for prevention and treatment of inflammation and cancer. *J. Ethnopharmacol.* **2007**, *109*, 177–206. [CrossRef]

9. Kalamara, E.; Goula, A.M.; Adamopoulos, K.G. An integrated process for utilization of pomegranate wastes-seeds. *Innov. Food Sci. Emerg.* **2015**, *27*, 144–153. [CrossRef]

10. Qu, W.; Pan, Z.; Ma, H. Extraction modeling and activities of antioxidants from pomegranate marc. *J. Food Eng.* **2010**, *99*, 16–23. [CrossRef]

11. Talekar, S.; Patti, A.F.; Singh, R.; Vijayraghavan, R.; Arora, A. From waste to wealth: High recovery of nutraceuticals from pomegranate seed waste using a green extraction process. *Ind. Crop. Prod.* **2018**, *112*, 790–802. [CrossRef]

12. Wang, L.; Weller, C.L. Recent advances in extraction of nutraceuticals from plants. *Trends Food Sci. Technol.* **2006**, *17*, 300–312. [CrossRef]

13. Sbihi, H.M.; Nehdi, I.A.; Mokbli, S.; Romdhani-Yeounes, M.; Al-Resayes, S.I. Hexane and ethanol extracted seed oils and leaf essential compositions from two castor plant (*Ricinus communis* L.) varieties. *Ind. Crop. Prod.* **2018**, *122*, 174–181. [CrossRef]
14. Citeau, M.; Slabi, S.A.; Joffre, F.; Carré, P. Improved rapeseed oil extraction yield and quality via cold separation of ethanol miscella. *OCL* **2018**, *25*, D207. [CrossRef]
15. Perrier, A.; Delsart, C.; Boussetta, N.; Grimi, N.; Citeau, M.; Vorobiev, E. Effect of ultrasound and green solvents addition on the oil extraction efficiency from rapeseed flakes. *Ultrason. Sonochem.* **2017**, *39*, 58–65. [CrossRef] [PubMed]
16. Tir, R.; Dutta, P.C.; Badjah-hadj-ahmed, A.Y. Effect of the extraction solvent polarity on the sesame seeds oil composition. *Eur. J. Lipid Sci. Technol.* **2012**, *114*, 1427–1438. [CrossRef]
17. Efthymiopoulos, I.; Hellier, P.; Ladommatos, N.; Russo-pro, A.; Eveleigh, A.; Aliev, A.; Kay, A.; Mills-lamptey, B. Influence of solvent selection and extraction temperature on yield and composition of lipids extracted from spent coffee grounds. *Ind. Crop. Prod.* **2018**, *119*, 49–56. [CrossRef]
18. Maskan, M. Microwave/air and microwave finish drying of banana. *J. Food Eng.* **2000**, *44*, 71–78. [CrossRef]
19. Đurđević, S.; Šavikin, K.; Živković, J.; Böhm, V.; Stanojković, T.; Damjanović, A.; Petrović, S. Antioxidant and cytotoxic activity of fatty oil isolated by supercritical fluid extraction from microwave pretreated seeds of wild growing *Punica granatum* L. *J. Supercrit. Fluid* **2018**, *133*, 225–232. [CrossRef]
20. Wroniak, M.; Rekas, A.; Siger, A.; Janowicz, M. Microwave pretreatment effects on the changes in seeds microstructure, chemical composition and oxidative stability of rapeseed oil. *LWT Food Sci. Technol.* **2016**, *68*, 634–641. [CrossRef]
21. Azadmard-Damirchi, S.; Habibi-Nodeh, F.; Hesari, J.; Nemati, J.; Achachlouei, B.F. Effect of pretreeatment with microwaves on oxidative stability and nutraceuticals content of oil from rapeseed. *Food Chem.* **2010**, *121*, 1211–1215. [CrossRef]
22. Zhang, W.; Jin, G. Microwave puffing-pretreated extraction of oil from Camellia oleifera seed and evaluation of its physicochemical characteristics. *Int. J. Food Sci. Technol.* **2011**, *46*, 2544–2549. [CrossRef]
23. Li, J.; Zu, Y.; Luo, M.; Gu, C.; Zhao, C.; Efferth, T.; Fu, Y. Aqueous enzymatic process assisted by microwave extraction of oil from yellow horn (*Xanthoceras sorbifolia Bunge.*) seed kernels and its quality evaluation. *Food Chem.* **2013**, *138*, 2152–2158. [CrossRef] [PubMed]
24. Da Porto, C.; Da Decorti, D.; Natolino, A. Microwave pretreatment of Moringa oleifera seed: Effect on oil obtained by pilot-scale supercritical carbon dioxide extraction and soxhlet apparatus. *J. Supercrit. Fluid* **2016**, *107*, 38–43. [CrossRef]
25. Güneşer, A.B.; Yilmaz, E. Effects of microwave roasting on the yield and composition of cold pressed orange seed oils. *Grasas Aceites* **2017**, *68*, 1–10.
26. Lin, D.; Xiao, M.; Zhao, J.; Li, Z.; Xing, B.; Li, X.; Kong, M.; Li, L.; Zhang, Q.; Liu, Y.; et al. An overview of plant phenolic compounds and their importance in human nutrition and management of type 2 diabetes. *Molecules* **2016**, *21*, 1–19. [CrossRef] [PubMed]
27. Górnaś, P.; Rudzińska, M.; Seglina, D. Lipophilic composition of eleven apple seed oils: A promising source of unconventional oil from industry by-products. *Ind. Crop. Prod.* **2014**, *60*, 86–91. [CrossRef]
28. Wang, H.; Hu, Z.; Wang, Y.; Chen, H.; Huang, X. Phenolic compounds and the antioxidant activities in litchi pericarp: Difference among cultivars. *Sci. Hortic.* **2011**, *129*, 784–789. [CrossRef]
29. Rekas, A.; Scibisz, I.; Siger, A.; Wroniak, M. The effect of microwave pretreatment of seeds on the stability and degradation kinetics of phenolics compounds in rapeseed oil during long time storage. *Food Chem.* **2017**, *222*, 43–54. [CrossRef]
30. Gaber, M.A.F.M.; Tujillo, F.J.; Mansour, M.P.; Juliano, P. Improving oil extraction from canola seeds by conventional and advanced methods. *Food Eng. Rev.* **2018**, *10*, 198–210. [CrossRef]
31. Fetzer, D.L.; Cruz, P.N.; Hamerski, F.; Corazza, M.L. Extraction of baru (*Dipteryx alata vogel*) seed oil using compressed solvents technology. *J. Supercrit. Fluid* **2018**, *137*, 23–33. [CrossRef]
32. Cruz, R.M.S.; Khmelinski, I.; Vieira, M.C. *Methods in Food Analysis*, 1st ed.; Taylor and Francis Group: New York, NY, USA, 2016; pp. 140–190.
33. ISO. Animal and Vegetable Fats and Oils. In *ISO 3656: Determination of Ultraviolet Absorbance Expressed as Specific UV Extinction*; International Organisation for Standardisation: Geneva, Switzerland, 2011; pp. 1–8. Available online: https://www.iso.org/standard/51008 (accessed on 27 May 2020).

34. AOCS. *In Official Methods and Recommended Practices of the American oil Chemists' Society*; Firestone, D., Ed.; AOCS Press: Champaign, IL, USA, 2003.

35. Samaram, S.; Mirhosseini, H.; Tan, C.P.; Ghazali, H.M. Ultrasound-assisted extraction (UAE) and solvent extraction of papaya seed oil: Yield, fatty acid composition and triacylglycerol profile. *Molecules* **2013**, *18*, 12474–12487. [CrossRef]

36. Ranjith, A.; Kumar, K.S.; Venugopalan, V.V.; Arumughan, C.; Sawhney, R.C.; Singh, V. Fatty acids, tocols, and carotenoids in pulp oil of three sea buckthorn species (*Hippophae rhamnoides, H. salicifolia, and H. tibetana*) grown in the Indian Himalayas. *J. Am. Oil Chem. Soc.* **2006**, *83*, 359–364. [CrossRef]

37. Abbasi, H.; Rezaei, K., Emamdjomeh, Z.; Ebrahimzadeh Mousavi, S.M. Effect of various extraction conditions on the phenolic contents of pomegranate seed oil. *Eur. J. Lipid Sci. Technol.* **2008**, *110*, 435–440. [CrossRef]

38. Siano, F.; Straccia, M.C.; Paolucci, M.; Fasulo, G.; Boscaino, F.; Volpe, M.G. Physico-chemical properties and fatty acid composition of pomegranate, cherry and pumpkin seed oils. *J. Sci. Food Agric.* **2015**, *96*, 1730–1735. [CrossRef] [PubMed]

39. Benzie, I.F.F.; Strain, J.J. The ferric reducing ability of plasma (FRAP) as a measure of antioxidant power: The FRAP Assay. *Anal. Biochem.* **1996**, *239*, 70–76. [CrossRef] [PubMed]

40. Mphahlele, R.R.; Fawole, O.A.; Makunga, N.P.; Opara, U.L. Functional properties of pomegranate fruit parts: Influence of packaging systems and storage time. *J. Food Meas. Charact.* **2017**, *11*, 2233–2246. [CrossRef]

41. Sarkhosh, A.; Zamani, Z.; Fatahi, R.; Ranjbar, H. Evaluation of genetic diversity among Iranian soft-seed pomegranate accessions by fruit characteristics and RAPD markers. *Sci. Hortic.* **2009**, *121*, 313–319. [CrossRef]

42. Luo, X.; Cao, D.; Zhang, J.; Chen, L.; Xia, X.; Li, H.; Zhao, D. Integrated microRNA and mRNA expression profiling reveals a complex network regulating pomegranate (*Punica granatum* L.) seed hardness. *Sci. Rep.* **2018**, *8*, 1–14. [CrossRef]

43. Fathi-Achachlouei, B.; Azadmard-damirchi, S.; Zahedi, Y.; Shaddel, R. Microwave pretreatment as a promising strategy for increment of nutraceutical content and extraction yield of oil from milk thistle seed. *Ind. Crop. Prod.* **2019**, *128*, 527–533. [CrossRef]

44. Uquiche, E.; Jeréz, M.; Ortíz, J. Effect of pretreatment with microwaves on mechanical extraction yield and quality of vegetable oil from Chilean hazelnuts (*Gevuina avellana Mol*). *Innov. Food Sci. Emerg. Technol.* **2008**, *9*, 495–500. [CrossRef]

45. Nikiforidis, C.V. Structure and function of oleosomes (oil bodies). *J. Colloid Interface Sci.* **2019**, *274*, 1–6. [CrossRef] [PubMed]

46. Davis, J.P.; Sweigart, D.S.; Price, K.M.; Dean, L.L.; Sanders, T.H. Refractive index and density measurements of peanut oil for determining oleic and linoleic acid contents. *J. Am. Oil Chem. Soc.* **2013**, *90*, 199–206. [CrossRef]

47. Costa, A.M.M.; Silva, L.O.; Torres, A.G. Chemical composition of commercial cold-pressed pomegranate (*Punica granatum*) seed oil from Turkey and Israel, and the use of bioactive compounds for samples' origin preliminary discrimination. *J. Food Compos. Anal.* **2019**, *75*, 8–16. [CrossRef]

48. Pathare, P.B.; Opara, U.L.; Al-Said, F.A. Colour measurement and analysis in fresh and processed foods: A review. *Food Bioproc. Technol.* **2013**, *6*, 36–60. [CrossRef]

49. Rekas, A.; Siger, A.; Wroniak, M.; Scibisz, I.; Derewiaka, D.; Anders, A. Dehulling and microwave pretreatment effects on the physicochemical composition and antioxidant capacity of virgin rapeseed oil. *J. Food Sci. Technol.* **2017**, *54*, 627–638. [CrossRef]

50. Codex Alimentarius. Standard for Named Vegetable Oils-Codex Stan 210-1999 Standard for Named Vegetable Oils-Codex Stan 210-1999. Codex Alimentarius. 1999. 1–15. Available online: http://www.fao.org/fao-who-codexalimentarius/codex-texts/list-standards (accessed on 7 September 2020).

51. Basiri, S. Evaluation of antioxidant and antiradical properties of pomegranate (*Punica granatum* L.) seed and defatted seed extracts. *J. Food Sci. Technol.* **2015**, *52*, 1117–1123. [CrossRef]

52. Choe, E.; Min, D.B. Mechanisms and factors for edible oil oxidation. *Compr. Rev. Food Sci. Food Saf.* **2006**, *5*, 169–186. [CrossRef]

53. Amri, Z.; Lazreg-Aref, H.; Mekni, M.; El-gharbi, S.; Dabbaghi, O.; Mechri, B.; Hammami, M. Oil characterization and lipids class composition of pomegranate seeds. *Biomed. Res. Int.* **2017**. [CrossRef]

54. Moghimi, M.; Farzaneh, V. The effect of ultrasound pretreatment on some selected physicochemical properties of black cumin (*Nigella Sativa*). *Nutrire* **2018**, *43*, 1–8. [CrossRef]

Foods **2020**, *9*, 1287

55. Young, A.J.; Lowe, G.M. Antioxidant and prooxidant properties of carotenoids. *Arch. Biochem. Biophys.* **2001**, *385*, 20–27. [CrossRef] [PubMed]
56. Mazaheri, Y.; Torbati, M.; Azadmard-Damirchi, S.; Savage, G.P. Effect of roasting and microwave pretreatments of Nigella sativa L. seeds on lipase activity and the quality of the oil. *Food Chem.* **2019**, *274*, 480–486. [CrossRef] [PubMed]
57. Kha, T.C.; Nguyen, M.H.; Roach, P.D.; Stathopoulous, C.E. Effect of galic aril microwave processing conditions on oil extraction efficiency and β-carotene and lycopene content. *J. Food Eng.* **2013**, *117*, 486–491. [CrossRef]
58. Kumar, N.; Goel, N. Phenolic acids: Natural versatile molecules with promising therapeutic applications. *Biotechnol. Rep.* **2019**, *24*, 1–10. [CrossRef]
59. Pande, G.; Akoh, C.C. Antioxidant capacity and lipid characterization of six Georgia-grown pomegranate cultivars. *J. Agric. Food Chem.* **2009**, *57*, 9427–9436. [CrossRef]
60. Xi, W.; Lu, J.; Qun, J.; Jiao, B. Characterization of phenolic profile and antioxidant capacity of different fruit part from lemon (*Citrus limon Burm.*) cultivars. *J. Food Sci. Technol.* **2017**, *54*, 1108–1118. [CrossRef]
61. De Melo, I.L.P.; de Carvalho, E.B.T.; Silva, A.M.O.; Yoshime, L.T.; Sattler, J.A.G.; Pavan, R.T.; Mancini-Filho, J. Characterization of constituents, quality and stability of pomegranate seed oil (*Punica granatum* L.). *Food Sci. Technol.* **2014**, *36*, 132–139. [CrossRef]
62. Caligiani, A.; Bonzanini, F.; Palla, G.; Cirlini, M.; Bruni, R. characterization of a potential nutraceutical ingredient: Pomegranate (*Punica granatum* L.) seed oil unsaponifiable fraction. *Plant. Foods Hum. Nutr.* **2010**, *65*, 277–283. [CrossRef]
63. Pieszka, M.; Migda, W.; Gdsior, R.; Rudziska, M.; Bederska-Aojewska, D.; Pieszka, M.; Szczurek, P. Native oils from apple, blackcurrant, raspberry, and strawberry seeds as a source of polyenoic fatty acids, tocochromanols, and phytosterols: A health implication. *J. Chem.* **2015**. [CrossRef]
64. Górnaś, P.; Rudzińska, M.; Raczyk, M.; Mišina, I.; Soliven, A.; Segliņa, D. Composition of bioactive compounds in kernel oils recovered from sour cherry (*Prunus cerasus L.*) by-products: Impact of the cultivar on potential applications. *Ind. Crop. Prod.* **2016**, *82*, 44–50. [CrossRef]
65. Uddin, M.S.; Ferdosh, S.; Haque Akanda, M.J.; Ghafoor, K.; Rukshana, A.H.; Ali, M.E.; Kamaruzzaman, B.Y.; Fauzi, M.B.; Hadijah, S.; Shaarani, S.; et al. Techniques for the extraction of phytosterols and their benefits in human health: A review. *Sep. Sci. Technol.* **2018**, *53*, 2206–2223.
66. Tian, Y.; Xu, Z.; Zheng, B.; Lo, Y.M. Ultrasonics sonochemistry optimization of ultrasonic-assisted extraction of pomegranate (*Punica granatum* L.) seed oil. *Ultrason. Sonochem.* **2013**, *20*, 202–208. [CrossRef] [PubMed]
67. Aruna, P.; Manohar, B.; Singh, R.P. Processing of pomegranate seed waste and mass transfer studies of extraction of pomegranate seed oil. *J. Food Process. Preserv.* **2018**, *42*, 1–11. [CrossRef]
68. Deniz Senyilmaz-Tiebe, D.; Pfaff, D.H.; Virtue, S.; Schwarz, K.V.; Fleming, T.; Altamura, S.; Muckenthaler, M.U.; Okun, J.G.; Vidal-Puig, A.; Nawroth, P.; et al. Dietary stearic acid regulates mitochondria in vivo in humans. *Nat. Commun.* **2018**, *9*, 1–10. [CrossRef] [PubMed]
69. Đurđević, S.; Šavikin, K.; Živković, J.; Böhm, V.; Stanojković, T.; Damjanović, A.; Petrović, S. Improvement of supercritical CO_2 and *n*-hexane extraction of wild growing pomegranate seed oil by microwave pretreatment. *Ind. Crop. Prod.* **2017**, *104*, 21–27. [CrossRef]
70. Shin, E.; Craft, B.D.; Pegg, R.B.; Phillips, R.D.; Etenmiller, R.R. Chemometric approach of fatty acids and profiles in Runneer-type peanut cultivar by principal component analysi (PCA). *Food Chem.* **2010**, *119*, 1262–1270. [CrossRef]

 *foods*

Article

Chemical Composition and Antioxidant Activity of Thyme, Hemp and Coriander Extracts: A Comparison Study of Maceration, Soxhlet, UAE and RSLDE Techniques

Sara Palmieri, Marika Pellegrini [†], Antonella Ricci [*], Dario Compagnone and Claudio Lo Sterzo

Faculty of Bioscience and Technology for Food, Agriculture and Environment, University of Teramo, Via R. Balzarini 1, 64100 Teramo, Italy; spalmieri@unite.it (S.P.); mpellegrini@unite.it (M.P.); dcompagnone@unite.it (D.C.); closterzo@unite.it (C.L.S.)

[*] Correspondence: aricci@unite.it; Tel.: +39-0861-266-904

[†] Present address: Department of Life, Health and Environmental Sciences, University of L'Aquila, Via Vetoio 9, 67010 Coppito, L'Aquila, Italy.

Received: 1 August 2020; Accepted: 27 August 2020; Published: 2 September 2020

Abstract: Appropriate and standardized techniques for the extraction of secondary metabolites with interesting biological activity from plants are required. In this work, a comparison of different conventional and unconventional extraction techniques (maceration—M, Soxhlet—S, ultrasound assisted extraction—UAE, and rapid solid-liquid dynamic extraction—RSLDE) was investigated. Bioactive compounds were extracted from *Thymus vulgaris* L. (thyme), *Cannabis sativa* L. (industrial hemp) and *Coriandrum sativum* L. (coriander) and chemically characterized for their volatile fraction and polyphenolic content by means of gas chromatography-mass spectrometry (GC-MS) and high performance liquid chromatography-ultraviolet (HPLC-UV). Linalool (48.19%, RSLDE) and carvacrol (21.30%, M) for thyme, caryophyllene (54.78%, S) and humulene (14.13%, S) for hemp, and linalool (84.16%, RSLDE) for coriander seeds were the main compounds among terpenes, while thyme was the richest source of polyphenols with rosmarinic acid (51.7 mg/g dry extract-S), apigenin (7.6 mg/g dry extract-S), and luteolin (4.1 mg/g dry extract-UAE) being the most abundant. In order to shed light on their potential as natural food preservatives, the biological activity of the extracts was assessed in terms of antioxidant activity (2,2′-azino-bis(3-ethylbenzothiazoline-6-sulphonic acid—ABTS$^{·+}$, ferric reducing antioxidant power—FRAP, 2,2-diphenyl-1-picrylhydrazyl—DPPH$^·$ assays) and phenolic content (Folin–Ciocâlteu method). For thyme, Soxhlet extracts showed best performances in FRAP and ABTS$^{·+}$ assays (74 mg TE/g dry extract and 134 mg TE/g dry extract, respectively), while Soxhlet and RSLDE extracts recorded similar activity in DPPH$^·$ (107–109 mg TE/g dry extract). For hemp and coriander, indeed, RSLDE extracts accounted for higher antioxidant activity as evidenced by FRAP (80 mg TE/g dry extract and 18 mg TE/g dry extract, respectively) and ABTS$^{·+}$ (557 mg TE/g dry extract and 48 mg TE/g dry extract, respectively) assays. With respect to DPPH$^·$, the best results were observed for UAE extracts (45 mg TE/g dry extract and 220 mg TE/g dry extract, respectively). Our findings suggest that all the investigated techniques are valid extraction methods to retain bioactive compounds and preserve their activity for application in food and pharmaceutical formulations. Among them, the innovative RSLDE stands out for the slightly higher antioxidant performances of the extracts, coupled with the facility of use and standardization of the extraction process.

Keywords: ultrasound assisted extraction—UAE; rapid solid-liquid dynamic extraction—RSLDE; gas chromatography-mass spectrometry—GC-MS; antioxidants; *C. sativa*; *T. vulgaris*; *C. sativum*

---

1. Introduction

Plant bioactive compounds are defined as secondary plant metabolites capable of exerting a positive effect on animal or human health. Secondary metabolites are produced within the plants beyond the primary biosynthetic and metabolic routes of compounds [1]. These components are not needed for plant basic metabolism, can be regarded as products of biochemical "sidetracks" in the plant cells, and can cover important functions in living plants. Polyphenols, for example, can protect plants against free radicals generated during photosynthesis. Terpenoids may attract pollinators or seed dispersers or inhibit competing plants, whereas alkaloids usually ward off herbivore animals or insect attacks.

Among the best-known bioactive compounds, polyphenols and terpenes can delay or inhibit the oxidation of lipids or other biomolecules, and, thus, prevent or repair the damage of human cells caused by oxygen [2,3]. The importance of these components has been emphasized in the last years. The ever-increasing consumer sensibility to the consumption of food with lower content of synthetic chemical products and the loss of efficacy of common preservatives, due to the development and diffusion of resistant bacteria, have led to increasing research activities regarding the extraction and the evaluation of the efficacy of natural antioxidants [4,5].

The use of plant bioactive compounds as antioxidants in different commercial sectors, such as the pharmaceutical, food, and chemical industries, needs an appropriate and standardized extraction technique [6]. Extraction is the first step of any plant chemical component study and plays a significant and crucial role. The efficiency of conventional and non-conventional extraction methods strongly depends on the input parameters, the nature of the plant matrix, the chemistry of bioactive compounds, and the operator expertise [7,8].

Traditional methods, like maceration, percolation, and Soxhlet, are known to have some limits such as time and solvent consumption, and decomposition of heat sensitivity bioactive compounds [8]. However, Soxhlet technique is still common in laboratories and industries being involved in a wide variety of official methods [9]. Recently, the need of enhancing the biological activity of plant extracts has led to the development of unconventional extraction methods. Among the latter, microwave assisted extraction (MAE), supercritical fluid extraction (SFE), ultrasound assisted extraction (UAE), and rapid solid–liquid dynamic extraction (RSLDE) are the most interesting [10–12].

In UAE, the propagation of ultrasonic waves through a liquid medium damages plant wall, resulting in an improvement in solvent penetration; thus, bioactive components can be extracted in minutes. Therefore, with respect to conventional methods, UAE has the advantage of reducing the extraction process time and energy consumption retaining high efficiency [13,14].

The RSLDE, performed by Naviglio Extractor®, can be considered among the "greenest" strategies, operating at room temperature, with a minimum waste of energy and solvents. Naviglio's principle is based on generating, with a suitable solvent, a negative pressure gradient between the internal and external sides of a solid matrix containing extractable material, followed by a sudden restoration of the initial equilibrium conditions. This process induces the forced extraction of the compounds not chemically linked to the main structure of the solid [15].

Scientific literature presents several works about RSLDE comparison with other extraction techniques. However, few records of this comparison are aimed at food preservation [16–19]. The present work focuses on the comparison of different conventional and unconventional extraction techniques (maceration, Soxhlet, UAE, and RSLDE), to obtain extracts suitable for food preservation. Three aromatic species were investigated: *Thymus vulgaris* L., *Cannabis sativa* L., and *Coriandrum sativum* L. The obtained extracts were chemically characterized, and their biological activity was assessed in terms of antioxidant activity.

2. Materials and Methods

2.1. Plant Material

Plants were open field cultivated in Abruzzo's territory starting from certified seeds. Dry inflorescences of *Cannabis sativa* 'Futura 75' (hemp), dry apical stems and leaves of *Thymus vulgaris* (thyme) and seeds of *Coriandrum sativum* (coriander) were obtained from a local farmer (Hemp Farm Italia, Tortoreto (TE), Azienda Agricola Luigi Barlafante, Roseto degli Abruzzi (TE), and Mediterranea Sementi, Sant'Atto (TE), respectively).

Inflorescences of hemp were collected during the flowering period (September), let dry in a dark room at room temperature (20–25 °C), with controlled relative humidity (45–55%), and stored in the same conditions until processing. Little branches of *T. vulgaris* were collected during the balsamic period (June), dried on the field, and stored in a dry and darkroom until processing. Seed heads of *C. sativum* were cut off when the plant began to turn brown, put in a paper bag, and hanged. After drying, seeds were collected and stored in sealed bags.

2.2. Chemicals

Ethanol absolute was obtained from Carlo Erba (Milan, Italy). Acetic acid, acetonitrile, methanol, and water (high performance liquid chromatography—HPLC grade) were purchased from VWR (Milan, Italy).

α-pinene, β-pinene, linalool, β-myrcene, terpinolene, caryophyllene, humulene, and β-bisabolene, gallic acid, *p*-OH benzoic acid, chlorogenic acid, vanillic acid, caffeic acid, syringic acid, ferulic acid, and rosmarinic acid (from Sigma-Aldrich, Darmstadt, Germany) standards were employed. Working standard mixtures were prepared by appropriate dilution of the standards in methanol. All solutions were stored at −20 °C in the dark.

Folin–Ciocâlteu's reagent, 6-hydroxy-2,5,7,8-tetramethylchroman-2-carboxylic acid (Trolox), 2,2-diphenyl-1-picrylhydrazyl (DPPH·), and 2,2'-azino-bis(3-ethylbenzothiazoline-6-sulphonic acid) diammonium salt (ABTS·+) were purchased from Sigma-Aldrich (Darmstadt, Germany). Sodium carbonate, potassium persulfate, potassium hexacyanoferrate(III), trichloroacetic acid, ferric chloride, and potassium phosphate monobasic were obtained from Carlo Erba (Milan, Italy).

2.3. Extractions

Before extraction, the samples were homogenized by trituration with a chopper (Kenwood Quad Blade CH580 Chopper, Kenwood Limited, Havant, UK) 3 times and then crushed with a mortar. Trituration time was as follows: hemp inflorescences, 10 s; thyme leaves and little stems, 20 s; coriander seeds, 15 s.

Extracts were obtained both with conventional methods, as maceration and Soxhlet, and using the unconventional UAE and RSLDE. RSLDE and Soxhlet extracts were produced with two commonly utilized total time extraction processes: 2 and 6 h.

The extracts were all collected in flasks, filtered, and brought to dry by Rotavapor Steroglass S.r.l. (Perugia, Italy).

The extraction yields were calculated according to the equation:

$$Yield\ (\%\ w/w) = \frac{mass\ dried\ extract\,(g)}{mass\ dried\ matrix\,(g)} \times 100$$

The results were expressed as the average of two replicates of the extraction.

2.3.1. RSLDE Extraction

RSLDE technique was performed using Naviglio Extractor® (Atlas Filtri, Padua, Italy), using the same quantitative for both extraction processes (at 2 and 6 h): 50 g of inflorescences for *C. sativa*, 20 g of leaves and stems for *T. vulgaris* and 106 g of seeds for *C. sativum*. 250 mL of ethanol were used as extraction solvent. The 2 hours process (N2h) was carried out by processing plant matrix for 30 cycles (with a maximum pressure of 8 bar); each cycle was composed by 12 hits in the dynamic phase (2 min duration) and a duration of the static phase of 2 min. The 6 h extracts (N6h) were obtained with the same conditions, but with a major number of cycles (i.e., 90).

2.3.2. Soxhlet Extraction

Soxhlet extracts were produced starting from the same quantitative for both extraction processes (at 2 and 6 h, S2h and S6h, respectively): 50 g of inflorescences for *C. sativa*, 20 g of leaves and stems for *T. vulgaris* and 106 g of seeds for *C. sativum* were used. The extractions were performed with 250 mL of ethanol at 100 °C.

2.3.3. Maceration

Macerations were performed using 9 g of inflorescences for *C. sativa*, 4 g of leaves and stems for *T. vulgaris*, and 21 g of seeds for *C. sativum*. The macerates (M) were obtained with 50 mL of ethanol as solvent for 30 days at room temperature without light exposure.

2.3.4. UAE Extraction

The UAE extractions were performed using 19 g of inflorescences for *C. sativa*, 8 g of leaves and little stems for *T. vulgaris* and 42 g of seeds for *C. sativum*. Plant matrices were extracted with 100 mL of ethanol in 250 mL flasks, sealed and immersed in an ultrasonic water bath (Argo Lab DU-45, Milan, Italy) for 15 min (40 kHz, 180 W).

2.4. SPME/GC–MS Characterization of Extracts Volatile Fraction

Chemical characterizations of extracts volatile fraction were performed by solid-phase microextraction/gas chromatography coupled to mass spectrometry (SPME/GC-MS). SPMEs were obtained by a Supelco-57299-U SPME DVB/CAR/PDMS (Divinylbenzene/Carboxen/Polydimethylsiloxane) fiber (Sigma Aldrich-Saint Louis, MO, USA). All extracts were processed as follows: 0.50 g of dry extract were put into a 20 mL capacity glass vial and sealed with a rubber septum and an aluminum. The vial was placed on a heated plate (50 °C) and the SPME needle was inserted into the vial. The grey fiber was exposed to the headspace for 20 min. After exposure, the fiber was retracted into a needle and loaded into the injection port of the gas chromatographer for fiber desorption at 250 °C for 15 min.

A Clarus 580 GC apparatus (PerkinElmer-Waltham, MA, USA) coupled to a Clarus SQ 8 S GC/MS (PerkinElmer-Waltham, MA, USA) was used for GC-MS analysis. Separations were achieved on a fused silica Zebron-ZB-SemiVolatile column (30 m × 250 μm × 0.25 μm—Phenomenex, Torrance, CA, USA). Analyses were carried following a different temperature gradient depending on samples.

The temperature gradient for hemp extracts was as follows: starting temperature 50 °C (hold 1 min), up to 145 °C at 7 °C/min (hold 5 min), up to 175 °C at 4 °C/min and up to 250 °C at 7 °C/min (hold 5 min). The carrier gas was Helium (flow rate 1 mL/min). The split of the injector was set to 1:50, while the injector and the transfer line temperature were set at 250 °C.

The temperature gradient for thyme and coriander extracts was as follows: starting temperature 45 °C (hold 10 min), up to 180 °C at 2.5 °C/min (hold 5 min). The carrier gas was Helium (flow 1 mL/min), while the injector and the transfer line temperature were set at 250 °C.

The semi-quantitative characterization was carried out through Turbomass 6.1.0.1963 software (PerkinElmer-Waltham, MA, USA). The unknown compounds were identified by matching the obtained spectra with the NIST Mass Spectral Library 2.0 (NIST-Gaithersburg, MD, USA) and confirmed by comparison of the retention index (RI) with those retrieved from http://webbook.nist.gov/chemistry/. A mix of *n*-alkanes, ranging from octane (C8) to triacontane (C30) was obtained from Supelco (Bellefonte, CA, USA) and injected using the analytical conditions above reported to determine the retention index (RI) as proposed by Lee et al. [20].

Semi-quantitative analysis was made by peak area normalization without response factors. Relative abundances (%) were the mean of two replicates.

2.5. HPLC-UV Characterization of the Phenolic Fraction

Phenolic compounds were determined by HPLC (Perkin-Elmer series 200, Monza, Italy) equipped with an autosampler and a UV-Vis detector (Perkin Elmer LC 240, Monza, Italy) set at 280 nm. For separation, a Phenomenex Kinetex C18 column was used (dimensions: 250 × 4.6 mm, particle size: 5 μm, pore size: 110 Å; Phenomenex, Bologna, Italy). The mobile phases used were: (A) 1% acetic acid in water and (B) acetonitrile. For analyte separation, the mobile phase gradient was programmed as follows: from 10% to 100% solvent B for 30 min and subsequent return to initial composition in 4 min, achieving mobile phase stabilization for 10 min.

40 mg of dry extracts sample was dissolved in 1 mL of water/methanol (50:50), vortexed for 3 min, centrifugated for 15 min, filtered with 0.2 μm PTFE filter and analyzed.

Quantification of polyphenols was carried out by the external standard method. Linear regression curves based on peak area were calculated for each phenolics compound after injection of mix phenolic standard solutions covering the sample range of concentrations (6-12-25-50-100 ppm).

For quantitative analysis, a calibration curve for each available phenolic standard were constructed based on the UV signal: gallic acid ($y = 36,255x - 26,062$; $R^2 = 0.9983$), p-OH-benzoic ($y = 31,711x - 16,966$; $R^2 = 0.9992$), vanillic acid ($y = 33,123x - 52,417$; $R^2 = 0.9974$), rosmarinic acid ($y = 34,344x - 40,066$; $R^2 = 0.9992$), ferulic acid ($y = 61,245x + 74,735$; $R^2 = 0.9912$), caffeic acid ($y = 44,841x - 416,813$; 0.9952), syringic acid ($y = 39,490x - 106,101$; $R^2 = 0.9987$), luteolin ($y = 7593,1x - 19,075$; $R^2 = 0.9991$), apigenin ($y = 83,755x - 7443,3$; $R^2 = 0.9979$), and chlorogenic acid ($y = 29,136x - 63,864$; $R^2 = 0.9971$).

2.6. Total Phenolic Content (TPC) and Antioxidant Capacity (AOC)

Total Phenolic Content (TPC) estimation was carried out by means of Folin-Ciocâlteu's reagent, following the Singelton and Rossi method [21]. The reference standard was gallic acid (GA). Results are expressed as mg GA equivalents (GAE)/g dry extract, mean value of two replicates.

The antioxidant activity (AOC) was investigated employing:

- DPPH· assay, following the method proposed Brand-Williams et al. [22];
- ABTS·+ assay, with the Gullon et al. method [23],
- FRAP assay, assessed by means of potassium ferricyanide-ferric chloride method described by Oyaizu [24].

For FRAP, DPPH·, and ABTS·+ assays, Trolox was used as a reference standard. Results are expressed as mg Trolox equivalents (TE)/g dry extract, mean value of two replicates.

2.7. Statistical Analysis

Results were expressed as means ± standard deviations. Yields, chemical, and biological characterization data were subjected to ANOVA (analysis of variance), followed by Tukey's HSD post-hoc test at a significance level of 5% ($p < 0.05$). Terpenes classes composition obtained by SPME/GC-MS were processed through principal component analysis (PCA) to observe the possible correlations within the extracts of the different matrices. Before applying the PCA algorithm, the data were linearized and automatically scaled (zero mean and unit variance) to eliminate the differences in the concentration range. The data set consisted of 18 × 4, in which rows represented the 18 extracts and columns the 4 terpenes classes. Data on terpenes classes were also treated using a hierarchical clustering method. Dendrograms were constructed using Euclidean distance measure and Ward's method of dissimilarity between clusters. Both statistical tests were performed with Microsoft Xlstat 2016 statistical software (Addinsoft, Paris, France).

3. Results and Discussion

3.1. Yields

The extracts yields obtained for the three plant matrices are reported in Table 1. The highest yield for thyme was obtained for 2 h Soxhlet extraction (S2h), while the lowest for maceration (M) and ultrasound assisted extraction (UAE) ($p < 0.05$). The best yield for hemp was achieved, indeed, for 6 h Soxhlet extraction (S6h), while the lowest for M and UAE extracts ($p < 0.05$). For this matrix, no significant differences were recorded among RSLDE extraction times (N2h and N6h) and S2h ($p > 0.05$). A totally different behavior was observed for coriander seeds extracts: the best yield was found for UAE and, then M; the lowest was N2h.

According to these data, Soxhlet seems to be the most suitable technique for the extraction of plants aerial parts in terms of yield. However, it should be pointed out that this extraction technique carried out at high temperature allows the co-extraction of the fibers [25–29]. These contribute to the dry extract weight. On the other hand, ultrasound assisted method seems to be the best extraction technique to process plant seeds.

Lower yields were obtained for coriander seeds with respect to hemp and thyme. This is common to other plant species. In fact, the best yields of extraction are usually recovered from stems and leaves [30,31]. In any case, our findings are in line with literature data, falling within the intervals normally reported in several works for the same species for some of these techniques [28,32].

Table 1. Yields of extracts (% *w*/*w*). N2h, RSLDE 2 h; N6h, RSLDE 6 h; S2h, Soxhlet 2 h; S6h, Soxhlet 6 h; UAE, ultrasound assisted extraction; M, maceration.

	N2h	N6h	S2h	S6h	UAE	M
Thyme	2.30 ± 0.06 d	2.45 ± 0.09 c	9.25 ± 0.02 a	8.65 ± 0.05 b	1.62 ± 0.02 e	1.78 ± 0.04 e
Hemp	6.00 ± 0.03 b	5.81 ± 0.05 b	5.70 ± 0.01 b	10.00 ± 0.07 a	0.71 ± 0.09 c	0.95 ± 0.04 c
Coriander seeds	0.57 ± 0.09 f	0.73 ± 0.08 e	1.18 ± 0.06 d	1.63 ± 0.09 c	2.36 ± 0.07 a	2.17 ± 0.08 b

Results followed by the same case-letter are not different according to Tukey's HSD post-hoc test ($p > 0.05$).

3.2. Chemical Composition of Extracts Volatile Fraction

The SPME/GC–MS characterization data of the volatile fraction of the extracts of the three plants are shown in Table 2.

In thyme extracts 22 compounds were found, 21 monoterpenes, and one sesquiterpene. Thymol has generally been reported to be the main component of *T. vulgaris*. However, this cultivar contains carvacrol, the isomer of thymol, that has the same biological activity. Linalool was the most abundant volatile compound in all the extracts, but using Soxhlet for two hours, a much lower quantity was found. Both carvacrol and linalool are natural effective antimicrobials used to control the growth of spoilage microorganisms in food as demonstrated in some studies in literature [33,34]. They have been reported to have also therapeutic properties (e.g., vs. Alzheimer's disease) [35].

A total of 25 compounds were identified in hemp extracts, 13 monoterpenes, and 12 sesquiterpenes. The predominant compounds were: β-myrcene and caryophyllene within the monoterpenes and sesquiterpenes, respectively. β-myrcene is known to possess anti-inflammatory, analgesic, and anxiolytic properties [36,37]. Caryophyllene has been reported as anti-inflammatory compound in some cannabis preparations because of the interaction with the cannabinoid receptors and a gastric cytoprotective activity has been also found [38–40]. Interestingly, caryophyllene oxide seems to be a multi-target molecule, known for its anticancer and analgesic properties [37].

Table 2. Solid-phase microextraction/gas chromatography-mass spectrometry (SPME/GC-MS) characterization of the volatile fraction of thyme, hemp and coriander seeds extracts. Results are expressed as relative abundances % (means ± sd).

ID-Thyme	Terpenes Class	RI	N2h	N6h	S2h	S6h	UAE	M
α-pinene	bicyclic monoterpenes	921	2.11 ± 0.73	2.34 ± 0.31	2.53 ± 0.19	2.26 ± 0.10	2.47 ± 0.35	1.57 ± 0.16
sabinene	bicyclic monoterpenes	967	2.49 ± 0.61	2.82 ± 0.27	2.36 ± 0.09	2.90 ± 0.22	3.04 ± 0.22	2.21 ± 0.25
1-octen-3-ol	acyclic monoterpenes	980	0.63 ± 0.02	0.31 ± 0.01	0.30 ± 0.00	0.61 ± 0.04	0.82 ± 0.13	0.22 ± 0.00
β-myrcene	acyclic monoterpenes	999	0.52 ± 0.02	0.56 ± 0.08	0.82 ± 0.23	0.60 ± 0.01	0.56 ± 0.03	0.43 ± 0.00
Y-3-carene	bicyclic monoterpenes	1010	3.52 ± 0.38	4.20 ± 0.94	3.26 ± 0.96	4.16 ± 0.04	4.10 ± 0.19	3.15 ± 0.15
o-cymene	monocyclic monoterpenes	1019	0.18 ± 0.01	0.90 ± 0.08	1.24 ± 0.54	0.45 ± 0.08	1.31 ± 0.31	0.16 ± 0.01
p-cymene	monocyclic monoterpenes	1023	1.18 ± 0.04	1.33 ± 0.20	7.81 ± 0.26	1.38 ± 0.10	1.34 ± 0.03	1.01 ± 0.04
Y-terpinene	monocyclic monoterpenes	1054	5.45 ± 0.49	3.21 ± 0.54	5.48 ± 0.16	7.17 ± 0.80	6.81 ± 0.19	5.03 ± 0.22
cis-sabinene hydrate	bicyclic monoterpenes	1066	5.93 ± 0.76	4.88 ± 0.26	2.29 ± 0.53	5.06 ± 0.42	5.96 ± 0.05	4.14 ± 0.09
cis-linalool oxide	monocyclic monoterpenes	1079	1.14 ± 0.14	1.30 ± 0.27	5.68 ± 0.33	1.33 ± 0.06	1.20 ± 0.04	1.03 ± 0.01
linalool	acyclic monoterpenes	1096	48.19 ± 1.95	46.32 ± 1.23	16.19 ± 2.88	42.59 ± 1.37	47.66 ± 3.79	44.95 ± 1.81
cis-p-menth-2-en-1-ol	monocyclic monoterpenes	1123	2.03 ± 0.65	0.81 ± 0.02	14.36 ± 0.19	0.37 ± 0.05	0.68 ± 0.01	0.19 ± 0.01
trans-limonene oxide	monocyclic monoterpenes	1138	2.20 ± 0.54	0.89 ± 0.11	1.24 ± 1.68	0.15 ± 0.02	0.19 ± 0.07	0.40 ± 0.00
β-pinene oxide	bicyclic monoterpenes	1155	0.01 ± 0.00	1.06 ± 0.01	6.92 ± 0.86	0.36 ± 0.04	0.79 ± 0.02	0.30 ± 0.00
trans-linalool oxide	monocyclic monoterpenes	1174	5.09 ± 0.01	5.00 ± 0.76	6.36 ± 0.67	5.25 ± 0.25	6.75 ± 0.24	3.24 ± 0.27
α-terpineol	monocyclic monoterpenes	1190	3.07 ± 0.46	3.46 ± 0.47	1.44 ± 0.28	3.57 ± 0.12	4.28 ± 1.17	3.54 ± 0.11
trans-piperitol	monocyclic monoterpenes	1208	0.71 ± 0.14	0.44 ± 0.03	4.76 ± 0.65	0.19 ± 0.00	0.21 ± 0.03	0.19 ± 0.15
6,7-epoxigeranial	acyclic monoterpenes	1232	0.76 ± 0.03	0.65 ± 0.01	0.85 ± 0.15	0.39 ± 0.11	0.26 ± 0.10	0.32 ± 0.01
carvone	monocyclic monoterpenes	1243	0.76 ± 0.01	0.3 ± 0.01	0.05 ± 0.01	0.66 ± 0.06	1.01 ± 0.42	0.45 ± 0.02
linalyl acetate	acyclic monoterpenes	1247	2.49 ± 0.20	1.71 ± 0.64	2.74 ± 0.70	3.35 ± 0.27	2.82 ± 0.69	2.56 ± 0.14
carvacrol	monocyclic monoterpenes	1296	9.18 ± 2.01	15.84 ± 1.09	6.52 ± 2.26	13.99 ± 2.68	5.49 ± 0.75	21.30 ± 0.72
β-bisabolene	sesquiterpenes	1508	1.29 ± 0.31	0.67 ± 0.01	2.00 ± 0.49	1.50 ± 0.23	1.21 ± 0.50	1.70 ± 0.08

D-Hemp	Terpenes Class	RI	N2h	N6h	S2h	S6h	UAE	M
α-thuyene	bicyclic monoterpenes	898	0.07 ± 0.00	0.06 ± 0.02	0.02 ± 0.0	0.02 ± 0.0	-	0.13 ± 0.01
α-pinene	bicyclic monoterpenes	915	0.09 ± 0.00	0.10 ± 0.04	0.40 ± 0.1	0.16 ± 0.1	-	0.22 ± 0.04
β-pinene	bicyclic monoterpenes	959	0.16 ± 0.01	0.15 ± 0.00	0.34 ± 0.0	0.14 ± 0.0	0.28 ± 0.03	0.20 ± 0.03
β-myrcene	acyclic monoterpenes	969	2.47 ± 0.29	0.78 ± 0.11	4.27 ± 0.04	3.03 ± 0.8	-	1.98 ± 0.14
D-limonene	monocyclic monoterpenes	1011	0.65 ± 0.34	0.26 ± 0.02	0.55 ± 0.01	0.36 ± 0.0	0.10 ± 0.03	0.61 ± 0.05
eucaliptol	bicyclic monoterpenes	1016	0.86 ± 0.41	0.46 ± 0.07	1.01 ± 0.0	0.74 ± 0.1	0.11 ± 0.02	2.14 ± 0.26
β-ocymene	acyclic monoterpenes	1027	0.95 ± 0.65	0.15 ± 0.04	2.27 ± 0.1	1.11 ± 0.3	2.25 ± 0.08	0.52 ± 0.01
Y-terpinene	monocyclic monoterpenes	1039	0.41 ± 0.15	0.89 ± 0.11	0.46 ± 0.00	0.25 ± 0.00	0.32 ± 0.05	1.66 ± 0.07
terpinolene	monocyclic monoterpenes	1067	3.75 ± 0.17	0.37 ± 0.04	3.79 ± 0.03	5.05 ± 0.9	0.03 ± 0.00	2.15 ± 0.72
linalool	acyclic monoterpenes	1079	4.93 ± 0.30	9.42 ± 1.67	2.69 ± 0.04	0.63 ± 0.0	0.26 ± 0.04	25.98 ± 1.73

Table 2. *Cont.*

		RI	N2h	N6h	S2h	S6h	UAE	M
L-*trans*-pinocarveol	bicyclic monoterpenes	1109	0.87 ± 0.36	0.75 ± 0.01	0.42 ± 0.1	0.60 ± 0.1	0.81 ± 0.15	0.60 ± 0.06
cis-p-mentha-2,8-dien-1-ol	monocyclic monoterpenes	1279	1.58 ± 0.39	1.23 ± 0.10	0.92 ± 0.2	1.35 ± 0.0	0.23 ± 0.04	1.20 ± 0.01
geranyl acetate	acyclic monoterpenes	1372	1.00 ± 0.16	0.72 ± 0.01	1.46 ± 0.59	1.53 ± 0.36	0.96 ± 0.63	1.50 ± 0.18
caryophyllene	sesquiterpenes	1390	51.85 ± 2.57	52.78 ± 2.61	52.44 ± 0.1	54.78 ± 0.1	40.00 ± 0.36	39.61 ± 3.31
α-bergamotene	sesquiterpenes	1400	5.86 ± 1.26	6.14 ± 0.51	4.83 ± 0.0	6.53 ± 0.2	4.79 ± 0.22	2.56 ± 0.52
cis-β-farnesene	sesquiterpenes	1416	3.82 ± 0.50	3.54 ± 0.42	2.70 ± 0.1	4.20 ± 0.3	4.03 ± 0.35	2.18 ± 0.18
humulene	sesquiterpenes	1423	13.60 ± 1.81	13.49 ± 0.14	12.11 ± 0.2	14.13 ± 0.4	12.25 ± 0.99	8.67 ± 0.05
aromadendrene	sesquiterpenes	1428	1.85 ± 0.18	2.86 ± 0.12	2.35 ± 0.0	2.09 ± 0.1	3.16 ± 0.19	2.55 ± 0.01
β-selinene	sesquiterpenes	1459	1.96 ± 0.79	2.81 ± 0.17	2.39 ± 0.1	0.96 ± 0.6	0.53 ± 0.10	3.37 ± 0.39
α-selinene	sesquiterpenes	1466	1.14 ± 0.86	1.91 ± 0.15	1.63 ± 0.0	1.12 ± 0.1	3.26 ± 0.59	2.30 ± 0.25
ɣ-cadinene	sesquiterpenes	1408	0.32 ± 0.11	0.20 ± 0.00	0.40 ± 0.0	0.41 ± 0.0	0.37 ± 0.07	0.27 ± 0.03
guaia-3–9-diene	sesquiterpenes	1413	1.08 ± 0.11	0.80 ± 0.09	1.60 ± 0.0	0.86 ± 0.1	0.00 ± 0.00	0.70 ± 0.09
selina-3,7(11)-diene	sesquiterpenes	1418	0.91 ± 0.08	0.48 ± 0.01	1.65 ± 0.1	0.80 ± 0.1	1.19 ± 0.21	0.47 ± 0.08
caryophyllene oxide	sesquiterpenes	1458	0.29 ± 0.04	0.27 ± 0.01	0.23 ± 0.0	0.29 ± 0.1	0.17 ± 0.03	0.20 ± 0.02
cis-α-bisabolol	sesquiterpenes	1589	0.05 ± 0.01	0.09 ± 0.07	0.04 ± 0.0	0.06 ± 0.00	0.32 ± 0.06	0.05 ± 0.00
ID-Coriander Seeds	**Terpenes Class**	**RI**	**N2h**	**N6h**	**S2h**	**S6h**	**UAE**	**M**
α-pinene	bicyclic monoterpenes	915	0.15 ± 0.04	0.33 ± 0.02	0.49 ± 0.07	0.15 ± 0.04	0.50 ± 0.01	0.34 ± 0.03
2-carene	bicyclic monoterpenes	920	0.04 ± 0.01	0.07 ± 0.02	0.21 ± 0.01	0.04 ± 0.01	0.39 ± 0.01	0.04 ± 0.01
p-mentha-1,3,8-triene	monocyclic monoterpenes	980	0.04 ± 0.01	0.05 ± 0.01	1.40 ± 0.03	0.04 ± 0.01	-	0.49 ± 0.01
β-terpinyl-acetate	monocyclic monoterpenes	1343	0.04 ± 0.01	0.10 ± 0.00	0.04 ± 0.01	0.04 ± 0.01	-	0.13 ± 0.02
eucaliptol	bicyclic monoterpenes	935	-	-	0.24 ± 0.03	-	0.33 ± 0.00	0.03 ± 0.00
β-myrcene	acyclic monoterpenes	987	0.14 ± 0.03	0.21 ± 0.00	2.21 ± 0.01	0.15 ± 0.03	0.98 ± 0.01	0.23 ± 0.02
β-ocymene	acyclic monoterpenes	1027	0.05 ± 0.01	0.09 ± 0.01	0.02 ± 0.01	0.05 ± 0.01	-	0.04 ± 0.01
ɣ-terpinene	monocyclic monoterpenes	1053	0.15 ± 0.02	0.08 ± 0.01	2.24 ± 0.03	0.16 ± 0.02	1.41 ± 0.00	1.12 ± 0.01
trans-linalool oxide	monocyclic monoterpenes	1083	0.74 ± 0.03	0.59 ± 0.01	7.50 ± 0.04	0.75 ± 0.03	0.55 ± 0.04	2.09 ± 0.01
terpinolene	monocyclic monoterpenes	1085	0.04 ± 0.01	0.06 ± 0.01	0.03 ± 0.01	0.04 ± 0.01	-	0.03 ± 0.01
borneol	bicyclic monoterpenes	1162	0.45 ± 0.03	0.39 ± 0.02	0.38 ± 0.06	0.45 ± 0.03	-	1.65 ± 0.01
linalool	acyclic monoterpenes	1079	84.16 ± 0.56	72.55 ± 0.03	54.73 ± 0.05	83.99 ± 0.56	82.95 ± 0.05	78.20 ± 0.02
canfora	bicyclic monoterpenes	1144	6.41 ± 0.05	3.26 ± 0.02	3.11 ± 0.45	6.47 ± 0.05	1.87 ± 0.03	7.90 ± 0.02
terpinen-4-ol	monocyclic monoterpenes	1173	0.66 ± 0.02	0.78 ± 0.00	6.05 ± 0.04	0.67 ± 0.02	0.47 ± 0.02	0.76 ± 0.01
α-terpineol	monocyclic monoterpenes	1190	1.05 ± 0.02	2.14 ± 0.03	13.30 ± 0.04	1.06 ± 0.02	5.65 ± 0.04	1.19 ± 0.01
cis-geraniol	acyclic monoterpenes	1248	3.96 ± 0.14	13.08 ± 0.02	0.25 ± 0.01	4.00 ± 0.14	0.67 ± 0.04	4.25 ± 0.04
lavandulyl acetate	acyclic monoterpenes	1270	1.42 ± 0.03	5.50 ± 0.01	0.77 ± 0.05	1.44 ± 0.03	1.44 ± 0.03	1.28 ± 0.02
geranyl acetate	acyclic monoterpenes	1372	0.48 ± 0.03	0.70 ± 0.02	2.77 ± 0.01	0.49 ± 0.03	0.58 ± 0.03	0.22 ± 0.01

In the table: ID, component name; RI, retention index; N2h, RSLDE 2 h; N6h, RSLDE 6 h; S2h, Soxhlet 2 h; S6h, Soxhlet 6 h; UAE, ultrasound assisted extraction; M, maceration.

Furthermore, 18 terpenoids were identified in coriander seeds extracts, all belonging to the monoterpenes class. The most abundant compound was linalool, which has antibacterial activity [33,34] and anti-tumorigenic potential [41]. Canfora and *cis*-geraniol were also present in smaller amounts; however, they have been reported to contribute to biological and antioxidant activity [37,42]. In literature, there are few studies of the chemical composition of extracts from coriander fruits; the terpenes profile found is similar to coriander seeds essential oils previously reported by Pellegrini et al. [43] and found in literature [44–46].

For each matrix, the SPME/GC-MS identified compounds were associated with four main classes of terpenes (class assignment of each compound is in Table 2). To explore potential correlations among the whole data set the PCA algorithm was used.

Figure 1 reports the PCA biplot obtained for the different terpenes classes (loadings), determined in extracts of thyme, hemp, and coriander (scores). The total variance explained was 70.51%, with the first component accounting for 42.94% and the second for 27.57%.

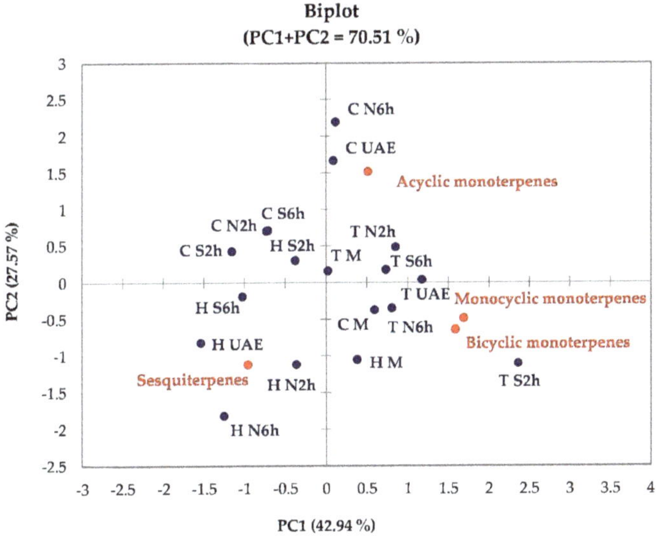

Figure 1. Biplot (scores and loadings) obtained from the PCA on data set of different extracts (rows) and terpenes classes analyzed (columns). In the Figure: T, thyme; H, hemp; C, coriander seeds; N2h, RSLDE 2 h; N6h, RSLDE 6 h; S2h, Soxhlet 2 h; S6h, Soxhlet 6 h; UAE, ultrasound assisted extraction; M, maceration.

From the biplot, it is evident that thyme extract obtained from Soxhlet at 2 h of extraction (T S2h) is separated from all extracts based on the major content of monocyclic and bicyclic monoterpenes. These two classes of terpenes are strongly correlated. Sesquiterpenes were the most abundant compounds in hemp 6 h RSLDE extract (H N6h), while acyclic monoterpenes represented mainly the volatile fractions of coriander ultrasounds-assisted (C UAE) and 6 h RSLDE (C N6h) extracts. Based on the studied variables, from PCA is also evident the presence of different clusters; in particular, on PC1 the extracts C M, T N6h, T S6h, and T N2h are well grouped (positive correlation with PC1). The same applies for C S6h, C S2h, C N2h, H S2h, and H S6h (negative correlation with PC1).

To evaluate the influence of the extraction techniques on each matrix, the dataset was also processed through cluster analysis. Cluster analysis is a valid tool of multivariate analysis that has been already used and is useful to underline the differences among extraction techniques and conditions for the isolation of compounds from plant matrices [47,48]. Clusters were formed to contain four components (acyclic monoterpenes, monocyclic monoterpenes, bicyclic monoterpenes, sesquiterpenes).

The dendrograms obtained from a cluster analysis of coriander, hemp, and thyme data are illustrated in Figures 2–4, respectively.

From the coriander dendrogram (Figure 2) is evident that the diagram is divided into three classes: one class comprising only M extract, and the other 2 constituted by N2h and Soxhlet at 2 and 6 h (S2h and S6h) extracts, and N6h and UAE extracts, respectively. A similar classification was achieved for hemp (Figure 3). In both dendrograms, the N6h/UAE class has large distance from the M class, meaning that these extraction techniques allowed for the isolation of different classes of terpenes. Indeed, N2h and Soxhlet at both 2 and 6 h of distillation time, have similar extraction patterns.

For thyme (Figure 4), a different distribution is obtained; one class consists of M, N2h, S6h, and UAE, the second and third consisting of only N6h and S2h, respectively. In this case, a larger distance of S2h from the first group was observed, because of the ability of this technique to extract more monocyclic and bicyclic monoterpenes, as already evidenced by PCA (Figure 1).

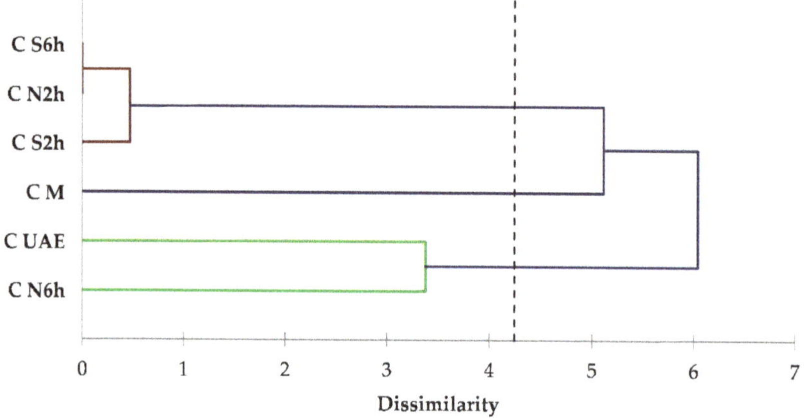

Figure 2. Dendrogram obtained from the cluster analysis based on terpenes classes data for coriander seeds extracts.

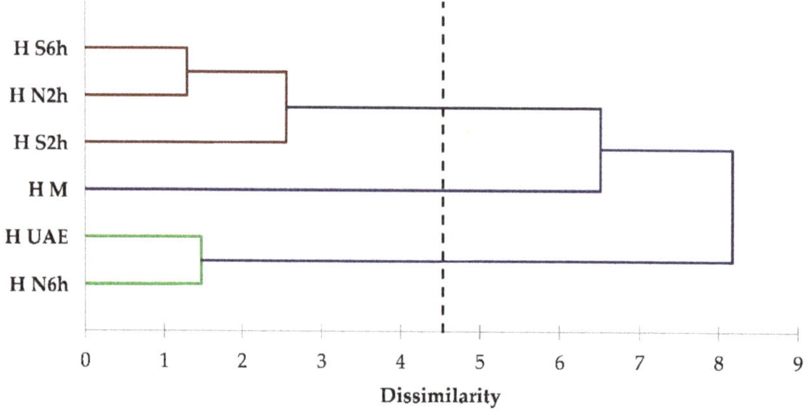

Figure 3. Dendrogram obtained from the cluster analysis based on terpenes classes data for hemp extracts.

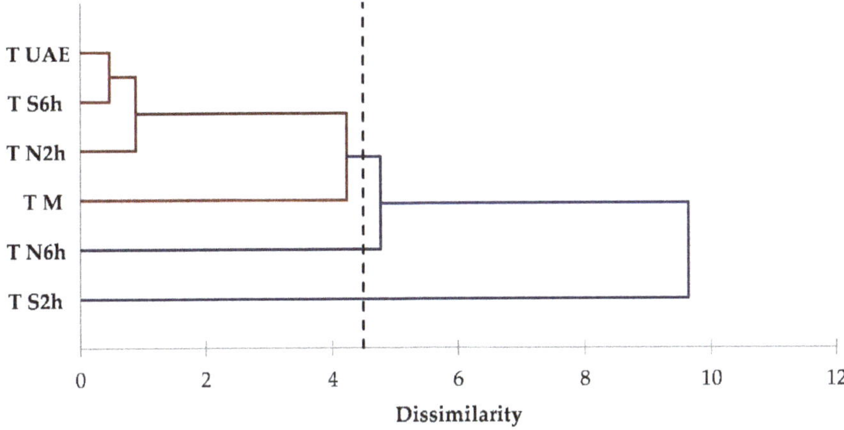

Figure 4. Dendrogram obtained from the cluster analysis based on terpenes classes data for thyme extracts.

The cluster analyses of terpenes allowed to underline that there were not clear patterns of extraction that can help in selecting a particular technique for the extraction of a certain class of terpenes. This is mainly related to the variables that occur before, during and after the extraction process and that influence the outcomes [8]. Anyhow, for all matrices the RSLDE at 2 h of extraction is always clustered with Soxhlet at 6 hours, indicating that, regardless of matrix nature, the extraction patterns are very similar for the two approaches.

3.3. Polyphenolic Composition

The HPLC-UV qualitative and quantitative analysis results of the extracts are presented in Table 3. Eight phenolic acids, (i.e., gallic acid, *p*-OH-benzoic acid, chlorogenic acid, vanillic acid, caffeic acid, syringic acid, ferulic acid, and rosmarinic acid), one phenolic monoterpene (carvacrol) and two flavonoids (i.e., luteolin and apigenin) were identified in thyme. Rosmarinic acid was the compound with the highest concentration in all extracts. Rosmarinic acid is known as one of the main constituents of thyme and it has been recognized for antioxidant, antiviral, anti-inflammatory, antibacterial, and immunostimulant activities [49,50]. UAE and M extracts were significantly poorer of polyphenols, except for the highest UAE luteolin content; however, they were the only extracts containing some phenolic acids (i.e., vanillic and caffeic acids in UAE, syringic acid in M). RSLDE exhibited improved extraction of *p*-OH-benzoic and chlorogenic acids. A non-univocal effect of the increase of the extraction time was observed, as some components have increased and others decreased. In line with our findings, the decrease in content of rosmarinic acid and luteolin at prolonged extraction time has been reported by other authors [51,52]. Besides the extraction operative conditions like solvent, temperature, and time, the stability of natural products in certain conditions is a variable that may influence the chemical composition of the extract. The bioactive compound, during the extraction procedure, are exposed to chemical reactions with solvent and/or other components in the solution that rearrange chemical structures. Chemical alterations occur also after the extraction process, due to manipulation (e.g., solvent removal) and/or conservation conditions (e.g., compounds breakdown by oxidation or light) [53,54].

Table 3. Contents of phenolic compounds (μg/g dry extract).

	N2h	N6h	S2h	S6h	UAE	M
Thyme						
Gallic acid	42.74 ± 0.50 d	58.73 ± 0.90 b	678.67 ± 0.40 a	48.78 ± 0.32 c	-	40.23 ± 0.89 e
p-OH-benzoic acid	190.47 ± 1.02 b	197.15 ± 0.89 a	-	148.68 ± 0.98 c	55.04 ± 0.95 d	17.97 ± 1.02 e
Chlorogenic acid	66.54 ± 0.96 c	120.44 ± 0.75 a	30.07 ± 0.89 d	97.21 ± 0.98 b	-	-
Vanillic acid	-	-	-	-	376.27 ± 0.56 a	-
Caffeic acid	-	-	-	-	145.92 ± 0.85 a	-
Syringic acid	-	-	-	-	-	159.69 ± 1.02 a
Ferulic acid	-	139.24 ± 0.96b	-	516.02 ± 0.84 a	-	-
Rosmarinic acid	34201.41 ± 1.05 b	33955.20 ± 1.02 c	51686.96 ± 0.95 a	31549.93 ± 1.25 d	15049.48 ± 1.09 e	261.55 ± 0.82 f
Luteolin	2671.96 ± 1.10 c	1554.86 ± 1.05 f	1931.34 ± 0.95 d	1704.21 ± 1.32 e	4143.43 ± 0.65 b	2099.47 ± 0.84 a
Apigenin	6608.97 ± 1.15 b	5309.40 ± 1.03 c	2909.13 ± 1.01 e	7618.77 ± 0.98 a	4416.70 ± 0.87 d	-
Carvacrol	3499.84 ± 1.15 b	885.03 ± 1.02 d	2873.81 ± 0.99 c	5595.41 ± 1.05 a	-	220.49 ± 0.98 e
Hemp						
Gallic acid	-	52.29 ± 0.98 c	118.07 ± 0.32 b	408.92 ± 0.63 a	35.10 ± 0.98 d	36.02 ± 0.65 d
p-OH-benzoic acid	-	47.70 ± 0.75 c	95.18 ± 0.95 a	36.42 ± 0.35 d	-	52.41 ± 0.96 b
Chlorogenic acid	-	-	-	-	-	-
Vanillic acid	-	-	-	-	-	-
Caffeic acid	-	-	36.98 ± 0.48 b	-	81.81 ± 0.91 a	-
Syringic acid	-	-	-	-	57.28 ± 0.64 a	-
Ferulic acid	-	247.77 ± 0.64 a	-	-	-	96.70 ± 0.93 b
Rosmarinic acid	259.56 ± 0.97 c	27.09 ± 0.85 f	206.30 ± 0.94 d	152.06 ± 0.65 e	514.33 ± 1.01 a	328.21 ± 1.10 b
Luteolin	1572.05 ± 1.04 a	304.37 ± 1.10 e	502.83 ± 0.95 d	753.01 ± 0.84 c	1384.09 ± 1.09 b	127.67 ± 1.03 f
Apigenin	72.99 ± 1.02 a	51.43 ± 0.48 c	35.77 ± 0.95 d	54.22 ± 1.06 b	-	-
Carvacrol	-	-	-	-	-	-
Coriander Seeds						
Gallic acid	42.37 ± 0.98 b	49.05 ± 0.65 a	22.45 ± 0.35 d	22.73 ± 0.36 d	-	31.02 ± 0.39 c
p-OH-benzoic acid	274.83 ± 0.95 a	31.80 ± 0.98 d	-	89.22 ± 0.84 b	-	46.61 ± 0.91 c
Chlorogenic acid	74.80 ± 0.67 e	480.49 ± 0.92 b	149.47 ± 0.41 c	490.56 ± 0.35 a	146.90 ± 0.83 d	27.82 ± 0.94 f
Vanillic acid	120.27 ± 0.87 e	208.65 ± 0.80 c	440.90 ± 1.25 a	273.23 ± 1.65 b	205.46 ± 0.96 d	-
Caffeic acid	-	208.66 ± 0.85 b	440.17 ± 0.75 a	54.21 ± 0.65 c	27.88 ± 0.35 d	-
Syringic acid	-	-	24.09 ± 0.35 b	65.20 ± 0.92 a	-	23.56 ± 0.24 b
Ferulic acid	188.92 ± 0.95 b	78.81 ± 0.85 c	20.64 ± 0.77 e	239.21 ± 0.8 7a	22.75 ± 0.64 d	23.37 ± 0.81 d
Rosmarinic acid	81.05 ± 1.05a	82.13 ± 0.97a	34.78 ± 0.91b	31.00 ± 0.85c	21.97 ± 0.98c	-
Luteolin	324.50 ± 0.94a	295.82 ± 1.12b	172.40 ± 1.18c	86.90 ± 0.98e	152.01 ± 1.32d	-
Apigenin	182.67 ± 1.20a	182.04 ± 1.15a	30.09 ± 1.06d	35.81 ± 1.14b	101.07 ± 1.23e	87.31 ± 0.98c
Carvacrol	-	-	-	-	-	-

Results followed by the same case-letter are not different according to Tukey's HSD post-hoc test ($p > 0.05$).

Three phenolic acids (i.e., gallic acid, *p*-OH-benzoic acid, and rosmarinic acid) and two flavonoids (i.e., luteolin and apigenin) were mainly found in hemp. The most abundant component was luteolin, a flavonoid with antioxidant [55], anti-inflammatory and antiallergic [56] activity. Among the few literature studies on the identification of polyphenols in hemp extracts, the presence in hemp essential oil of gallic acid and *p*-OH benzoic acid has been already reported [57]. Considering the different extraction techniques, N2h was the extract richest in the three main compounds: rosmarinic acid, luteolin, and apigenin. N6h indicated that the increase in extraction time allowed the recovery of gallic acid, *p*-OH-benzoic acid, and ferulic acid, with a loss in the main components. Indeed, the increase in Soxhlet extraction time from 2 h to 6 h led to an increase in the concentration of gallic acid and of the two flavonoids luteolin and apigenin, accompanied by the decrease of *p*-OH-benzoic acid and rosmarinic acid. M was the poorest extract compared to other techniques.

Similarly to thyme, eight phenolic acids (i.e., gallic acid, *p*-OH-benzoic acid, chlorogenic acid, vanillic acid, caffeic acid, syringic acid, ferulic acid, and rosmarinic acid) and two flavonoids (i.e., luteolin and apigenin) were detected in coriander seeds. Barros and collaborators [58] reported a similar profile in the analysis of polyphenols of coriander seeds. However, gallic acid, luteolin, and apigenin were not detected. Msaada and collaborators [59] described a similar composition, with a higher number of flavonoids components (quercetin, rutin, luteolin, apigenin, and kaempferol). Among the different extraction techniques, RSLDE extracts were richer in gallic, *p*-OH-benzoic, and rosmarinic phenolic acids and in luteolin and apigenin, with a positive effect of longer extraction time on their concentration, except for *p*-OH-benzoic acid and luteolin. On the contrary, Soxhlet extracts showed higher concentrations of chlorogenic, vanillic, and caffeic acids, in addition to the presence of syringic acid, that was not detected in N2h and N6h.

3.4. Total Phenolic Content and Antioxidant Activity

Total phenolic content (TPC) and antioxidant activity (AOC-FRAP, DPPH·, ABTS·+) data are reported in Table 4. The different extraction techniques were compared using an analysis of variance (ANOVA).

Table 4. Total phenolic content (TPC) and antioxidant activity. Ferric Reducing Antioxidant Power (FRAP), and Trolox Equivalent Antioxidant Capacity with 2,2-diphenyl-1-picrylhydrazyl (DPPH·), and 2,2′-azinobis-(3-ethylbenzothiazoline-6-sulfonic acid (ABTS·+) of extracts of different techniques.

	TPC (mg GAE/g Dry Extract)	FRAP (mg TE/g Dry Extract)	DPPH· (mg TE/g Dry Extract)	ABTS·+ (mg TE/g Dry Extract)
Thyme				
N2h	141.86 ± 7.85 b	63.86 ± 0.29 b	108.17 ± 0.08 a	23.62 ± 8.20 d
N6h	179.67 ± 12.71 a	66.95 ± 5.74 b	108.52 ± 0.16 a	52.12 ± 0.11 bc
S2h	157.86 ± 1.71 b	68.48 ± 1.23 ab	108.57 ± 0.08 a	49.07 ± 0.11 c
S6h	154.26 ± 11.23 b	73.53 ± 1.29 a	107.19 ± 1.22 a	134.46 ± 0.22 a
UAE	107.91 ± 9.01 c	40.60 ± 1.04 c	98.50 ± 1.71 b	53.72 ± 0.34 bc
M	142.08 ± 1.69 b	40.25 ± 2.93 c	92.60 ± 4.33 c	57.56 ± 0.34 b
Hemp				
N2h	140.25 ± 2.56 a	81.073 ± 2.31ab	34.02 ± 1.86 b	557.16 ± 6.57 a
N6h	139.52 ± 2.49 a	80.21 ± 1.76 abc	33.76 ± 0.89 b	485.10 ± 4.16 b
S2h	120.55 ± 5.42 b	74.66 ± 0.81 d	29.92 ± 0.32 d	394.39 ± 3.41 c
S6h	124.25 ± 3.61 b	78.69 ± 1.88 bc	31.54 ± 0.08 cd	433.22 ± 19.81 bc
UAE	110.30 ± 3.71 c	77.22 ± 0.92 cd	45.04 ± 1.23 a	381.26 ± 9.05 c d
M	125.12 ± 3.54 b	83.14 ± 1.63 a	32.43 ± 0.32 bc	502.16 ± 5.62 ab
Coriander Seeds				
N2h	19.87 ± 1.52 b	18.43 ± 0.15 a	147.60 ± 3.97 b	48.05 ± 1.60 a
N6h	17.67 ± 0.47 c	12.69 ± 0.39 d	163.55 ± 2.94 b	46.48 ± 1.78 a
S2h	15.54 ± 0.57 d	15.22 ± 0.39 bc	128.84 ± 3.94 b	21.90 ± 2.00 b
S6h	24.36 ± 1.01 a	16.91 ± 2.34 ab	150.61± 4.92 b	23.85 ± 1.60 b
UAE	3.01 ± 0.61 e	13.45 ± 1.26 cd	219.95 ± 2.44 a	1.12 ± 0.14 d
M	18.95 ± 0.11 bc	14.40 ± 0.09 cd	177.23 ± 1.46 ab	5.64 ± 1.04 c

Results followed by the same case-letter are not different according to Tukey's HSD post-hoc test ($p > 0.05$).

The comparison with literature indicates that our data are in accordance with those reported by different authors [60–64].

The influence of the extraction time on TPC was more evident for thyme RSLDE extracts and coriander Soxhlet extracts, with improved extraction for longer times (6 h with respect to 2 h). In the comparison of the different techniques, the highest values were registered for RSLDE extracts for thyme and hemp, at 6 h and 2 h, respectively. Indeed, in the case of coriander seeds, Soxhlet extraction at 6 h was more efficient. For all plant matrices, UAE extracts had the lowest TPC values.

Concerning AOC (Table 4) of hemp and coriander seeds, a higher activity for the 2 h RSLDE extracts in all spectrophotometric assays was achieved, except for the DPPH·. The highest antiradical activity was obtained for UAE extracts. Moreover, in the case of hemp, M demonstrated good activity, similarly to RSLDE extracts.

It is known that the differences in the antioxidant activity might be related to the different availability of extractable components, resulting from the varied chemical composition of plants [65]. The amount of the antioxidant components that can be extracted from a plant material is mainly affected by the strength of the extraction procedure and may vary from sample to sample. Usually, the TPC and AOC of extracts obtained by reflux extraction technique (e.g., Soxhlet) are lower than other methods [27], in contrast to the trends noted for extraction yields. This decrease can be attributed to the thermal decomposition of some antioxidants at the temperatures used in the process. Several studies reported that thermal processing conditions might result in the loss of natural antioxidants because heat may accelerate oxidation and other degradation reactions [27,66,67].

In this work, a different behavior was observed; for thyme, in fact, Soxhlet at 6 h was the best extract in terms of antioxidant activity. This indicates that for this aromatic plant, hot solvent systems under reflux state are more efficient for the recovery of antioxidant components. Dutra et al. [68] reported that among different extraction techniques (i.e., reflux, maceration, ultrasound, heating plate), extraction made under reflux using ethanol/water (70:30, *v*/*v*) offered the highest polyphenol levels in *Pterodon emarginatus* vogel seeds. This was attributed to the effective extraction under reflux conditions, leading to a higher release of some bound phenolics and with an increase of antioxidant activity [69].

To highlight the comparison of the different extraction techniques, for each plant matrix, the best performing extracts in TPC and antioxidant activity assays are collected in Table 5.

Table 5. Extracts with the highest value of total phenolic content and antioxidant activity

	TPC	FRAP	DPPH$^\cdot$	ABTS$^{\cdot+}$
Thyme	N6h	S6h	N2h/N6h/S2h/S6h	S6h
Hemp	N2h/N6h	M	UAE	N2h
Coriander seeds	S6h	N2h	UAE	N2h/N6h

Table 5 clearly shows that all the extraction techniques are very efficient methods for the recovery of bioactive compounds from plant matrices, particularly to produce extracts retaining good antioxidant capacity. It is worth to notice, however, that more than half of the cases are carried out using unconventional techniques, (N2h/N6h and UAE), thus reducing solvent and energy consumption.

4. Conclusions

In this study, extracts have been obtained from three plant species *T. vulgaris*, *C. sativa*, and *C. sativum* cultivated in the Abruzzo region. Four different extraction methods were applied to recover the plants' bioactive components, two conventional, namely the maceration and Soxhlet technique, and two unconventional, namely the ultrasound assisted extraction and the rapid solid–liquid dynamic extraction, performed by Naviglio Extractor®. Moreover, for the Soxhlet and RSLDE techniques, the effect of the extraction time was also investigated.

All the obtained extracts are rich in bioactive compounds and display good antioxidant properties. Although the results do not show univocal trends, slightly higher performances are observed for the extracts obtained by the unconventional RSLDE and UAE techniques.

Given the limited effect of the increase in extraction time, the RSLDE technique performed by Naviglio Extractor® at 2 h of extraction time can be considered a good standardized method to obtain extracts with interesting in vitro antioxidant activity and potential candidates as natural preservatives in food.

Author Contributions: Conceptualization A.R. and C.L.S.; methodology and investigation, S.P. and M.P.; data curation, S.P., A.R., and M.P.; writing—original draft preparation, S.P., and M.P.; writing—review and editing, A.R. and D.C.; supervision, A.R., C.L.S., and D.C. All authors have read and agreed to the published version of the manuscript.

Funding: This research has been supported by the grant of Regione Abruzzo, for the project "PSR 2017-2013 Misura 1.2.4" cod. CUA 2446850691, title "Sviluppo di sistemi convenzionali e innovazioni per la produzione di composti bioattivi da materie prime vegetali per l'impiego nel settore alimentare", and by the FARDIB Research Project of the University of Teramo.

Conflicts of Interest: The authors declare no conflict of interest.

References

1. Sasidharan, S.; Chen, Y.; Saravanan, D.; Sundram, K.; Latha, L.Y. Extraction, isolation and characterization of bioactive compounds from plants' extracts. *Afr. J. Tradit. Complement Altern. Med.* **2011**, *8*, 1–10. [CrossRef] [PubMed]
2. Tajkarimi, M.; Ibrahim, S.A.; Cliver, D. Antimicrobial herb and spice compounds in food. *Food Control* **2010**, *21*, 1199–1218. [CrossRef]
3. Gonzalez-Burgos, E.; Gomez-Serranillos, M. Terpene compounds in nature: A review of their potential antioxidant activity. *Curr. Med. Chem.* **2012**, *19*, 5319–5341. [CrossRef] [PubMed]
4. Briskin, D.P. Medicinal plants and phytomedicines. Linking plant biochemistry and physiology to human health. *Plant Physiol.* **2000**, *124*, 507–514. [CrossRef] [PubMed]
5. Tomaino, A.; Cimino, F.; Zimbalatti, V.; Venuti, V.; Sulfaro, V.; De Pasquale, A.; Saija, A. Influence of heating on antioxidant activity and the chemical composition of some spice essential oils. *Food Chem.* **2005**, *89*, 549–554. [CrossRef]
6. Altemimi, A.; Lakhssassi, N.; Baharlouei, A.; Watson, D.G.; Lightfoot, D.A. Phytochemicals: Extraction, isolation, and identification of bioactive compounds from plant extracts. *Plants* **2017**, *6*, 42. [CrossRef]
7. Smith, R.M. Before the injection—Modern methods of sample preparation for separation techniques. *J. Chromatogr. A* **2003**, *1000*, 3–27. [CrossRef]
8. Azmir, J.; Zaidul, I.; Rahman, M.; Sharif, K.; Mohamed, A.; Sahena, F.; Jahurul, M.; Ghafoor, K.; Norulaini, N.; Omar, A. Techniques for extraction of bioactive compounds from plant materials: A review. *J. Food Eng.* **2013**, *117*, 426–436. [CrossRef]
9. De Castro, M.L.; Priego-Capote, F. Soxhlet extraction: Past and present *Panacea. J. Chromatogr. A* **2010**, *1217*, 2383–2389. [CrossRef]
10. Belwal, T.; Ezzat, S.M.; Rastrelli, L.; Bhatt, I.D.; Daglia, M.; Baldi, A.; Prasad Devkota, H.; Erdogan Orhan, I.; Kumar Patra, J.; Das, G.; et al. A critical analysis of extraction techniques used for botanicals: Trends, priorities, industrial uses and optimization strategies. *Trends Anal. Chem.* **2018**, *100*, 82–102. [CrossRef]
11. Chemat, F.; Rombaut, N.; Sicaire, A.-G.; Meullemiestre, A.; Fabiano-Tixier, A.-S.; Abert-Vian, M. Ultrasound assisted extraction of food and natural products. Mechanisms, techniques, combinations, protocols and applications A Review. *Ultrason. Sonochem.* **2017**, *34*, 540–560. [CrossRef] [PubMed]
12. Naviglio, D.; Scarano, P.; Ciaravolo, M.; Gallo, M. Rapid Solid-Liquid Dynamic Extraction (RSLDE): A powerful and greener alternative to the latest solid-liquid extraction techniques. *Foods* **2019**, *8*, 245. [CrossRef] [PubMed]
13. Skenderidis, P.; Petrotos, K.; Giavasis, I.; Hadjichristodoulou, C.; Tsakalof, A. Optimization of ultrasound assisted extraction of of goji berry (*Lycium barbarum*) fruits and evaluation of extracts' bioactivity. *J. Food Process Eng.* **2017**, *40*, e12522. [CrossRef]
14. Medina-Torres, N.; Ayora-Talavera, T.; Espinosa-Andrews, H.; Sánchez-Contreras, A.; Pacheco, N. Ultrasound Assisted Extraction for the Recovery of Phenolic Compounds from Vegetable Sources. *Agronomy* **2017**, *7*, 47. [CrossRef]
15. Naviglio, D. Naviglio's principle and presentation of an innovative solid–liquid extraction technology: Extractor Naviglio®. *Anal. Lett.* **2003**, *36*, 1647–1659. [CrossRef]
16. Gallo, M.; Vitulano, M.; Andolfi, A.; Della Greca, M.; Conte, E.; Ciaravolo, M.; Naviglio, D. Rapid Solid-Liquid Dynamic Extraction (RSLDE): A new rapid and greener method for extracting two steviol glycosides (*Stevioside* and *Rebaudioside* A) from stevia leaves. *Plant Foods Hum. Nutr.* **2017**, *72*, 141–148. [CrossRef]
17. Ferrara, L.; Naviglio, D.; Gallo, M. Extraction of bioactive compounds of saffron (*Crocus sativus* L.) by ultrasound assisted extraction (UAE) and by rapid solid-liquid dynamic extraction (RSLDE). *Eur. Sci. J.* **2014**, *10*, 1–13.
18. Cozzolino, I.; Vitulano, M.; Conte, E.; D'Onofrio, F.; Aletta, L.; Ferrara, L.; Andolfi, A.; Naviglio, D.; Gallo, M. Extraction and curcuminoids activity from the roots of *Curcuma longa* by RSLDE using the Naviglio extractor. *Eur. Sci. J. (Special Edition)* **2016**, *12*, 119–127.
19. Gallo, M.; Formato, A.; Ciaravolo, M.; Langella, C.; Cataldo, R.; Naviglio, D. A water extraction process for lycopene from tomato waste using a pressurized method: An application of a numerical simulation. *Eur. Food Res. Technol.* **2019**, *245*, 1767–1775. [CrossRef]

20. Lee, K.-G.; Shibamoto, T. Antioxidant properties of aroma compounds isolated from soybeans and mung beans. *J. Agric. Food Chem.* **2000**, *48*, 4290–4293. [CrossRef]
21. Singleton, V.L.; Rossi, J.A. Colorimetry of total phenolics with phosphomolybdic-phosphotungstic acid reagents. *Am. J. Enol. Vitic.* **1965**, *16*, 144–158.
22. Brand-Williams, W.; Cuvelier, M.-E.; Berset, C. Use of a free radical method to evaluate antioxidant activity. *LWT-Food Sci. Technol.* **1995**, *28*, 25–30. [CrossRef]
23. Gullon, B.; Pintado, M.E.; Barber, X.; Fernández-López, J.; Pérez-Álvarez, J.A.; Viuda-Martos, M. Bioaccessibility, changes in the antioxidant potential and colonic fermentation of date pits and apple bagasse flours obtained from co-products during simulated in vitro gastrointestinal digestion. *Food Res. Int.* **2015**, *78*, 169–176. [CrossRef] [PubMed]
24. Oyaizu, M. Studies on products of browning reaction. *Jpn. J. Nutr. Diet.* **1986**, *44*, 307–315. [CrossRef]
25. Zekovic, Z.P. Analysis of thyme (*Thymus vulgaris* L.) extracts. *Acta Period. Technol. Yugosl.* **2000**, *31*, 617–622.
26. Zhang, Z.-M.; Li, G.-K. A preliminary study of plant aroma profile characteristics by a combination sampling method coupled with GC–MS. *Microchem. J.* **2007**, *86*, 29–36. [CrossRef]
27. Sultana, B.; Anwar, F.; Ashraf, M. Effect of extraction solvent/technique on the antioxidant activity of selected medicinal plant extracts. *Molecules* **2009**, *14*, 2167–2180. [CrossRef] [PubMed]
28. Wei, J.-N.; Liu, Z.-H.; Zhao, Y.-P.; Zhao, L.-L.; Xue, T.-K.; Lan, Q.-K. Phytochemical and Bioactive Profile of *Coriandrum sativum* L. *Food Chem.* **2019**, *286*, 260–267. [CrossRef]
29. Devi, V.; Khanam, S. Comparative study of different extraction processes for hemp (*Cannabis sativa*) seed oil considering physical, chemical and industrial-scale economic aspects. *J. Clean. Prod.* **2019**, *207*, 645–657. [CrossRef]
30. Guillen, M.; Manzanos, M. Study of the composition of the different parts of a Spanish *Thymus vulgaris* L. plant. *Food Chem.* **1998**, *63*, 373–383. [CrossRef]
31. Cappelletto, P.; Brizzi, M.; Mongardini, F.; Barberi, B.; Sannibale, M.; Nenci, G.; Poli, M.; Corsi, G.; Grassi, G.; Pasini, P. Italy-grown hemp: Yield, composition and cannabinoid content. *Ind. Crops Prod.* **2001**, *13*, 101–113. [CrossRef]
32. Butt, A.S.; Nisar, N.; Ghani, N.; Altaf, I.; Mughal, T.A. Isolation of thymoquinone from *Nigella sativa* L. and *Thymus vulgaris* L., and its anti-proliferative effect on HeLa cancer cell lines. *Trop. J. Pharm. Res.* **2019**, *18*, 37–42. [CrossRef]
33. Karam, L.; Roustom, R.; Abiad, M.G.; El-Obeid, T.; Savvaidis, I.N. Combined effects of thymol, carvacrol and packaging on the shelf-life of marinated chicken. *Int. J. Food Microbiol.* **2019**, *291*, 42–47. [CrossRef] [PubMed]
34. Flores, P.I.G.; Valenzuela, R.B.; Ruiz, L.D.; López, C.M.; Cháirez, F.E. Antibacterial activity of five terpenoid compounds: Carvacrol, limonene, linalool, α-terpinene and thymol. *Trop. Subtrop. Agroecosyst.* **2019**, *22*, 443–450.
35. Agatonovic-Kustrin, S.; Kustrin, E.; Morton, D.W. Essential oils and functional herbs for healthy aging. *Neural Regen. Res.* **2019**, *14*, 441–445. [CrossRef] [PubMed]
36. Andre, C.M.; Hausman, J.-F.; Guerriero, G. *Cannabis sativa*: The plant of the thousand and one molecules. *Front. Plant Sci.* **2016**, *7*, 1–17. [CrossRef]
37. Nuutinen, T. Medicinal properties of terpenes found in *Cannabis sativa* and *Humulus lupulus*. *Eur. J. Med. Chem.* **2018**, *157*, 198–228. [CrossRef]
38. Tambe, Y.; Tsujiuchi, H.; Honda, G.; Ikeshiro, Y.; Tanaka, S. Gastric Cytoprotection of the Non-Steroidal Anti-Inflammatory Sesquiterpene, β-Caryophyllene. *Planta Med.* **1996**, *62*, 469–470. [CrossRef]
39. Pellati, F.; Brighenti, V.; Sperlea, J.; Marchetti, L.; Bertelli, D.; Benvenuti, S. New methods for the comprehensive analysis of bioactive compounds in *Cannabis sativa* L.(hemp). *Molecules* **2018**, *23*, 2639. [CrossRef]
40. Stenerson, K.K.; Halpenny, M.R. Analysis of Terpenes in Cannabis Using Headspace Solid-Phase Microextraction and GC–MS. *LCGC North Am.* **2017**, *35*, 28.
41. Jana, S.; Patra, K.; Sarkar, S.; Jana, J.; Mukherjee, G.; Bhattacharjee, S.; Mandal, D.P. Antitumorigenic potential of linalool is accompanied by modulation of oxidative stress: An in vivo study in sarcoma-180 solid tumor model. *Nutr. Cancer* **2014**, *66*, 835–848. [CrossRef] [PubMed]
42. Chen, W.; Viljoen, A. Geraniol—A review of a commercially important fragrance material. *South Afr. J. Bot.* **2010**, *76*, 643–651. [CrossRef]

43. Pellegrini, M.; Rossi, C.; Palmieri, S.; Maggio, F.; Chaves, C.; Sterzo, C.L.; Paparella, A.; de Medici, D.; Ricci, A.; Serio, A. *Salmonella enterica* Control in Stick Carrots through Incorporation of Coriander Seeds Essential Oil in Sustainable Washing Treatments. *Front. Sustain. Food Syst.* **2020**, *4*, 1–9. [CrossRef]

44. Gębarowska, E.; Pytlarz-Kozicka, M.; Nöfer, J.; Łyczko, J.; Adamski, M.; Szumny, A. The Effect of *Trichoderma* spp. on the Composition of Volatile Secondary Metabolites and Biometric Parameters of Coriander (*Coriandrum sativum* L.). *J. Food Qual.* **2019**, *2019*. Article ID 5687032, 7p. [CrossRef]

45. Pellegrini, M.; Ricci, A.; Serio, A.; Chaves-López, C.; Mazzarrino, G.; D'Amato, S.; Lo Sterzo, C.; Paparella, A. Characterization of essential oils obtained from Abruzzo autochthonous plants: Antioxidant and antimicrobial activities assessment for food application. *Foods* **2018**, *7*, 19. [CrossRef]

46. Silva, F.; Ferreira, S.; Duarte, A.; Mendonca, D.I.; Domingues, F.C. Antifungal activity of *Coriandrum sativum* essential oil, its mode of action against *Candida* species and potential synergism with amphotericin B. *Phytomedicine* **2011**, *19*, 42–47. [CrossRef]

47. Rodrigues, V.H.; de Melo, M.M.R.; Portugal, I.; Silva, C.M. Extraction of Eucalyptus leaves using solvents of distinct polarity. Cluster analysis and extracts characterization. *J. Supercrit. Fluids* **2018**, *135*, 263–274. [CrossRef]

48. Vural, N.; Algan-Cavuldak, Ö.; Akay, M.A.; Anli, R.E. Determination of the various extraction solvent effects on polyphenolic profile and antioxidant activities of selected tea samples by chemometric approach. *J. Food Meas. and Charact.* **2020**, *14*, 1286–1305. [CrossRef]

49. Armatu, A.; Colceru-Mihul, S.; Bubueanu, C.; Draghici, E.; Pirvu, L. Evaluation of antioxidant and free scavenging potential of some *Lamiaceae* species growing in Romania. *Rom. Biotechnol. Lett.* **2010**, *15*, 5274–5280.

50. Chaouche, T.M.; Haddouchi, F.; Ksouri, R.; Medini, F.; El-Haci, I.A.; Boucherit, Z.; Sekkal, F.Z.; Atik-Bekara, F. Antioxidant Potential of Hydro-methanolic Extract of *Prasium majus* L: An in vitro Study. *Pak. J. Biol. Sci.* **2013**, *16*, 1318–1323. [CrossRef]

51. Ngo, Y.L.; Lau, C.H.; Chua, L.S. Review on rosmarinic acid extraction, fractionation and its anti-diabetic potential. *Food Chem. Toxicol.* **2018**, *121*, 687–700. [CrossRef] [PubMed]

52. Manzoor, M.F.; Ahmad, N.; Ahmed, Z.; Siddique, R.; Zeng, X.-A.; Rahaman, A.; Aadil, R.M.; Wahab, A. Novel extraction techniques and pharmaceutical activities of luteolin and its derivatives. *J. Food Biochem.* **2019**, *43*, e12974. [CrossRef] [PubMed]

53. Fierascu, R.C.; Fierascu, I.; Avramescu, S.M.; Sieniawska, E. Recovery of Natural Antioxidants from Agro-Industrial Side Streams through Advanced Extraction Techniques. *Molecules* **2019**, *24*, 4212. [CrossRef] [PubMed]

54. Maltese, F.; van der Kooy, F.; Verpoorte, R. Solvent Derived Artifacts in Natural Products Chemistry. *Nat. Prod. Commun.* **2009**, *4*, 447–454. [CrossRef]

55. Vaya, J.; Mahmood, S.; Goldblum, A.; Aviram, M.; Volkova, N.; Shaalan, A.; Musa, R.; Tamir, S. Inhibition of LDL oxidation by flavonoids in relation to their structure and calculated enthalpy. *Phytochemistry* **2003**, *62*, 89–99. [CrossRef]

56. Bazylko, A.; Strzelecka, H. A HPTLC densitometric determination of luteolin in *Thymus vulgaris* and its extracts. *Fitoterapia* **2007**, *78*, 391–395. [CrossRef]

57. Zengin, G.; Menghini, L.; di Sotto, A.; Mancinelli, R.; Sisto, F.; Carradori, S.; Cesa, S.; Fraschetti, C.; Filippi, A.; Angiolella, L. Chromatographic analyses, in vitro biological activities, and cytotoxicity of *Cannabis sativa* L. Essential oil: A multidisciplinary study. *Molecules* **2018**, *23*, 3266. [CrossRef]

58. Barros, L.; Duenas, M.; Dias, M.I.; Sousa, M.J.; Santos-Buelga, C.; Ferreira, I.C. Phenolic profiles of in vivo and in vitro grown *Coriandrum sativum* L. *Food Chem.* **2012**, *132*, 841–848. [CrossRef]

59. Msaada, K.; Jemia, M.B.; Salem, N.; Bachrouch, O.; Sriti, J.; Tammar, S.; Bettaieb, I.; Jabri, I.; Kefi, S.; Limam, F. Antioxidant activity of methanolic extracts from three coriander (*Coriandrum sativum* L.) fruit varieties. *Arab. J. Chem.* **2017**, *10*, CS3176–CS3183. [CrossRef]

60. Tohidi, B.; Rahimmalek, M.; Trindade, H. Review on essential oil, extracts composition, molecular and phytochemical properties of *Thymus* species in Iran. *Ind. Crops Prod.* **2019**, *134*, 89–99. [CrossRef]

61. Sena, S.R.; Dantas, T.R.; Pereira, C.G. Extracts from *Thymus vulgaris* and *Origanum vulgare* L. obtained by different separation processes: Global yield and functional profile. *Trends Phytochem. Res.* **2018**, *2*, 13–20.

62. Sriti, J.; Wannes, W.A.; Talou, T.; Jemia, M.B.; Kchouk, M.E.; Marzouk, B. Antioxidant properties and polyphenol contents of different parts of coriander (*Coriandrum sativum* L.) fruit. *La Rivista Italiana Delle Sostanze Grasse* **2012**, *49*, 253–262.

63. Moccia, S.; Siano, F.; Russo, G.L.; Volpe, M.G.; La Cara, F.; Pacifico, S.; Piccolella, S.; Picariello, G. Antiproliferative and antioxidant effect of polar hemp extracts (*Cannabis sativa* L., *Fedora* cv.) in human colorectal cell lines. *Int. J. Food Sci. Nutr.* **2020**, *71*, 410–423. [CrossRef] [PubMed]

64. Smeriglio, A.; Galati, E.M.; Monforte, M.T.; Lanuzza, F.; D'Angelo, V.; Circosta, C. Polyphenolic Compounds and Antioxidant Activity of Cold-Pressed Seed Oil from Finola Cultivar of *Cannabis sativa* L. *Phytother. Res.* **2016**, *30*, 1298–1307. [CrossRef]

65. Hsu, B.; Coupar, I.M.; Ng, K. Antioxidant activity of hot water extract from the fruit of the Doum palm, *Hyphaene Thebaica*. *Food Chem.* **2006**, *98*, 317–328. [CrossRef]

66. Van der Sluis, A.A.; Dekker, M.; van Boekel, M.A.J.S. Activity and concentration of polyphenolic antioxidants in apple juice. 3. Stability during storage. *J. Agric. Food Chem.* **2005**, *53*, 1073–1080. [CrossRef]

67. Cheng, Z.; Su, L.; Moore, J.; Zhou, K.; Luther, M.; Yin, J.-J.; Yu, L. Effects of postharvest treatment and heat stress on availability of wheat antioxidants. *J. Agric. Food Chem.* **2006**, *54*, 5623–5629. [CrossRef]

68. Dutra, R.; Leite, M.; Barbosa, N. Quantification of phenolic constituents and antioxidant activity of *Pterodon Emarginatus* vogel seeds. *Int. J. Mol. Sci.* **2008**, *9*, 606. [CrossRef]

69. Antolovich, M.; Prenzler, P.; Robards, K.; Ryan, D. Sample preparation in the determination of phenolic compounds in fruits. *Analyst* **2000**, *125*, 989–1009. [CrossRef]

 *foods*

Article

Organic Selenium as Antioxidant Additive in Mitigating Acrylamide in Coffee Beans Roasted via Conventional and Superheated Steam

Ahmad K. Alafeef, Fazilah Ariffin * and Musfirah Zulkurnain

Food Technology Division, School of Industrial Technology, Universiti Sains Malaysia, 11800 Minden, Penang, Malaysia; afeef@student.usm.my (A.K.A.); musfirah.z@usm.my (M.Z.)
* Correspondence: fazilah@usm.my; Tel.: +6-04-653-5208

Received: 3 August 2020; Accepted: 24 August 2020; Published: 29 August 2020

Abstract: Selenium is an essential micronutrient with significant antioxidant activity promising in mitigating the formation of acrylamide during high-temperature roasting. In this study, green coffee beans pretreated with selenium (Se-coffee) were investigated on their selenium uptake, selenium retention in green and roasted beans, antioxidant activities, and formation of acrylamide during conventional and superheated steam roasting. Comparisons were made with positive (pretreated without selenium) and negative (untreated) controls. The acrylamide formation was significantly inhibited in Se-coffee (108.9–165.3 µg/kg) compared to the positive and negative controls by 73.9% and 52.8%, respectively. The reduction of acrylamide by superheated steam roasting only observed in the untreated coffee beans (negative control) by 32.4% parallel to the increase in its antioxidant activity. Selenium pretreatment significantly increased antioxidant activity of the roasted Se-coffee beans after roasting although soaking pretreatment significantly reduced antioxidant activity in the green beans. Acrylamide reduction in the roasted coffee beans strongly correlated with the change in antioxidant capacities after roasting (ΔFRAP, 0.858; ΔDPPH, 0.836). The results indicate that the antioxidant properties of the organic selenium suppressed acrylamide formation during coffee roasting.

Keywords: selenomethionine; green coffee; acrylamide; Arabica; Robusta; Maillard reaction; selenium uptake; pretreatment

---

1. Introduction

Coffee is consumed widely around the world, mainly due to its refreshing and stimulating effects. Arabica (*Coffea arabica*) and Robusta (*Coffea canephora*) are two coffee species most widely cultivated with the former known for better organoleptic characteristics, while Robusta coffee possesses higher antioxidant activity but less favorable flavor. The potential benefits of coffee consumption have been well studied to associate with its rich phytochemicals and antioxidant properties originating from the green beans and complex bioactive compounds formed during coffee roasting [1]. Main bioactive compounds in coffee brew are chlorogenic acids, caffeine, pentacyclic diterpenes (cafestol and kahweol), trigonelline, and melanoidins which have been associated with reduced incidences of developing neurodegenerative diseases, several types of cancer, cardiovascular diseases, and type 2 diabetes [2,3]. Antioxidative properties of coffee brews are attributed to phenolic compounds predominantly from melanoidins, chlorogenic acids, and caffeine [4–6]. The contribution of coffee to the daily intake of dietary antioxidants for many people is more than other food sources such as fruit, vegetables, and herbs [7,8].

Roasting of coffee beans at high temperatures involves a series of reactions including caramelization, Maillard reaction, Strecker degradation, and pyrolytic reaction responsible for the development of desirable organoleptic characteristics and antioxidant capacity of coffee [9,10]. Maillard reaction products (MRP) formed during roasting are important flavor and color components in roasted coffee, chiefly Melanoidins, strong antioxidants that account for ~29% of coffee brew dry matter [6]. However, the Maillard reaction also contributes to the formation of undesired toxic components that may counteract the health benefits of coffee, such as acrylamide. Acrylamide is classified as carcinogenic compounds for human health (Group 2A) due to neural, reproductive, and genetic toxicities [11]. Roasted coffee contributes to significant levels of dietary exposure of acrylamide, leading to the recent establishment of the benchmark level of acrylamide in roasted ground coffee at 400 µg/kg by the European Commission [12].

Acrylamide forms in coffee during roasting via reaction of carbonyl group mainly from degradation products of sugars and polysaccharides with amino group of asparagine in Maillard reaction. Mitigation of acrylamide during coffee roasting should target for limiting the source of carbonyl and asparagine content of the coffee beans. Several studies have demonstrated the ability of antioxidants to inhibit the formation of acrylamide in several food matrixes by limiting the source of carbonyl pool, reacting with key Maillard reaction intermediates (e.g., 3-aminopropionamide), and with the acrylamide itself [13]. However, some antioxidants showed contradictory observations by increasing acrylamide levels. Different structures and functional groups of antioxidants can induce the formation of reactive carbonyl pool at high temperatures and low moisture conditions [13]. In particular, antioxidants that contain carbonyl groups such as chlorogenic acids and caffeine were shown to accelerate the formation of acrylamide [14–16]. Studies using several model-systems simulating coffee showed that chlorogenic acid triggered the decomposition of sugars to form 3,4-dideoxyosone and 5-hydroxymethylfurfural (HMF), which then able to react with asparagine via Maillard reaction and promote the formation of acrylamide [14–16]. Alternatively, organic selenium, such as L(+)-selenomethionine, has a high melting properties above 250 °C and remains stable at high temperatures [17]. Therefore, it can be utilized as an antioxidant additive and supplemented to green beans for reducing the formation of acrylamide during coffee roasting.

Selenium is essential to human health with a recommended daily allowance (RDA) value of 55 µg/day and a tolerable upper intake of 400 µg/day for adults [18]. Selenium plays a role in several major metabolic pathways such as immune functions, antioxidants defense systems, and thyroid hormone metabolism [19]. Simultaneously, the deficiency of selenium is associated with cardiovascular and inflammatory diseases, cancer, cirrhosis, diabetes, asthma, and other free radical related problems such as premature aging. Recently, there has been intense interest in selenium supplementation due to its role in protecting the immune system, improve cardiovascular functions, and protection against cancer [20,21]. Coffee contains a trace amount of selenium in the organic form, which is not significant for selenium supplementation [22]. The infusion of selenium via raw material pretreatment has been shown to contribute to aroma profile of roasted coffee and enrichment of its natural selenium level [23] but no impact on the fate of acrylamide is available.

The mitigation strategies of acrylamide in coffee have taken multiple approaches from the modification of raw material to remove potential precursors (e.g., steam and enzyme pretreatments of green beans) and modification of roasting process (e.g., vacuum and steam roasting) to the addition of external additives (e.g., adding amino acid) [24–27]. However, the impact on the sensorial and nutritional properties of coffee are among the limitations of the intervention studies. Superheated steam roasting demonstrated improvement in sensorial quality, increase antioxidant capacity, and reduce lipid oxidation of roasted coffee beans due to the generation of oxygen absent environment at higher bed temperature [9,28,29]. Nevertheless, no information is available on the effect of superheated steam roasting on the fate of acrylamide in coffee. The present study aimed to investigate the effects of selenium pretreatment on the acrylamide formation in Arabica and Robusta coffee during roasting in relation to their antioxidant activities, comparing the conventional and superheated steam roasting. The positive and negative controls were employed to evaluate the true effects of organic selenium on acrylamide formation in the coffee beans during roasting.

2. Materials and Methods

2.1. Materials

Single-origin green Arabica (Santos, Brazil) and Robusta (Laos) coffees beans were obtained from Rasta Brew Ent. (Penang, Malaysia). Selenomethionine, acrylamide (99.3% purity), and 2,2-diphenyl -1-picrylhydrazyl (DPPH) were supplied by Sigma Aldrich (St. Louis, MO, USA).

2.2. Selenium Pretreatment

The green beans were cut using laboratory mills (ZM 200, Retsch GmbH, Haan, Germany) to one third in size to increase surface area for selenium uptake. The crushed green beans (100 g) were soaked overnight in different concentrations of selenomethionine solutions (0, 200, 400 µg/L) at the ratio of 1:5 (*w/v*) at 25 °C. The Se-coffee beans were then subjected to drying using the hot air dryer at 45 °C until moisture content reached ~10% prior to roasting. The selenium pretreatment was investigated by comparing it with a positive and negative control sample. The positive control was the sample pretreated with similar procedures in the absence of selenium, while the negative control sample was the untreated green coffee beans.

2.3. Coffee Roasting

The coffee roasting was done in a laboratory drum roaster hosted in a superheated steam oven (Healsio, AV-1500V, Sharp, Japan). The perforated metal drum is equipped with a motor (Buiacs, JWD Motor Co., Ltd., Zhejiang, China) to spin the drum, short fins at the sides of the drums to allow the beans to fall to the center of the drum, and thermocouple at the center of the drum for actual temperature monitoring. The oven is functional for both conventional roasting and roasting with superheated steam at a pressure of approximately 1 bar, steam generation capacity of 16 cm^3/min, and steam engine heater of 900 W. The oven was preheated to 240 °C before roasting 100 g green beans for 20 min. The roasted coffee beans were then cooled to room temperature and stored in the air-tight container for subsequent analyses.

2.4. Determination of Acrylamide in Coffee

The acrylamide determination was conducted according to PerkinElmer (2004) and Ku Madihah et al. (2013) [30,31]. Ground roasted coffee beans (5 g) was mixed with 50 mL distilled water and heated at 50 °C under continuous stirring for 10 min. The mixture was filtered using filter paper (No. 4, Whatman, Darmstadt, Germany) and loaded to SPE Bond Elut C18 (Agilent, Santa Clara, CA, USA) under gravity flow. The SPE cartridge (500 mg) was conditioned with 3 mL acetone followed by 3 mL 0.1% formic acid. After extraction, the SPE cartridge was washed with 2 mL deionized water and applied vacuum to remove excess water. The acrylamide was eluted using 3 mL of acetone and passed through a nylon syringe filter (0.45 µm, Whatman, Darmstadt, Germany). The sample was kept at 4 °C until further GC-FID analysis.

Acrylamide was determined using a gas chromatograph with a flame ionization detector (Shimadzu Corp, Kyoto, Japan) using the Elite-Wax ETR column (15 m, 0.53 ID, 0.50 µm film thickness) (PerkinElmer Inc., Waltham, MA, USA). Working conditions were as follows: Carrier gas helium (1 mL/min at constant flow); injector, 260 °C; oven temperature: From 100 °C (0.5 min) to 200 °C at 15 °C/min.

Quantification of acrylamide was carried out by preparing a standard curve (r^2 = 0.996) with eight different concentration levels of acrylamide (0.02–2000 µg). Data are mean values of at least two experiments with coefficients of variation at the different concentrations lower than 10%.

2.5. Determination of Selenium

The selenium determination was conducted according to Choi et al. (2009) with slight modifications [32]. An amount of 5 g of finely ground coffee samples was weighed accurately and digested with 40 mL of nitric acid in a digestion tube for 4 h at 150 °C until the darkness of the solution disappeared. After cooling, the digested solution was diluted with deionized water and transferred to a 50 mL volumetric flask. The diluted solution was filtered through filter paper (No. 4, Whatman, Darmstadt, Germany) into a plastic tube designed for the autosampler analysis. Total selenium levels were measured in the aqueous selenium solutions before and after the beans were removed. The difference in selenium levels correspond to the selenium uptake by the coffee beans (µg/kg) [23].

Selenium was determined using NexION™ 300Q Perkin Elmer ICP–MS (PerkinElmer Inc., Waltham, MA, USA) with analysis conditions as follows: Forward power, 1000 W; plasma gas flow rate (Ar), 17.0 L/min; auxiliary gas flow rate (Ar), 1.20 L/min; nebulizer gas flow rate (Ar), 0.93 L/min; Baffled Cycronic Spray Chamber; Ni/Ni sampling cone/skimmer cone; dwell time, 0.50 s per isotope; isotope monitored, Se^{82}. Selenium concentrations of the samples were quantified using the selenium standard curve ($r^2 = 0.999$).

2.6. Determination of Antioxidants Activity

The ground green and roasted coffee beans (1 g) were extracted with 40 mL of methanol for two hours and the extract was filtered using filter paper (No. 1, Whatman, Darmstadt, Germany).

2.6.1. FRAP Assay

The FRAP values were determined according to Moreira et al. (2005) with slight modification [33]. The FRAP reagent was prepared by adding a 10 mM tripyridyl triazine (TPTZ) solution in 40 mM HCl and 20 mM of $FeCl_3$. $6H_2O$ to 300 mM sodium acetate buffer (pH 3.6) in a ratio 1:1:10 (*v/v/v*) and incubated in a water bath at 37 °C for 30 min. An amount of 100 µL of each sample extract was mixed with 900 µL FRAP reagent. The mixture was incubated for 30 min at 37 °C and the absorbance at 593 nm was recorded using an ultraviolet-visible spectrophotometer (UV-1650PC, Shimadzu Corp., Kyoto, Japan). The FRAP values were obtained compared to the calibration curve for Fe^{+2} ($r^2 = 0.998$), expressed as mmol of Fe^{2+} equivalents per gram of soluble solids.

2.6.2. DPPH Assay

The DPPH assay for antioxidant activity was conducted according to Kwak et al. (2017) [34]. A methanol solution of DPPH (2,2-diphenyl-1-picrylhydrazyl) (50 mg/100 mL) at 0.2 mL was mixed with the extract, and the mixture was brought to a total volume of 4.0 mL. The mixture was mixed and kept in the dark for 45 min. Absorbance reading was taken at 515 nm (A_S), along with a blank prepared by substituting the coffee extract with methanol (A_c). DPPH radical scavenging activity was calculated using the following equation:

$$\text{DPPH radical scavenging ratio (\%)} = ((A_c - A_S)/A_c) \times 100 \tag{1}$$

The concentration of samples required to reduce the absorbance of DPPH by 50% (EC_{50}) expressed as mg/mL was determined from the regression of the curve of the methanolic solution of the coffee extract in five different concentrations.

2.7. Color Evaluation

Color measurements of ground roasted coffee were carried out using the CIE Lab-scale spectrophotometer (CM-3500D, Minolta Co., Ltd., Osaka, Japan). The measurement time was set at 2.5 s, the reflectance used was d/8 (diffuse illumination/viewing angle) geometry, and pulsed xenon arc lamp was used the light source. The color values of the ground roasted coffee were expressed as L^* (luminosity component), a^* (redness/greenness), and b^* (blueness/yellowness). The browning index (BI) was calculated using Equations (2) and (3) to estimate the purity of brown color as an indicator of the extend of non-enzymatic browning reaction according to Virgen-Navarro et al. (2016) and Buera et al. (1986) [35,36].

$$\text{Browning index, BI} = (z - 0.31)/(0.172 \times 100), \tag{2}$$

where

$$\text{Chromaticity coordinate of a color, } z = (a^* + 1.75L^*)/(5.645L^* + a^* - 3.012b^*) \tag{3}$$

2.8. Statistical and Data Analysis

All data expressed as means ± standard deviation were analyzed using MINITAB statistical software version 19 (Minitab, Inc., State College, PA, USA) for the one-way analysis of variance (ANOVA) with post-hoc Tukey's tests and Pearson correlation at a 0.05 significance level. Furthermore, two-way ANOVA was applied to determine the significant effects of linear as well as interaction of independent variables. A significance level of $p < 0.05$ was used throughout the study. All measures were taken in triplicate.

3. Results and Discussion

3.1. Selenium Uptake by Green Coffee Beans and Retention in Se-Coffee Beans

Figure 1 shows the selenium retention in the green and roasted Se-coffee beans pretreated at high and low selenium concentrations. The selenium uptake by the crushed Arabica green beans after overnight soaking was at ~71% and 77% from the 200 and 400 µg/L selenium solutions, respectively. An increase in selenium concentration significantly increased selenium uptake by the green beans as a higher concentration gradient is expected to increase the rate of selenium diffusion into the beans. The actual selenium content in the green Se-coffee beans was determined after the drying process found at 774.0 and 1446.0 µg/kg for low and high selenium concentrations, respectively. The loss of selenium after drying process was at ~46 and 53% for low and high selenium concentrations, respectively.

Roasting of the Arabica and Robusta Se-coffee beans showed no significant loss of selenium at lower selenium concentration with a final selenium content of 770 µg/kg in Arabica Se-coffee bean. For higher selenium concentration, ~30% selenium loss was observed after roasting with final selenium content at 1012.0 µg/kg. The loss of selenium was higher since the rate of chemical reactions depends on the concentration of reactants. Volatile selenium species have been reported in roasted coffee beans pretreated with selenomethionine and selenocysteine at a higher concentration (10 mg/L) that may suggest the loss of selenium during coffee roasting [23]. Between different coffee varieties, Robusta coffee showed significantly lower final selenium content after roasting at 638.0 and 655.0 µg/kg for both low and high selenium concentrations, respectively.

The Se-coffee pretreated with high selenium concentration can contribute to selenium content at about 13.6 and 11.5 µg per serving from "double shot" espresso (20 g ground beans) of Arabica and Robusta varieties assuming 67% recovery which equivalent to 24.7 and 20.8% of selenium RDA, respectively [37]. The possibility of exceeding the UL of selenium (400 µg/day) at 30 cups of "double shot" espresso is very unlikely. Thus, the higher selenium concentration was selected for the acrylamide mitigation study.

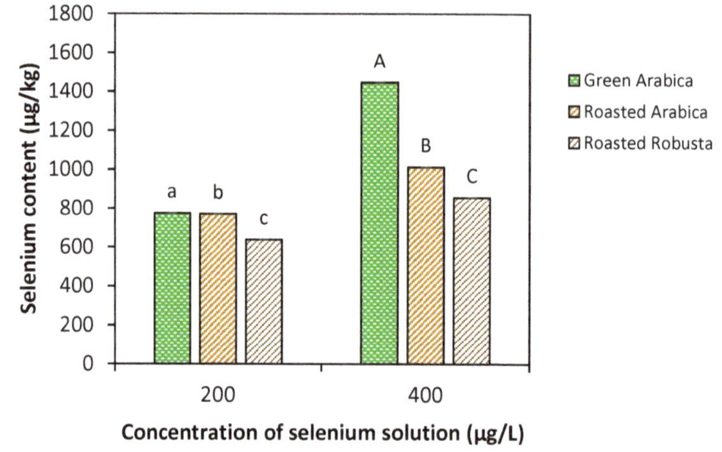

Figure 1. Selenium content of green Arabica, and roasted Arabica and Robusta Se-coffee beans (µg/kg) pretreated at different selenium concentrations and roasted via conventional roasting. Lowercase and uppercase letters represent significant differences ($p < 0.05$) between samples at low and high selenium concentrations, respectively.

3.2. Effects of Selenium Pretreatment on Acrylamide Formation

Figure 2 shows the acrylamide levels in Se-coffee compared to the positive control (pretreated without selenium) and negative control (untreated) roasted via conventional and superheated steam roasting for Arabica and Robusta. The two-way ANOVA of the data showed stronger influence of the selenium pretreatment on the acrylamide formation (F-value = 54.61; p-value = 0.000, Figure 2) compared to the roasting method (F-value = 7.31; p-value = 0.011, Figure 2) with insignificant interaction effect between them. The selenium pretreatment significantly ($p < 0.05$) reduced the acrylamide levels in Se-coffee (108.9–134.4 µg/kg) compared to the negative control (306.2–359.2 µg/kg) by ~64% in Arabica coffee. Robusta coffee showed higher acrylamide levels at 118.4–165.3 µg/kg and significantly ($p < 0.05$) larger acrylamide reduction compared to the negative control (427.3–632.4 µg/kg) at ~73%.

The pretreatment alone without selenium (positive control) significantly ($p < 0.05$) reduced acrylamide levels to 187.7–217.9 µg/kg and 250.8–276.3 µg/kg for Arabica and Robusta coffee, respectively with ~39 and 40% reduction compared to the negative controls. The pretreatment involving overnight soaking of coffee beans in water could facilitate the leaching of water-soluble precursors of Maillard reaction. Soaking green coffee beans in the wet processing method of coffee production has reported the loss of sugars (sucrose, glucose, fructose, mannitol), amino acids (asparagine, arginine, and aspartic acid), chlorogenic acids, and alkaloids (trigonelline and caffeine) through leaching [38]. The concentrations of sucrose, raffinose, and stachyose reduced to half after 40 h soaking, with a higher final concentration of sugars reported in Robusta coffee [39], which may elucidate the higher acrylamide levels of Robusta compared to Arabica coffee in the positive and negative controls.

Selenium supplementation resulted in about two-fold acrylamide reduction (~40–52.8%) compared to the pretreatment alone. These results suggest the role of selenium as an antioxidant additive in suppressing acrylamide formation during coffee roasting by reacting further with the precursors of Maillard reaction and Maillard reaction intermediates. Organic selenium can participate in Maillard reaction to form organoselenium of mono-, di-, and triselenium compounds such as methyl selenoacetate, 2- and 3-methylfuranyl methyl selenide, dimethyl diselenide, and 1,2,4-selenotrithiolane studied in a selenomethionine-glucose model system [40]. Large part of the volatile organoselenium generated is dimethyl diselenide, demonstrating a higher reactivity towards the Maillard reaction than selenomethionine [41].

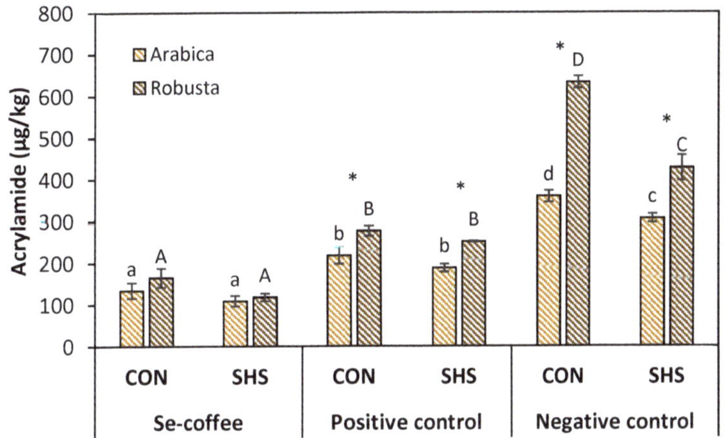

Figure 2. Acrylamide levels in Arabica and Robusta Se-coffee (pretreated with selenium), positive control (pretreated without selenium), and negative control (untreated) roasted via conventional (CON) and superheated steam (SHS) roasting. Lowercase letter, uppercase letter, and star symbol represent significant differences ($p < 0.05$) between different treatments for Arabica, Robusta coffees, and significant differences ($p < 0.05$) between the coffee varieties, respectively.

Roasting methods showed significant influence on the acrylamide formation only in the untreated coffee beans (negative control). The superheated steam roasting significantly ($p < 0.05$) reduced acrylamide formation by 14.8% from 359.2 to 306.2 µg/kg for Arabica coffee and at 32.4% from 632.4 to 427.3 µg/kg for Robusta coffee, respectively. These results suggest that chemical pretreatment is more effective than physical treatment in mitigating acrylamide in coffee, attributing to the removal of associated precursors. The lack of oxygen condition is the most important characteristic of the superheated steam roaster, because the air in the system is replaced by superheated steam, thus, the sample heated under environment lack of oxygen is not oxidized [9,28]. It has been reported that the coffee beans roasted under superheated steam had lower pH and higher sugar content [9,29]. The superior thermal properties of superheated steam to hot air have been reported to result in the lower formation of sugar degradation products in coffee which can limit the formation of Maillard reaction intermediates and formation of acrylamide [9].

Furthermore, the superheated steam roasting has been shown to reduce lipid oxidation under the environment of very small amount of oxygen (~2–3%) during roasting of high lipid products [42,43]. Roasting coffee under superheated steam has been reported to decrease the unsaturated aldehydes content generated during roasting from lipid oxidation reaction [29]. Several researchers suggested the role of lipid oxidation products contributing to carbonyls pool at a significant amount during coffee roasting as an alternative route for acrylamide formation [14,44,45]. Linoleic acid as predominant lipid in coffee beans is likely to oxidize under roasting conditions into corresponding esters containing reactive $\alpha,\beta,\gamma,\delta$-diunsaturated carbonyl group which shown increased reactivity by 1.6 fold towards asparagine in the presence of sugar in a binary system [14,45,46]. In the untreated green coffee beans, lipid (8–18%) and soluble carbohydrates (6–12%) of mainly sucrose occur at almost equal concentration, thus the synergistic effect is highly likely to result in high acrylamide formation in the conventional roasting setting in which the lipid oxidation is the preliminary step [45,46]. The significant acrylamide reduction in the superheated steam roasted coffee may be attributed to the lower Maillard reaction precursors from sugar and lipid oxidation products due to lack of oxygen under superheated steam conditions [47]. Results in this study suggest that superheated steam roasting can be an effective physical method to mitigate acrylamide.

3.3. Effects of Selenium Pretreatment on Antioxidant Activity of Green and Roasted Coffee Beans

3.3.1. Green Beans

The antioxidant activity of green coffee beans is attributed to their complex bioactive constituents mainly caffeine and chlorogenic acid of hydrophilic nature and trigonelline, cafestol, and kahweol of hydrophobic nature [4,5]. Because of different chemical and physical characteristics of different antioxidants and radical sources, no single assay will accurately reflect antioxidant activity in a complex matrix like coffee [48]. The antioxidant activity of green coffee beans is widely measured by means of ferric reducing antioxidant power (FRAP) assay based on the reduction of transition metal ions by hydrophilic antioxidants and 2,2-diphenyl-1-picrylhydrazyl (DPPH) assay of stable lipophilic radical based on radical quenching which also permits the determination of hydrophobic antioxidants [5,49]. The DPPH radical scavenging activity of the coffee extract is represented as an effective concentration necessary to reduce 50% of the DPPH radicals (EC_{50}) [5]. Two major reaction mechanisms of antioxidant with free radicals are hydrogen atom transfer (HAT) or single electron transfer (SET), or the combination of both mechanisms [48]. FRAP is a SET based assay that evaluate the participation of single electron transfer from nucleophile to substrate to produce radical cation that has the ability to stops radical chains [48]. However, FRAP cannot detect compounds with free radical scavenging activity. Thus, DPPH that represents both HAT and SET reaction mechanisms, can be used in combination with FRAP to distinguish different reaction mechanisms of antioxidants in coffee. The antioxidant activities of different organic and inorganic selenium species have been reported by Sentkowska and Pyrzyńska (2019) with selenomethionine showing highest antioxidant activities measured using DPPH, cupric reducing antioxidant capacity (CUPRAC) and Folin–Ciocalteu (FC) assays [50]. However, the author reported reduction in their radical scavenging activities when mixed with tea extracts due to alteration in antioxidant reaction mechanism in the presence of active components such as phenolic compounds. Therefore, in this study the comparison of antioxidant activities of the Se-coffee was not made with the pure selenomethionine, rather with coffee samples of different pretreatments.

The antioxidant activities of green Se-coffee, positive, and negative controls measured by FRAP and DPPH EC_{50} values are shown in Table 1. The FRAP values for untreated green coffee beans (negative control) were at 1.86 and 2.07 mmol Fe^{2+} eq/g for Arabica and Robusta coffees, respectively. The pretreatments involving soaking significantly ($p < 0.05$) reduced FRAP values of the green Se-coffee (0.58 mmol Fe^{2+} eq/g) and the positive control (0.47 mmol Fe^{2+} eq/g) by ~74 and 69%, respectively compared to the untreated Arabica beans. For the Robusta beans, the pretreatments resulted in a lower reduction of FRAP value with Se-coffee (1.00 mmol Fe^{2+} eq/g) significantly higher than the positive control (0.84 mmol Fe^{2+} eq/g).

Table 1. Ferric reducing antioxidant power (FRAP) reducing activity and 2,2-diphenyl-1-picrylhydrazyl (DPPH) scavenging activity expressed as EC_{50} for green Se-coffee (pretreated with selenium), positive control (pretreated without selenium), and negative control (untreated).

Pretreatments	Arabica Coffee		Robusta Coffee	
	FRAP [1] (mmol Fe^{2+} eq/g)	DPPH EC_{50} [1] (mg/mL)	FRAP [1] (mmol Fe^{2+} eq/g)	DPPH EC_{50} [1] (mg/mL)
Se-Coffee	0.58 ± 0.02 [a]	3.19 ± 0.02 [b]	1.00 ± 0.03 [b]	3.20 ± 0.02 [b]
Positive Control	0.74 ± 0.01 [a]	3.38 ± 0.02 [c]	0.84 ± 0.02 [a]	3.15 ± 0.05 [b]
Negative Control	1.86 ± 0.07 [b]	1.93 ± 0.01 [a]	2.07 ± 0.02 [c]	1.92 ± 0.03 [a]

[1] Values are mean (n = 3) ± s.d. Mean values within pretreatment followed by different uppercase letter are significantly different at $p < 0.05$.

The radical scavenging activity of the green beans was at 1.93 and 1.92 mg/mL for Arabica and Robusta coffees, respectively. The lower amount required to give EC_{50} indicates higher radical scavenging activity. The pretreatments resulted in an increment of EC_{50} values of the Se-coffee

(3.19 mg/mL) and the positive control (3.38 mg/mL) by 65 and 75%, respectively. Robusta coffee with higher antioxidant capacity also showed comparable trends at 67 and 64% increment for the Se-coffee (3.20 mg/mL) and the positive control (3.15 mg/mL). The selenium supplementation showed no significant decrease in the EC_{50} values of Se-coffee compared to the positive control samples. These observations are in agreement with the previous finding, which observed no significant effects of selenium infusion on the radical scavenging activity of black and green tea extracts [50]. This observation may suggest the formation of complex interactions between selenomethionine and other antioxidant and phenolic compounds in the green coffee beans with different antioxidant activity.

Polyphenols compounds of predominantly chlorogenic, ferulic, and caffeic acids account for 6–10% of green coffee bean's dry weight may largely contribute to the total antioxidant activity of untreated green beans. However, soaking of the crushed green coffee beans overnight is expected to result in leaching of water-soluble constituents with antioxidant properties into the soaking mediums. Several studies on extra soaking step in wet processing method of coffee beans reported leaching of soluble portion of the green beans, including diterpenes, polyphenols, and tannins with antioxidant properties such as caffeine, trigonelline and chlorogenic acids from the green coffee beans [38,51]. The loss of chlorogenic acids from green coffee beans was reported up to 50% of caffeoylquinic acids (CQAs) notably the 3,4-diCQA and 3,5-diCQA isomers due to soaking and degradation during drying [38]. In this study, further breaking of the beans and drying can expect a higher loss of these antioxidants from the pretreated samples. The selenium pretreatment significantly increased the antioxidant activity of Se-coffee by means of FRAP values but not EC_{50} because the FRAP assay measures the total antioxidant activity of the sample, representing the combined single-electron transfer reductive ability of all redox-active antioxidants present in the samples compared to the DPPH assay [52].

3.3.2. Roasted Beans

Roasting coffee at high temperatures results in changes in coffee compositions mainly on the degradation of polyphenol components and the formation of Millard reaction products that contributes to changes in its final antioxidant activities [53]. The FRAP and DPPH EC_{50} values of roasted coffee beans as affected by the selenium pretreatment and roasting methods are shown in Table 2. It is expected that the antioxidant activity of the negative control (untreated samples) significantly ($p < 0.05$) reduced after roasting. This is especially true for dark roasting degree as the thermal degradation of polyphenols is not counterbalanced by the formation of non-phenolic Millard reaction products such as melanoidins [54,55]. Interestingly, the antioxidant activities of the roasted Se-coffee and positive control were comparable to the negative control with the Se-coffee showed higher values for both antioxidant assays. Roasting of the Se-coffee significantly ($p < 0.05$) increased the antioxidant activity of the roasted Se-coffee, although lower initial antioxidant capacity was observed in the green Se-coffee. The participation of selenomethionine in the Maillard reaction during roasting enhances the formation of organoselenium of different structures and reactivity which can contribute to the increase in antioxidant activity of the Se-coffee [40,41].

The change of antioxidant capacity of coffee beans after roasting measured from the difference of FRAP values (ΔFRAP) and DPPH EC_{50} values (ΔDPPH) between green and roasted coffees showed a strong correlation with acrylamide levels of the roasted coffee as shown in Figure 3. The acrylamide reduction increased with increase in positive values of the ΔFRAP and ΔDPPH with significant Pearson's correlation of −0.858 ($p < 0.05$) and −0.836 ($p < 0.05$), respectively. These results suggest that high antioxidant capacity of Se-coffee attributed to the presence of organoselenium may acted as an inhibitor to the formation of acrylamide. Cheng et al. (2015) reported similar observations that the reduction of acrylamide levels in a Maillard reaction model strongly correlated with the addition of flavonoids at a moderate level without significantly contributing to the final antioxidant capacity of the Millard reaction products [56]. Moreover, the reduction in the selenium content of the roasted Se-coffee in comparison to the green Se-coffee beans (Figure 1) suggests the participation of selenomethionine in the Maillard in mitigating acrylamide in the Se-coffee.

Table 2. FRAP reducing activity and DPPH scavenging activity (expressed as EC_{50}) of Se-coffee, positive control and negative control roasted via conventional (CON) and superheated steam (SHS) roasting.

	Arabica Coffee		Robusta Coffee	
	FRAP [1] (mmol Fe^{2+} eq/g)	DPPH EC_{50} [1] (mg/mL)	FRAP [1] (mmol Fe^{2+} eq/g)	DPPH EC_{50} [1] (mg/mL)
Se-coffee				
CON	1.42 ± 0.01 [Ba]	2.59 ± 0.03 [Aa]	1.45 ± 0.05 [Ba]	2.33 ± 0.08 [Aa]
SHS	1.49 ± 0.01 [Bb]	2.57 ± 0.03 [Aa]	1.71 ± 0.03 [Bb]	2.30 ± 0.05 [Aa]
Positive control				
CON	1.23 ± 0.03 [Aa]	3.03 ± 0.02 [Ca]	1.36 ± 0.04 [Aa]	2.96 ± 0.05 [Ca]
SHS	1.40 ± 0.04 [Ab]	3.18 ± 0.04 [Bb]	1.45 ± 0.07 [Ab]	2.93 ± 0.01 [Ba]
Negative control				
CON	1.37 ± 0.02 [Ba]	2.80 ± 0.04 [Bb]	1.39 ± 0.01 [Aa]	2.69 ± 0.04 [Bb]
SHS	1.46 ± 0.02 [Bb]	2.64 ± 0.03 [Aa]	1.65 ± 0.03 [Bb]	2.45 ± 0.04 [Aa]
Two-way ANOVA analysis (*F*-value)				
Pretreatment	34.8 ***	295.5 ***	15.8 ***	215.5 ***
Roasting	59.4 ***	NS	62.4 ***	8.6 ***
Pretreatment × Roasting	4.6 ***	24.9 ***	4.8 ***	14.2 ***

[1] Values are mean (n = 3) ± s.d. Mean values within pretreatment followed by different uppercase letter are significantly different at $p < 0.05$. Mean values within roasting methods followed by different lowercase letter are significantly different at $p < 0.05$. *** Significance at $p < 0.05$; NS, non-significant.

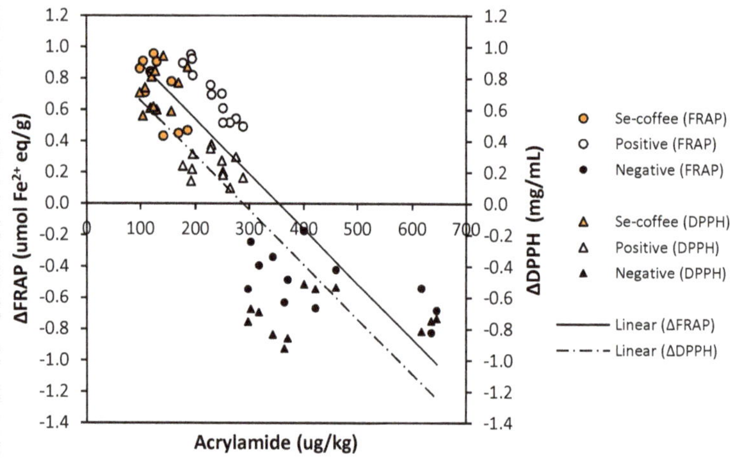

Figure 3. Correlation between acrylamide level and change of antioxidant capacities of Se-coffee (pretreated with selenium), positive control (pretreated without selenium), and negative control (untreated) measured from the difference of FRAP values (ΔFRAP) and DPPH EC_{50} values (ΔDPPH) between green and roasted coffees.

On the other hand, the negative values of the ΔFRAP and ΔDPPH represented by the negative control samples (untreated coffee beans) strongly correlated with the increase in acrylamide levels. High concentration of antioxidants with carbonyl compounds (predominantly chlorogenic acids) in the green negative control can trigger the decomposition of sugars and further reaction with Maillard reaction intermediates such as 3-aminopropionamide to promote the formation of acrylamide [14,56,57]. Pretreatment of the green beans may suggest lower contribution of phenolic compounds in the formation of α-dicarbonyl compounds from sugars and carbohydrates present in the green coffee beans that lead to lower acrylamide formation compared to the negative control.

Two-way ANOVA analysis showed that roasting method has a larger contribution to the reducing activity of the roasted coffee compared to the selenium pretreatment with significant interaction effect (Table 2). However, their radical scavenging activity was greatly influenced by selenium pretreatment with significant interaction effect. Compared to the conventional roasting, the superheated steam roasting significantly increased antioxidant activity of the negative control samples. The thermal degradation of phenolic antioxidants, predominantly chlorogenic acids and their degradation products (caffeic, ferulic, and coumaric acids) during roasting were reported at a lower degree under oxygen-free condition of superheated steam roasting compared to the conventional roasting [42,55].

3.4. Effects of Selenium Pretreatment on Color of Roasted coffee

Color of roasted coffee develops from Maillard reaction, caramelization, and polyphenols oxidation, also determines degree of roasting of the coffee beans. The roasting degree of coffee has been proposed by Sacchetti et al. (2009) from the L^* value as follow: Light roasted ($L^* > 35$), medium roasted ($25 < L^* < 35$), and dark roasted ($L^* < 25$) [54]. The effects of pretreatment and different roasting methods on the formation of browning pigments during coffee roasting were assessed using browning index (BI) and color parameters (L^*, a^*, and b^*), as shown in Figure 4. All samples were categorized as dark roast with L^* values lower than 25. The L^* values of Se-coffee (23.6–24.2) and positive control (23.7–24.2) were higher compared to the negative control (22.1–23.6) inferring formation of lighter coffee beans in the pretreated beans. The formation of brown color pigments from sugar caramelization and Maillard reaction can be impacted by the pretreatments due to the removal of soluble sugar components during soaking. The two-way ANOVA analysis of color and BI obtained significant main effects of both pretreatment and roasting methods and significant interaction effect with stronger influence of pretreatment on L^* and roasting method on BI.

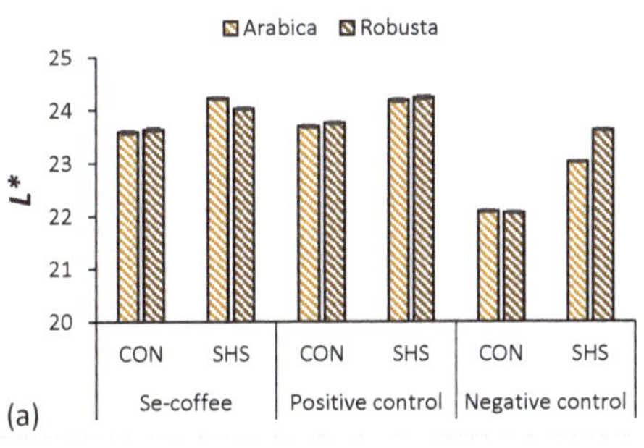

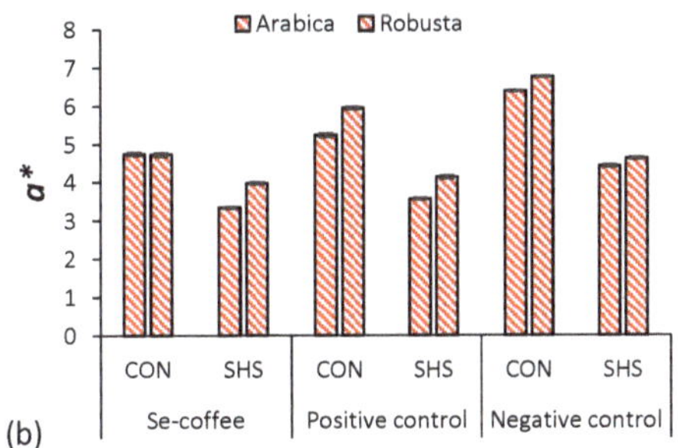

Figure 4. *Cont.*

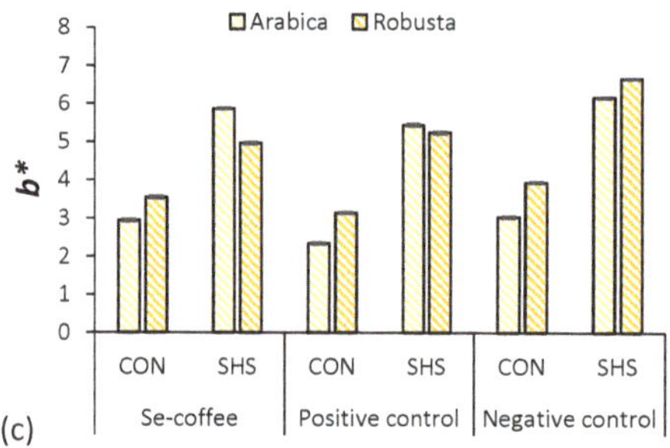

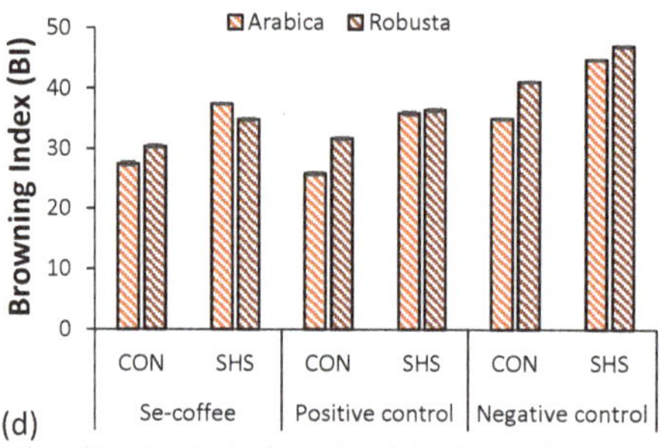

Figure 4. Chromatic parameters (**a**) L^*, (**b**) a^*, (**c**) b^*, and (**d**) browning index (BI) of Arabica and Robusta ground Se-coffee (pretreated with selenium), positive control (pretreated without selenium), and negative control (untreated) roasted via conventional (CON) and superheated steam (SHS) roasting.

The Se-coffee showed lower a^* value compared to the positive and negative controls, in agreement with the correlation between acrylamide and a^* value in roasted coffee beans [58]. The pretreatments also reduced b^* value and BI which can be associated with lower Maillard reaction products. Although Se-coffee showed lower L^*, a^*, and b^* values compared to the positive control, it has higher BI signifying preferred overall characteristic of the brown color of roasted coffee beans. On the other hand, the superheated steam roasting increased L^*, b^*, BI, and decreased a^* value of the SHS samples compared to the conventional roasted samples which associated with the ability of superheated steam to maintain the melanoidin and phenolic compounds from thermal degradation [9].

4. Conclusions

Organic selenium as antioxidant additive supplemented via pretreatment of green coffee beans was effective in reducing acrylamide formation by 73% to a level 70% below the benchmark levels established by the European Commission for roasted coffee (400 μg/kg). Although pretreatments of green coffee reduced the antioxidant activity of green beans, roasting resulted in a comparable antioxidant activity between samples of different pretreatments. The increase in antioxidation capacity from selenium fortification and removal of water-soluble precursors of the Maillard reaction may explain the acrylamide reduction mechanism of Se-coffee. Superheated steam roasting significantly reduced acrylamide levels up to 32% and increased antioxidant activity which was only noticed in the untreated coffee beans. This study suggests that the superheated steam roasting is an effective physical method in mitigating acrylamide in coffee. However, the chemical pretreatment of selenium supplementation is more effective attributing to the removal of associated precursors and an increase in antioxidation activity which also provides supplementation of selenium to coffee beverages. Besides costing, sensory properties of the Se-coffee remain the most important consumer parameter warrants study.

Author Contributions: Conceptualization, A.K.A., M.Z. and F.A.; methodology, A.K.A., M.Z. and F.A.; data analysis, A.K.A., and M.Z.; investigation, A.K.A.; resources, F.A.; data curation, M.Z.; writing—original draft preparation, A.K.A.; writing—review and editing, M.Z. and F.A.; visualization, M.Z.; supervision, F.A. and M.Z.; project administration, F.A.; funding acquisition, F.A. and M.Z. All authors have read and agreed to the published version of the manuscript.

Funding: This research was funded by Universiti Sains Malaysia through Research University Grant (1001.PTEKIND.8011042) and Short-Term Grant (304/PTEKIND/631161).

Acknowledgments: The authors gratefully acknowledge the financial support provided by Universiti Sains Malaysia.

Conflicts of Interest: The authors declare no conflict of interest.

References

1. Hu, G.L.; Wang, X.; Zhang, L.; Qiu, M.H. The sources and mechanisms of bioactive ingredients in coffee. *Food Funct.* **2019**, *10*, 3113–3126. [CrossRef] [PubMed]

2. Higashi, Y. Coffee and endothelial function: A coffee paradox? *Nutrients* **2019**, *11*, 2104. [CrossRef] [PubMed]

3. Gökcen, B.B.; Şanlier, N. Coffee consumption and disease correlations. *Crit. Rev. Food Sci. Nutr.* **2019**, *59*, 336–348. [CrossRef] [PubMed]

4. Acidri, R.; Sawai, Y.; Sugimoto, Y.; Handa, T.; Sasagawa, D.; Masunaga, T.; Yamamoto, S.; Nishihara, E. Phytochemical profile and antioxidant capacity of coffee plant organs compared to green and roasted coffee beans. *Antioxidants* **2020**, *9*, 93. [CrossRef]

5. Liang, N.; Kitts, D.D. Antioxidant property of coffee components: Assessment of methods that define mechanism of action. *Molecules* **2014**, *19*, 19180–19208. [CrossRef]

6. Liu, Y.; Kitts, D.D. Confirmation that the Maillard reaction is the principle contributor to the antioxidant capacity of coffee brews. *Food Res. Int.* **2011**, *44*, 2418–2424. [CrossRef]

7. Lee, B.H.; Nam, T.G.; Kim, S.Y.; Chun, O.K.; Kim, D.O. Estimated daily per capita intakes of phenolics and antioxidants from coffee in the Korean diet. *Food Sci. Biotechnol.* **2019**, *28*, 269–279. [CrossRef]

8. Torres, T.; Farah, A. Coffee, maté, açaí and beans are the main contributors to the antioxidant capacity of Brazilian's diet. *Eur. J. Nutr.* **2017**, *56*, 1523–1533. [CrossRef]

9. Chindapan, N.; Soydok, S.; Devahastin, S. Roasting Kinetics and Chemical Composition Changes of Robusta Coffee Beans During Hot Air and Superheated Steam Roasting. *J. Food Sci.* **2019**, *84*, 292–302. [CrossRef]

10. Opitz, S.; Smrke, S.; Goodman, B.; Keller, M.; Schenker, S.; Yeretzian, C. Antioxidant Generation during Coffee Roasting: A Comparison and Interpretation from Three Complementary Assays. *Foods* **2014**, *3*, 586–604. [CrossRef]

11. IARC. *Acrylamide. Monographs on the Evaluation of Carcinogenic Risks to Humans: Some Industrial Chemicals*; IARC: Lyon, France, 1994; Volume 60.

12. Europen Commission COMMISSION REGULATION (EU) 2017/2158 of 20 November 2017 establishing mitigation measures and benchmark levels for the reduction of the presence of acrylamide in food. *Off. J. Eur. Union* **2017**, *L 304*, 24–44.

13. Jin, C.; Wu, X.; Zhang, Y. Relationship between antioxidants and acrylamide formation: A review. *Food Res. Int.* **2013**, *51*, 611–620. [CrossRef]

14. Kocadagli, T.; Göncüoglu, N.; Hamzalioglu, A.; Gökmen, V. In depth study of acrylamide formation in coffee during roasting: Role of sucrose decomposition and lipid oxidation. *Food Funct.* **2012**, *3*, 970–975. [CrossRef] [PubMed]

15. Cai, Y.; Zhang, Z.; Jiang, S.; Yu, M.; Huang, C.; Qiu, R.; Zou, Y.; Zhang, Q.; Ou, S.; Zhou, H.; et al. Chlorogenic acid increased acrylamide formation through promotion of HMF formation and 3-aminopropionamide deamination. *J. Hazard. Mater.* **2014**, *268*, 1–5. [CrossRef] [PubMed]

16. Gökmen, V.; Kocadağli, T.; Göncüoğlu, N.; Mogol, B.A. Model studies on the role of 5-hydroxymethyl-2-furfural in acrylamide formation from asparagine. *Food Chem.* **2012**, *132*, 168–174. [CrossRef]

17. Kápolna, E.; Gergely, V.; Dernovics, M.; Illés, A.; Fodor, P. Fate of selenium species in sesame seeds during simulated bakery process. *J. Food Eng.* **2007**, *79*, 494–501. [CrossRef]

18. Institute of Medicine. *Dietary Reference Intakes for Vitamin C, Vitamin E, Selenium, and Carotenoids*; The National Academies Press: Washington, DC, USA; ISBN 9780309069496.

19. Kieliszek, M.; Błazejak, S. Current knowledge on the importance of selenium in food for living organisms: A review. *Molecules* **2016**, *21*, 609. [CrossRef] [PubMed]

20. Kuršvietienė, L.; Mongirdienė, A.; Bernatonienė, J.; Šulinskienė, J.; Stanevičienė, I. Selenium anticancer properties and impact on cellular redox status. *Antioxidants* **2020**, *9*, 80. [CrossRef] [PubMed]

21. Bodnar, M.; Szczyglowska, M.; Konieczka, P.; Namiesnik, J. Methods of Selenium Supplementation: Bioavailability and Determination of Selenium Compounds. *Crit. Rev. Food Sci. Nutr.* **2016**, *56*, 36–55. [CrossRef]

22. Messaoudi, M.; Begaa, S.; Hamidatou, L.; Salhi, M. Determination of selenium in roasted beans coffee samples consumed in Algeria by radiochemical neutron activation analysis method. *Radiochim. Acta* **2018**, *106*, 141–146. [CrossRef]

23. Meija, J.; Bryson, J.M.; Vonderheide, A.P.; Montes-Bayón, M.; Caruso, J.A. Studies of selenium-containing volatiles in roasted coffee. *J. Agric. Food Chem.* **2003**, *51*, 5116–5122. [CrossRef] [PubMed]

24. Anese, M.; Nicoli, M.C.; Verardo, G.; Munari, M.; Mirolo, G.; Bortolomeazzi, R. Effect of vacuum roasting on acrylamide formation and reduction in coffee beans. *Food Chem.* **2014**, *145*, 168–172. [CrossRef] [PubMed]

25. Guenther, H.; Anklam, E.; Wenzl, T.; Stadler, R.H. Acrylamide in coffee: Review of progress in analysis, formation and level reduction. *Food Addit. Contam.* **2007**, *24*, 60–70. [CrossRef] [PubMed]

26. Narita, Y.; Inouye, K. Decrease in the acrylamide content in canned coffee by heat treatment with the addition of cysteine. *J. Agric. Food Chem.* **2014**, *62*, 12218–12222. [CrossRef] [PubMed]

27. Lynglev, G.B.; Schoesler, S. Method for producing roasted coffee beans. U.S. Patent US20160295877A1, 13 October 2016.

28. Devahastin, S.; Mujumdar, A.S. Superheated Steam Drying of Foods and Biomaterials. *Modern Drying Technol.* **2014**, *5*, 57–84.

29. Maki, Y.; Haruyama, T. Process for roasting coffee beans with steam. U.S. Patent US5681607A, 28 October 1997.

30. Perkin, E. Acrylamide Analysis by Gas Chromatography. *USA PerkinElmer Life Anal. Sci.* **2004**, *20*, 5–7.

31. Ku Madihah, K.Y.; Zaibunnisa, A.H.; Norashikin, S.; Rozita, O.; Misnawi, J. Optimization of roasting conditions for high-quality Arabica coffee. *Int. Food Res. J.* **2013**, *20*, 1623–1627.

32. Choi, Y.; Kim, J.; Lee, H.S.; Kim, C.; Hwang, I.K.; Park, H.K.; Oh, C.H. Selenium content in representative Korean foods. *J. Food Compos. Anal.* **2009**, *22*, 117–122. [CrossRef]

33. Moreira, D.P.; Monteiro, M.C.; Ribeiro-Alves, M.; Donangelo, C.M.; Trugo, L.C. Contribution of chlorogenic acids to the iron-reducing activity of coffee beverages. *J. Agric. Food Chem.* **2005**, *53*, 1399–1402. [CrossRef]

34. Kwak, H.S.; Ji, S.; Jeong, Y. The effect of air flow in coffee roasting for antioxidant activity and total polyphenol content. *Food Control* **2017**, *71*, 210–216. [CrossRef]

35. Virgen-Navarro, L.; Herrera-López, E.J.; Corona-González, R.I.; Arriola-Guevara, E.; Guatemala-Morales, G.M. Neuro-fuzzy model based on digital images for the monitoring of coffee bean color during roasting in a spouted bed. *Expert Syst. Appl.* **2016**, *54*, 162–169. [CrossRef]

36. Buera, M.P.; Lozano, R.D.; Petriella, C. Definition of Color in the Non-enzymatic Browning Process. *Die Farbe* **1986**, *33*, 316–326.

37. Ludwig, I.A.; Mena, P.; Calani, L.; Cid, C.; Del Rio, D.; Leand, M.E.; Crozier, A.; Lean, J. Variations in caffeine and chlorogenic acid contents of coffees: What are we drinking? *Food Funct.* **2014**, *5*, 1718–1726. [CrossRef] [PubMed]

38. Zhang, S.J.; De Bruyn, F.; Pothakos, V.; Contreras, G.F.; Cai, Z.; Moccand, C.; Weckx, S.; De Vuyst, L. Influence of Various Processing Parameters on the Microbial Community Dynamics, Metabolomic Profiles, and Cup Quality During Wet Coffee Processing. *Front. Microbiol.* **2019**, *10*, 1–24. [CrossRef] [PubMed]

39. Shadaksharaswamy, M.; Ramachandra, G. Changes in the oligosaccharides and the α-galactosidase content of coffee seeds during soaking and germination. *Phytochemistry* **1968**, *7*, 715–719. [CrossRef]

40. Tsai, J.H.; Hiserodt, R.D.; Ho, C.T.; Hartman, T.G.; Rosen, R.T. Determination of Volatile organic Selenium Compounds from the Maillard Reaction in a Selenomethionine - Glucose Model System. *J. Agric. Food Chem.* **1998**, *46*, 2541–2545. [CrossRef]

41. Wei, G.J.; Ho, C.T.; Huang, A.S. Determination of volatile compounds formed in a glucose-selenomethionine model system by gas chromatography-atomic emission detector and gas chromatography-mass spectrometry. *Food Chem.* **2009**, *116*, 774–778. [CrossRef]

42. Zzaman, W.; Bhat, R.; Yang, T.A. Effect of superheated steam roasting on the phenolic antioxidant properties of cocoa beans. *J. Food Process. Preserv.* **2014**, *38*, 1932–1938. [CrossRef]

43. Yodkaew, P.; Chindapan, N.; Devahastin, S. Influences of Superheated Steam Roasting and Water Activity Control as Oxidation Mitigation Methods on Physicochemical Properties, Lipid Oxidation, and Free Fatty Acids Compositions of Roasted Rice. *J. Food Sci.* **2017**, *82*, 69–79. [CrossRef]

44. Lim, P.K.; Jinap, S.; Sanny, M.; Tan, C.P.; Khatib, A. The influence of deep frying using various vegetable oils on acrylamide formation in sweet potato (*Ipomoea batatas* L. Lam) chips. *J. Food Sci.* **2014**, *79*. [CrossRef]

45. Capuano, E. *Lipid Oxidation Promotes Acrylamide Formation in Fat-Rich Systems*; Elsevier Inc.: Amsterdam, The Netherlands, 2016; ISBN 9780128028759.

46. Zamora, R.; Hidalgo, F.J. Contribution of lipid oxidation products to acrylamide formation in model systems. *J. Agric. Food Chem.* **2008**, *56*, 6075–6080. [CrossRef] [PubMed]

47. Hidalgo, F.J.; Delgado, R.M.; Zamora, R. Degradation of asparagine to acrylamide by carbonyl-amine reactions initiated by alkadienals. *Food Chem.* **2009**, *116*, 779–784. [CrossRef]

48. Prior, R.L.; Wu, X.; Schaich, K. Standardized methods for the determination of antioxidant capacity and phenolics in foods and dietary supplements. *J. Agric. Food Chem.* **2005**, *53*, 4290–4302. [CrossRef] [PubMed]

49. Masek, A.; Latos-Brozio, M.; Kałuzna-Czaplińska, J.; Rosiak, A.; Chrzescijanska, E. Antioxidant properties of green coffee extract. *Forests* **2020**, *11*, 557. [CrossRef]

50. Sentkowska, A.; Pyrzyńska, K. Investigation of antioxidant activity of selenium compounds and their mixtures with tea polyphenols. *Mol. Biol. Rep.* **2019**, *46*, 3019–3024. [CrossRef] [PubMed]

51. De Bruyn, F.; Zhang, S.J.; Pothakos, V.; Torres, J.; Lambot, C.; Moroni, A.V.; Callanan, M.; Sybesma, W.; Weckx, S.; De Vuysta, L. Exploring the Impacts of Postharvest Processing on the Microbiota and Metabolite Profiles during Green Coffee Bean Production. *Appl. Environ. Microbiol.* **2017**, *83*, 1–16. [CrossRef]

52. Benzie, I.F.F.; Devaki, M. The ferric reducing/antioxidant power (FRAP) assay for non-enzymatic antioxidant capacity: Concepts, procedures, limitations and applications. In *Measurement of Antioxidant Activity & Capacity*; Apak, R., Capanoglu, E., Shahidi, F., Eds.; John Wiley & Sons Ltd.: Hoboken, NJ, USA, 2018; pp. 77–106. ISBN 9781119135357.

53. Vignoli, J.A.; Viegas, M.C.; Bassoli, D.G.; de Benassi, M.T. Roasting process affects differently the bioactive compounds and the antioxidant activity of arabica and robusta coffees. *Food Res. Int.* **2014**, *61*, 279–285. [CrossRef]

54. Sacchetti, G.; Di Mattia, C.; Pittia, P.; Mastrocola, D. Effect of roasting degree, equivalent thermal effect and coffee type on the radical scavenging activity of coffee brews and their phenolic fraction. *J. Food Eng.* **2009**, *90*, 74–80. [CrossRef]

55. Budryn, G.; Nebesny, E.; Oracz, J. Correlation between the stability of chlorogenic acids, antioxidant activity and acrylamide content in coffee beans roasted in different conditions. *Int. J. Food Prop.* **2015**, *18*, 290–302. [CrossRef]

56. Cheng, J.; Chen, X.; Zhao, S.; Zhang, Y. Antioxidant-capacity-based models for the prediction of acrylamide reduction by flavonoids. *Food Chem.* **2015**, *168*, 90–99. [CrossRef]

57. Hamzalıoğlu, A.; Gökmen, V. 5-Hydroxymethylfurfural accumulation plays a critical role on acrylamide formation in coffee during roasting as confirmed by multiresponse kinetic modelling. *Food Chem.* **2020**, *318*, 126467. [CrossRef] [PubMed]

58. Summa, C.A.; de la Calle, B.; Brohee, M.; Stadler, R.H.; Anklam, E. Impact of the roasting degree of coffee on the in vitro radical scavenging capacity and content of acrylamide. *LWT Food Sci. Technol.* **2007**, *40*, 1849–1854. [CrossRef]

Article

Antioxidant Compounds for the Inhibition of Enzymatic Browning by Polyphenol Oxidases in the Fruiting Body Extract of the Edible Mushroom *Hericium erinaceus*

Seonghun Kim [1,2]

[1] Jeonbuk Branch Institute, Korea Research Institute of Bioscience and Biotechnology, 181 Ipsin-gil, Jeongeup 56212, Korea; seonghun@kribb.re.kr; Tel.: +82-63-570-5113; Fax: +82-63-570-5109

[2] Department of Biosystems and Bioengineering, KRIBB School of Biotechnology, University of Science and Technology (UST), 217 Gajeong-ro, Daejeon 34113, Korea

Received: 22 June 2020; Accepted: 16 July 2020; Published: 17 July 2020

Abstract: Mushrooms are attractive resources for novel enzymes and bioactive compounds. Nevertheless, mushrooms spontaneously form brown pigments during food processing as well as extraction procedures for functional compounds. In this study, the dark browning pigment in the extract derived from the edible mushroom *Hericium erinaceus* was determined to be caused by the oxidation of endogenous polyphenol compounds by the polyphenol oxidase (PPO) enzyme family. These oxidized pigment compounds were measured quantitatively using a fluorospectrophotometer and, through chelation deactivation and heat inactivation, were confirmed to be enzymatic browning products of reactions by a metalloprotein tyrosinase in the PPO family. Furthermore, a transcript analysis of the identified putative PPO-coding genes in the different growth phases showed that tyrosinase and laccase isoenzymes were highly expressed in the mushroom fruiting body, and these could be potential PPOs involved in the enzymatic browning reaction. A metabolite profiling analysis of two different growth phases also revealed a number of potential enzymatic browning substances that were grouped into amino acids and their derivatives, phenolic compounds, and purine and pyrimidine nucleobases. In addition, these analyses also demonstrated that the mushroom contained a relatively high amount of natural antioxidant compounds that can effectively decrease the browning reaction via PPO-inhibitory mechanisms that inhibit tyrosinase and scavenge free radicals in the fruiting body. Altogether, these results contribute to an understanding of the metabolites and PPO enzymes responsible for the enzymatic browning reaction of *H. erinaceus*.

Keywords: enzymatic browning; antioxidant compounds; *Hericium erinaceus*; mushroom metabolites; polyphenol oxidase; tyrosinase; laccase; natural inhibitor

---

1. Introduction

Fungi, including yeasts, molds, and mushrooms, are attractive resources for biotechnological applications. Of these organisms, many mushroom species have historically been used for food or medicinal purposes. Approximately 140,000 mushroom species belonging to the phyla Ascomycetes and Basidomycetes have been identified [1]. Over 2000 mushroom species are edible, and at least 10% of these species have been used traditionally for food or medicine [1]. Moreover, mushrooms produce small bioactive molecules such as secondary metabolites and polysaccharides that can be used for their medicinal properties in antitumor, immunomodulatory, antibacterial, and antiviral treatments [1–4].

Hericium erinaceus (also called lion's mane mushroom, bearded tooth mushroom, satyr's beard, bearded hedgehog mushroom, pom pom mushroom, or bearded tooth fungus) is an edible and medicinal mushroom found in East Asia (China, Korea, and Japan) that belongs to the tooth fungus group [5]. This mushroom species has a non-typical mushroom shape, with no cap and no stem. It is found on hardwoods and identifiable as a single clump within long dangling spines. The mushroom fruiting body is mainly white, although it becomes brown or yellow with age. In East Asia, *Hericium* species mushrooms are highly valued for their medicinal properties and have been used in traditional Chinese medicine [4]. In the past few decades, there has been considerable research interest in the bioactive properties of this species [5]. Several compounds isolated from *H. erinaceus* have been found to have biological activities, including antitumor, antimicrobial, antioxidant, and cytotoxic activities [5–8].

The mushroom fruiting body is usually white initially, but, when processed for food, the mushroom sometimes develops brown to black pigments. Indeed, in the initial purification process for both proteins and bioactive compounds, the solution extracted from the mushroom lysate is pale yellow, and the extract darkens over time during these processes [9]. The oxidation reaction of polyphenols and other oxidizable substances can potentially be prevented by antioxidant compounds in the mushroom fruiting body. However, oxidizable and aromatic-ring-containing substances in the mushroom fruiting-body extract can be exposed, resulting in their oxidization under aerobic conditions. Similarly to well-known plant substances [10], a number of aromatic compounds, including hericines, erinacines, alkalonides, lactones, and their derivatives [5], can presumably be converted to pigment-forming molecules. These mushroom metabolite substances are possibly oxidized products that are formed by polyphenol oxidases (PPOs) in the mushroom fruiting body. However, there are no reports on these pigment-forming compounds or the potential enzymes in *H. erinaceus*.

PPOs, including tyrosinase and laccase, are enzymes well-known for their activities related to post-harvest browning pigments in agricultural products [11]. Although these pigment compounds, such as certain secondary metabolites, might be useful for enhancing bioactive compounds in fermentation processes or for helping to preserve protein in forage crops, the browning reactions are generally viewed negatively in food processing or in fresh products [10–12]. The dark color reaction occurs via PPOs, which can be induced by wounding or pathogen attacks as part of the defense response in plants and fungi [10,12]. Indeed, it has been shown that PPOs play such a role in the white button mushroom (*Agaricus bisporus*) [13]. Nonetheless, the exact kinds of substances in the mushroom that are converted to brown products by some PPOs are unknown in mushroom species.

In our previous study, products of the dark browning reaction were observed as interfering substances during the purification process for a lectin protein from the fruiting-body lysate of *H. erinaceus* [9]; the chromatography fractions were dark brown with a gradient in the upper part. However, there is a lack of information available about this coloration event in the mushroom fruiting-body extract. Therefore, the aim of this study was to identify the cause of the dark brown pigmentation via the endogenous enzymatic reaction of the PPO family with oxidation activity, and to reduce the occurrence of this pigmentation. Furthermore, the putative PPO-coding genes and their transcripts in both the fruiting body and the mycelium were analyzed as putative proteins involved in the enzymatic browning reaction in the mushroom. In addition, a mushroom metabolite analysis was performed to identify natural compound families in the fruiting body and mycelium of *H. erinaceus* as potential antioxidant substances, or as the browning-reaction substrates of the PPOs.

2. Results and Discussion

2.1. Observation of Dark Brown Pigment in the Fruiting-Body Lysate

The fruiting body of the edible mushroom *H. erinaceus* is normally white. Its color becomes yellow or brown over time. In the preparation of the fruiting-body lysate for protein purification, the color of the mushroom crude lysate was initially pale yellow. However, over time, the color of the DEAE-sepharose column flow-through fractions collected during protein purification became dark brown (Figure 1). The dark brown color was observed in Fractions 3 (F3) and 4 (F4). Interestingly, the color pigments localized to the upper part of the test tubes, where the surface was exposed to air, which created a gradient from top to bottom. The yellow color remained at the bottom of the F3 and F4 tubes, but the dark pigmentation progressed over time. This dark brown pigmentation pattern implied that the pigmentation process involves oxygen and certain enzymes or proteins found in F3 and F4.

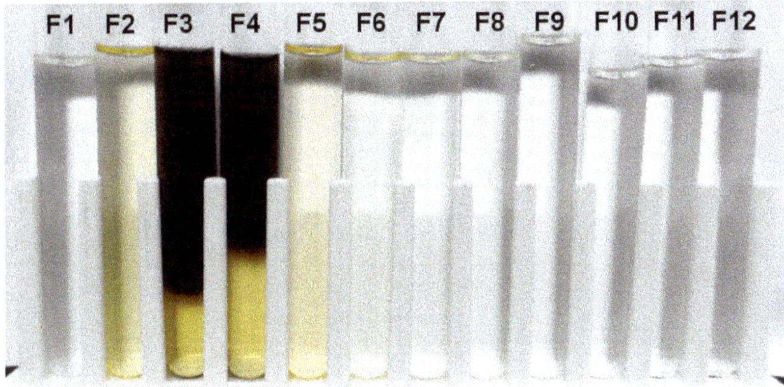

Figure 1. The fruiting body extract fractions (F1–F12) eluted from an ion-exchange column, DEAE-Sepharose, in the edible mushroom *Hericium erinaceus*, with dark brown pigments in F3 and F4.

2.2. Measurement of Dark Brown Pigment

To measure the relative amount of the brown pigment, the absorbance of the fractions was scanned by ultraviolet-visible (UV-vis) spectroscopy in the wavelength range of 210–750 nm. However, the detection wavelength for the pigment was less than 210 nm (data not shown). Therefore, UV-vis spectrophotometry was not useful for measuring the pigmentation, because a mixture of proteins as well as pigment products were present in the fraction (data not shown). The proteins were interfering substances to detect the pigment in the mushroom extract at 210 nm.

An alternative method, fluorescence spectroscopy, was used to detect the brown pigmentation. The fractions with dark brown pigment were scanned. To determine the optimum emission and excitation wavelength for the pigment, the emission wavelength (*Em*) was fixed at 460 nm and the excitation (*Ex*) spectra ranged from 250 to 420 nm was scanned. The highest fluorescence was obtained at 380 nm as the optimal emission wavelength. The excitation wavelength was then fixed at 380 nm, and the emission spectra were scanned from 420 to 700 nm. The maximum intensity signal was detected at the emission wavelength at 444 nm (Figure 2a) setting the excitation wavelength at 380 nm. The emission intensity of each sample corresponded to the relative amount of brown pigmentation, which formed in a time-dependent manner (Figure 2b).

The fluorescence intensity of fraction pools collected was measured at $Ex_{380nm} \sim Em_{444nm}$. All fractions exhibited positive fluorescence intensity readings. The visible intensity of the dark brown pigmentation seemed to correlate with the measured fluorescence intensity for all fractions (Figure 3a). The maximum fluorescence intensity was observed in the pooled crude (CL) fraction (Figure 3b). The fluorescence intensity in the CL fraction was 1.4-fold higher than that in the pooled flow-through (FT) fraction; the CL fraction showed higher fluorescence intensity than the other fractions eluted from the DEAE-sepharose column. The fluorescence intensities of both the CL and FT fractions were 2.7–3.6 times higher than those in the NaCl elution fractions. Although the elution fraction (E250) with 250 mM NaCl was a light yellow color (Figure 3a), the fluorescence intensity of this fraction was lower than that of the CL or FT fractions. However, the E250 fraction showed an approximately 1.2–1.3-fold higher intensity than the other NaCl elution fractions (E50, E500, E1000, and E5000). All of the NaCl elution fractions, except for the E250 fraction, appeared clear, but a fluorescence signal from the dark pigmented compounds was still detected.

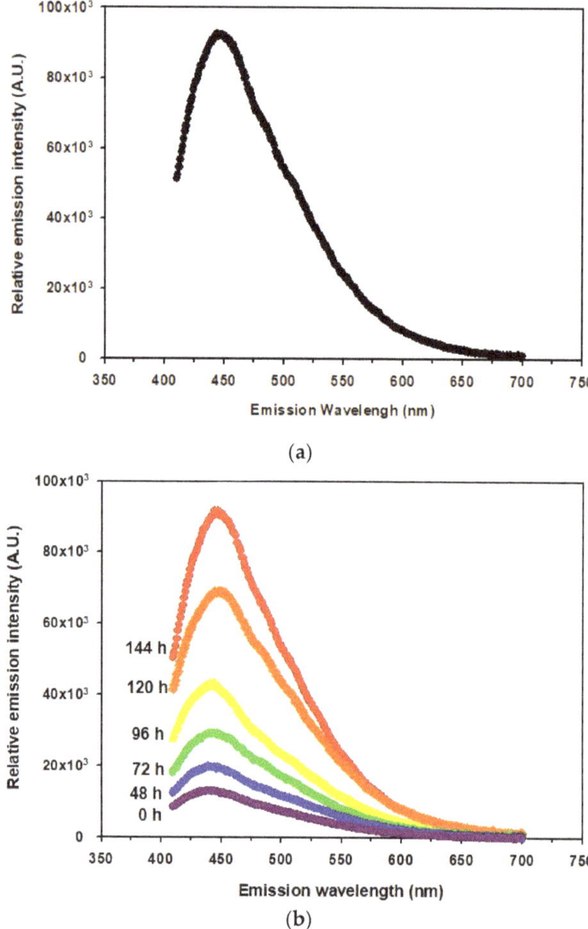

Figure 2. Fluorescence emission spectra of the dark brown pigment in the protein fraction. (**a**) Emission spectrum of the dark pigment fraction scanned by a fluorospectrophotometer with an excitation wavelength of 380 nm. (**b**) Fluorescence spectra of the pigment fractions for relative quantification of the brown pigment, which increased in a time-dependent manner under the same conditions. The fluorescence intensity of every fraction was measured with the following time points: initial 0 h, purple; 48 h, blue; 72 h, green; 96 h, yellow; 120 h, orange; 144 h, red.

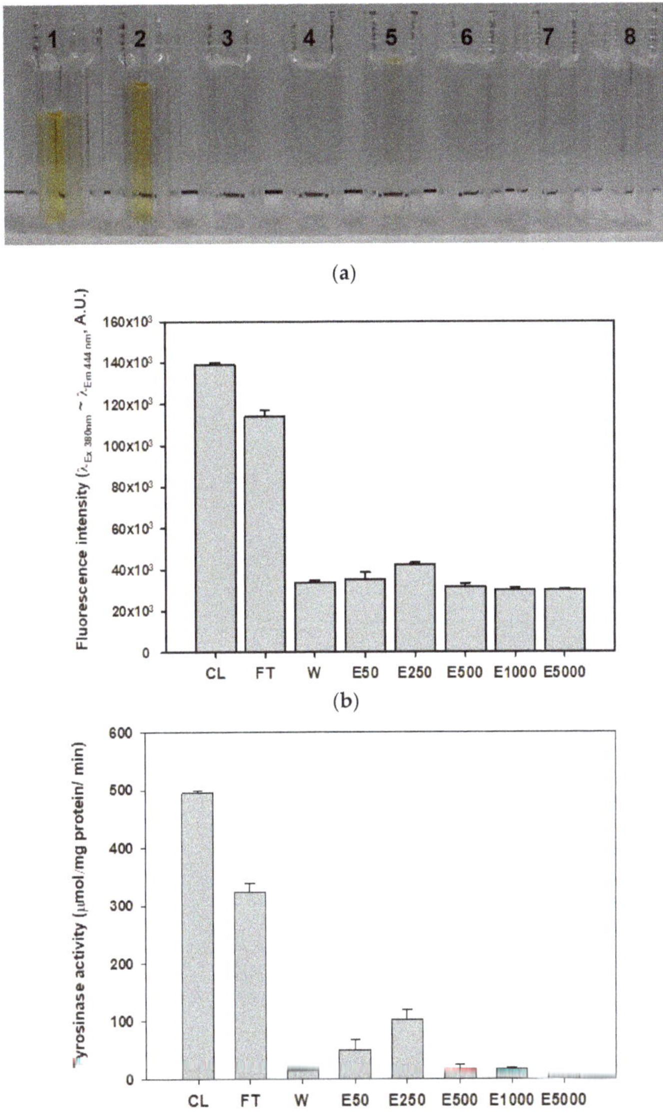

(a)

(b)

(c)

Figure 3. Measurement of the browning pigment substances and tyrosinase activities in the fraction pools collected in the DEAE-Sepharose column. (**a**) The protein fraction pools eluted from the DEAE-Sepharose column: 1, crude extract fraction pool (CL); 2, flow-through fraction pool (FT); 2, wash fraction pool (W); 4–8, elution fraction pools with 50 mM (E50), 250 mM (E250), 500 mM (E500), 1000 mM (E1000), and 5000 mM NaCl (E5000), respectively, in 50 mM Tris-HCl buffer. (**b**) Fluorescence intensities (Arbitrary Unit, A.U.) of these fraction pools for the measurement of brown pigment under the optimum conditions with $\lambda_{Ex380\,nm} \sim \lambda_{Em460nm}$. (**c**) Tyrosinase-specific activities of these fraction pools under the enzyme assay conditions described in the material and method section.

The fluorescence spectra of the dark brown pigment suggested that the compounds responsible for the pigmentation might be aromatic or heterocyclic aromatic compounds [14]. Mushrooms contain a considerable number of phenolic compounds, such as flavonoids [5,7,8]. These compounds are antioxidant substances in mushrooms, but are also good substrates for polyphenol oxidases (PPOs) [15]. The mechanisms by which the dark pigment is formed in *H. erinaceus* are not clearly understood, but browning in cultivated button mushrooms (*Agaricus bisporus*), an extensively cultivated mushroom, is induced by bacterial infection and oxidation after harvesting [16]. It has reported that the brown pigments form as a defense mechanism when mushroom tissue becomes damaged or infected [16]. This pigmentation is mainly caused by the oxidation of endogenous phenolic substances, which are enzymatically catalyzed by PPOs, including tyrosinases and laccases.

2.3. Measurement of Tyrosinase Activity and its Deactivation in the Mushroom Extract

To test whether PPOs were responsible for the pigmentation occurring in *H. erinaceus*, the enzymatic activity of the tyrosinase belonging to the PPO family was measured in the fractions eluted from the DEAE-sepharose column (Figure 3c). The CL fraction showed the highest tyrosinase activity per unit protein in all of the fractions tested. In addition, the FT fraction had higher enzyme activity than other fractions. However, tyrosinase activities in the NaCl elution fractions were much lower than those in the CL or FT fractions. The tyrosinase activity values correlated with the fluorescence intensity values in each fraction described above (Figure 3b). These results could suggest that the dark brown pigmentation that occurs in *H. erinaceus* could be due to enzymatic reactions catalyzed by tyrosinases. It seems the CL and FT fractions could contains tyrosinases with natural substrates such as phenolic compounds to convert the brown pigment. However, the NaCl eluted fractions contained low amount of the enzyme as well as less amount of the low molecule substances, which were already pass through the column. Therefore, the intensity of the brown pigment in the elution fractions would lower than the CL and FT fractions.

In addition, to confirm that the dark brown pigment was an enzymatic reaction product, the endogenous tyrosinase in the mushroom lysate was immediately deactivated or inactivated. The chelating agent ethylenediaminetetraacetic acid (EDTA) can eliminate metal ions in a copper-containing metalloprotein to deactivate tyrosinase [15,16]. EDTA was added in concentrations ranging from 1 to 25 mM into the mushroom lysate. At the initial time-point, the fluorescence intensities in EDTA-treated samples (black circles) and untreated samples were slightly different (Figure 4a). There was a slight trend towards decreasing fluorescence intensity, depending on the concentration of EDTA. After 14 days, the fluorescence intensity (white circles) did not increase significantly (less than 10%). With the chelation agent added to the lysate, tyrosinase activity was strongly inhibited (Figure 4b). Even low concentrations (1 mM) of EDTA abolished the tyrosinase activity by up to 90%.

For the inactivation of tyrosinase activity, the mushroom lysate was treated at 100 °C for 1 to 30 min. The fluorescence intensity and the tyrosinase activity in the heat-treated lysates were monitored. The fluorescence intensity (black squares) gradually increased in the heat-treated samples, depending on the incubation time (Figure 4c). The fluorescence intensity of the lysate heated for 30 min increased 1.3-fold compared with that of non-heated samples. The heat treatment may have acted to enhance the browning pigment responsible for the fluorescence intensity in these samples. Heat treatment reportedly changed the composition of polyphenolic compounds in shiitake mushroom extracts [17]. The polyphenolic content of shiitake extracts was enhanced with heating temperature and treatment time. However, after 14 days, the fluorescence intensity (white squares) of the heat-treated lysates had decreased by less than 10–13%. Moreover, heat treatment was effective with respect to tyrosinase activity, which decreased by 58.0% after 1 min, 75.2% after 5 min, and 76.8% after 10 min at 100 °C (Figure 4d). The enzymes in the samples treated for 30 min at 100 °C were almost completely deactivated, and the residual enzyme activity was 5.5%. However, the enzyme activity in the heat-treated samples was higher than that in EDTA-treated samples. The heat treatment did not completely inactivate the enzyme. In addition, the heat treatment could have influence on increasing spontaneously brown coloring by the chemical reactions such as sugar-amine reactions in the endogenous substance of the mushroom extract. These results indicate that the brown coloring pigmentation in *H. erinaceus* extracts could be the product of an enzymatic reaction by a tyrosinase belonging to the PPO family of enzymes.

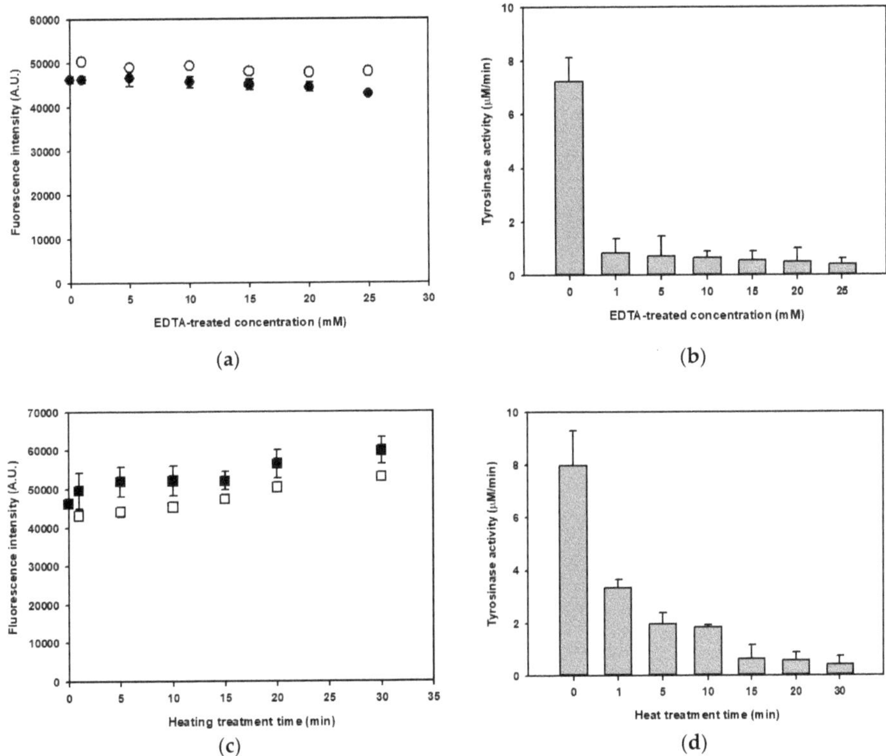

Figure 4. Measurement of fluorescence intensities and tyrosinase activities in the mushroom lysate treated with EDTA or heat. (**a**) Fluorescence intensities of the EDTA-treated lysates at the initial time (black circles) and 14 days later (white circles). (**b**) Total tyrosinase activities of EDTA-treated lysates. (**c**) Fluorescence intensities of the heat-treated lysates at the initial time (black squares) and 14 days later (white squares). (**d**) Total enzyme activities of the heat-treated lysates. Fluorescence intensities (Arbitrary Unit, A.U.) and tyrosinase activities of all fractions were measured under the optimum conditions with λE × 380 nm~λEm 460 nm and the enzyme assay conditions described, respectively.

2.4. Analysis of the Transcription Levels of PPO-Coding Genes in the Different Growth Phases of the Mushroom

To identify the correlation of the PPO-coding cDNAs with the dark browning event, the expression levels of the tyrosinase and the laccase genes in the RNA-seq data of the fruiting body and the mycelia were analyzed (Figure 5). The isotranscript genes for tyrosinase and laccase were differentially expressed in the growth phases; among seven putative tyrosinase-coding genes identified from expressed transcripts in the mushroom genome, three of them, *ust*Q, *tyr*1, and *asq*I, were observed in both the fruiting body and mycelium. Interestingly, the expression of *ust*Q was approximately 20-fold higher in the fruiting body than in the mycelium, whereas the expression of the *tyr*1 gene was about 1.1-fold higher. However, the expression of another tyrosinase gene (*asq*I) was 1.4-fold lower in the fruiting body than in the mycelium. The *ust*Q-encoded tyrosinase could be a major PPO enzyme involved in the enzymatic browning reaction in the *H. erinaceus* fruiting body.

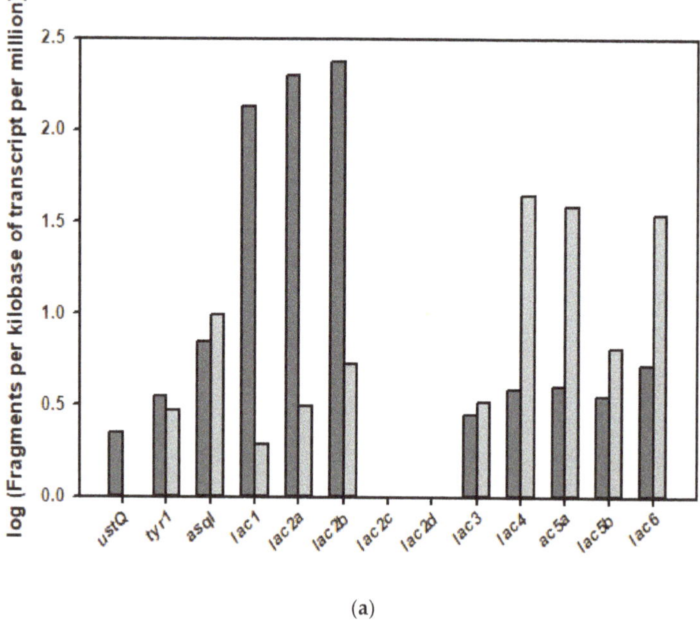

(**a**)

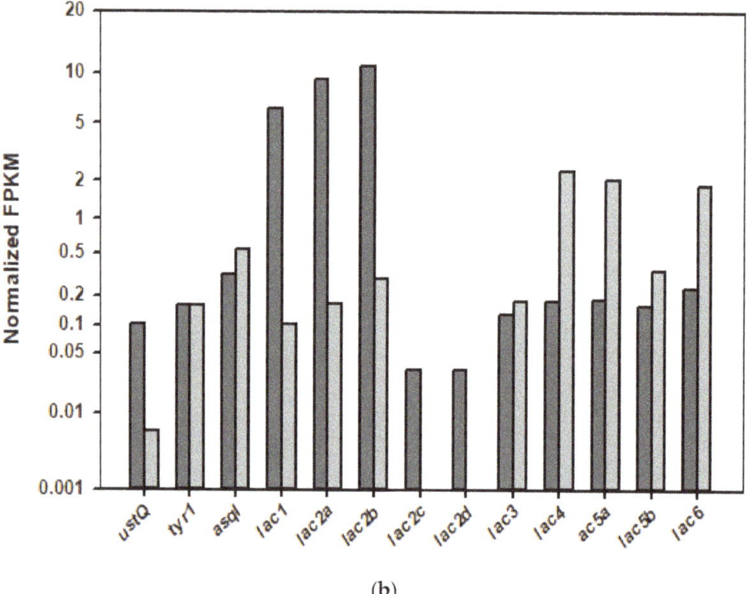

(**b**)

Figure 5. Transcription levels of tyrosinase- and laccase-coding genes involved in the enzymatic browning reaction in the fruiting body (dark gray) and the mycelium *(gray-white) of H. erinaceus*. (**a**) Log scale for transcription levels expressed as fragments per kilobase of transcript per million (FPKM). The log (FPKM) values of *lac2c* and *lac2d* were less than −0.1. (**b**) Normalized FPKM values relative to a suite of endogenous actin genes in each growth phase.

Thirteen putative laccase-coding genes were identified in the genome, but only 10 transcripts were expressed in the fruiting body and mycelium. These transcripts were differentially regulated in different growth phases (Figure 5). In the fruiting body, five laccase transcripts were upregulated, and four other genes were downregulated. Interestingly, the *lac1*, *lac2a*, and *lac2b* transcripts had significantly higher isotranscript expression in the fruiting body; the transcription levels of *lac1*, *lac2a*, and *lac2b* genes were upregulated 59-, 55-, and 38-fold in the fruiting body compared with the mycelium. Among these more highly expressed transcripts, *lac2* genes were expressed as four isotranscripts (*lac2a*, *lac3b*, *lac2c*, and *lac2d*), and the expression levels of all of these transcripts were higher in the fruiting body. On the other hand, five other laccase-coding transcript isoform genes were downregulated in the fruiting body, while these transcripts were highly expressed in the mycelium. Two transcripts, *lac4* and *lac5a*, among the five genes that were upregulated in the mycelium, had about 10-fold higher expression in the mycelium growth stage. The transcripts of *lac3*, *lac5b*, and *lac6* had 1.4-, 2.1-, and 7.7-fold higher expression in the mycelium.

These transcriptional analyses showed that the laccase transcripts were also highly expressed and regulated in the growth stages. Moreover, the laccase isoenzymes translated from the upregulated transcripts could be potential PPOs involved in the enzymatic browning reaction in the fruiting body.

2.5. Analysis of the Main Metabolites in the Mushroom Fruiting Body and Mycelium

To analyze the potential substances involved in the enzymatic browning reaction or compounds with protective roles in the mushroom fruiting body and mycelium, a metabolomic analysis was performed in two different growth phases using CE-TOF-MS and LC-TOF in two modes for cationic and anionic metabolites with molecular weights ranging from 50 to 1000 *m/z*. In the fruiting body, CE-TOF-MS detected 198 metabolites (138 metabolites in cation mode and 60 metabolites in anion mode), and LC-TOF-MS identified 63 metabolites (26 metabolites in positive mode and 37 metabolites in negative mode). On the other hand, in the mycelium, CE-TOF-MS detected 171 metabolites (107 metabolites in cation mode and 64 metabolites in anion mode), and LC-TOF-MS revealed 38 metabolites (12 metabolites in positive mode and 26 metabolites in negative mode).

Among these metabolites, well-known natural compounds with tyrosinase-inhibitory activity and antioxidant activity (Table 1) and low-molecular-weight substances potentially involved in the enzymatic browning reaction (Table 2) were identified, respectively, in the mushroom fruiting body and the mycelium. Interestingly, the antioxidant and enzymatic browning substances presented in a growth-stage-dependent manner. For protection against the endogenous browning reaction in the mushroom, 18 putative antioxidants substances were detected and categorized into tyrosinase inhibitors, antioxidants/free-radical scavengers, and polyamines (Table 1). A large number of metabolites with antioxidant activity were found in the fruiting body compared with in the mycelium (17 vs. 9), with an overlap of eight compounds. Of these antioxidant substances, ascorbic acid, caffeic acid, chlorogenic acid, curcumin, apyridoxal, and quinic acid—natural compounds with tyrosinase-inhibitor activity—were detected in the mushroom fruiting body, while vallinic acid was only found in the mycelium. However, terephalic acid was detected in both mushroom development stages. As tyrosinase inhibitors, the levels of ascorbic acid and quinic acid were significantly higher than those of other compounds in the fruiting body.

Table 1. Putative metabolites for the natural polyphenol oxidase inhibitors in *H. erinaceus* fruiting body and mycelium.

Compound	Relative Peak Area [1]		m/z [2]	MT/RT [3]
	Fruiting Body	**Mycelium**		
Antioxidant				
Tyrosinase inhibitor				
Ascorbic acid	640	-	175.025	8.34
Caffeic acid	88	-	179.035	8.88
Chlorogenic acid	92	-	355.107	5.51
Cucumin	45	-	269.134	10.32
Pyridoxal	6	-	166.066	8.52
Quinic acid	6400	-	191.058	8.07
Terephtalic acid	110	7.4	165.020	16.38
Vanillic acid	-	4.0	167.035	8.71
Antioxidant, free radical scavenger				
Adenosine	510	560	268.105	9.60
Ergosterol	1300	1300	379.339	15.63
Ergothioneine	2900	100	230.096	17.29
Glutathione (GSH)	12	240	308.092	12.98
5-Hydroxytryptophan	20	3.7	221.094	11.20
Hypotaurine	33	-	110.027	18.06
Tartaric acid	180	-	149.011	22.05
Polyamine				
1,3-Diaminopropane	19	-	75.091	4.31
Putrescine	16	1.3	89.107	4.58
Spermidine	190	2.4	146.165	4.40

[1] Peak area × 100,000 divided by the internal standard peak area of metabolic compounds in the two growth stages (the mushroom fruiting body and mycelium) [2] MS scan range: m/z 50~1000 [3] MT (migration time) in CE/RT (retention time) in LC.

Table 2. Potential substrates in the enzymatic browning reaction in the *H. erinaceus* fruiting body and mycelium.

Compound	Relative Peak Area		m/z [2]	MT/RT [3]
	Fruiting Body [1]	**Mycelium** [1]		
Enzymatic browning substrates				
Amino acid and its derivative				
Indole-3-carboxyaldehyde	-	170	146.062	6.91
Kynurenine	100	5.5	209.093	9.32
Phenylalanine	9400	1500	166.086	10.44
Quinolinic acid	160	-	166.016	15.47
Tryptamine	6.9	-	161.107	8.15
Tryptophan	520	110	205.097	10.69
Tyrosine	4900	280	182.081	11.09
Phenolic compounds				
o-Aminophenol	-	2.7	110.060	7.4
o-Hydroxybenzoic acid	-	2.1	137.025	10.10
2-Phenylethylamine	19	-	122.097	7.52
3-Phenylpropionic acid	-	5.6	149.062	8.60
p-, *m*-, *o*-Toluic acid	-	33	135.046	8.91
Purine and pyrimidine nucleobases				
Purine base				
1-Methyladenosine	4.8	3.0	282.119	9.39
2′-, 5′-Deoxyadenosine	-	2.1	252.111	9.09
3′-AMP	-	2.9	346.059	9.42
5′-Deoxy-5′-methylthioadenosine	-	2.2	298.097	9.49

Table 2. *Cont.*

Compound	Relative Peak Area		m/z[2]	MT/RT [3]
	Fruiting Body [1]	Mycelium [1]		
AMP	290	11	346.058	9.02
ADP	260	33	426.024	10.52
ATP	270	33	505.991	11.36
Adenine	-	9.5	136.062	7.20
GMP	33	-	362.051	8.97
GDP	-	3.3	442.020	10.40
GTP	41	-	521.986	11.24
Guanosine	100	380	284.100	12.27
Hypoxanthine	23	14	137.046	10.45
IMP	-	1.8	347.038	9.28
Inosine	-	25	269.090	18.37
Xanthine	-	13	153.042	18.35
Pyrimidine base				
Cytidine	32	33	244.093	9.12
UMP	35	3.6	323.032	9.51
UDP	-	6.7	402.998	11.22
UTP	-	7.5	482.962	12.10
Uracil	-	17	113.036	20.08
Uridine	29	210	245.078	20.08

[1] Peak area × 100,000 divided by the internal standard peak area of metabolic compounds in the two growth stages (the mushroom fruiting body and mycelium) [2] MS scan range: m/z 50~1000 [3] MT (migration time) in CE/RT (retention time) in LC.

Moreover, adenosine, ergosterol, erothioneine, glutathione, and 5-hydroxytryptophan were commonly observed as other antioxidants/free-radical scavengers in both the fruiting body and the mycelium (Table 1). In terms of the differences among these compounds, the amount of erothioneine was 29 times higher, whereas the content of glutathione was 20 times lower in the fruiting body than in the mycelium. These data may indicate that different compounds in different mushroom growth stages are synthesized as antioxidant substances for the inhibition of the enzymatic browning reaction or for the scavenging of free radicals.

On the other hand, for the enzymatic browning products, free amino acids and their derivatives, phenolic compounds, and purine and pyrimidine nucleobases were identified as potential substrates (Table 2); these compounds were less abundant in the fruiting body than in the mycelium (18 vs. 29), with an overlap of 12 compounds. Interestingly, the compounds that were grouped as free amino acids and their derivatives were predominantly found in the fruiting body, while phenolic compounds and purine/pyrimidine nucleobases were mainly detected in the mycelium. Phenylalanine and tyrosine were the main substances in terms of the relative amounts of these compounds. These amino acids could be used as substrates for the enzymatic browning reaction by PPOs, including tyrosinase and laccase. Both phenylalanine and tyrosine are favorable substrates and initial precursors for the dark pigment in the melanin synthesis pathway [18–20]. In addition, purine and pyrimidine compounds that contain a pentose sugar—ribose or deoxyribose—with an amine group in their backbone structures are also known to be potential substrates for browning reactions via the Maillard reaction under high temperatures [21]. Adenosine nucleoside phosphate derivatives—AMP, ADP, and ATP—were the main compounds in the nucleobase family. They could be putative browning substances, since their levels were approximately 10 times higher in the fruiting body than in the mycelium. These metabolite analysis results showed that the fruiting body and the mycelium can biosynthesize endogenous antioxidant compounds to reduce the browning reaction, whereas synthesized metabolites such as aromatic amino acids, nucleobases, and their derivatives might be prospective substrates of PPOs for the enzymatic browning reaction.

In this study, to identify the browning reaction in the mushroom extract through the endogenous enzymatic reaction of the PPO-family enzymes with oxidation, a method was established to quantitatively measure dark brown pigmentation in mushroom fruiting-body lysate using a fluorospectrophotometer. Excitation and emission wavelengths of 380 nm and 444 nm, respectively, were found to be the most efficient for measuring this pigmentation. Compounds that fluoresce between these wavelengths generally have aromatic functional groups, aliphatic or alicyclic carbonyl structures, or highly conjugated double-bond structures with low energy-transition levels [22]. Although both the pH and solvent affected the fluorescence readings in these samples, the wavelengths used in this study quantitatively detected the dark brown pigments. Interestingly, the wavelengths used ($\lambda_{Ex380nm}$ ~ $\lambda_{Em460nm}$) were close to those used to detect cyclized o-quinone oxidation products (aminochrome, dopachrome, and furanoquinone), generated from the tyrosinase-catalyzed oxidation of dopamine, L-DOPA, and 3,4-dihydroxyphenylacetic acid [23]. These cyclized o-quinone oxidation products are eventually polymerized to melanin [23].

The pigment products were estimated to be oxidized polyphenol compounds similar to melanin, derived from PPOs in the mushroom lysate [24]. PPOs catalyze enzymatic browning by oxidizing phenolic compounds to their respective o-quinones, which subsequently undergo nonenzymatic oxidation to brown melanin pigments or participate in addition–polymerization reactions with protein functional groups to form cross-linked polymers [25]. This phenomenon is a major problem in the processing of both fruits and vegetables [26]. Moreover, both tyrosinase and laccase are well-known oxidase enzymes that cause brown pigmentation in mushroom species [15,16].

Tyrosinases (monophenol, o-diphenol:oxygen oxidoreductase, EC 1.14.18.1) are type-3 copper proteins that are involved in the initial step of melanin synthesis [15]. These enzymes catalyze both the o-hydroxylation of monophenols and the subsequent oxidation of the resulting o-diphenols into reactive o-quinones, which evolve spontaneously to produce intermediates that are associated with dark brown pigments [15]. In mushrooms, tyrosinases are endogenous biological factors that are associated with the formation and stability of spores in defense and virulence mechanisms, and in browning and pigmentations [27–29]. Although the catalytic mechanisms of these enzymes are still unknown, the deactivation and inactivation of the metalloprotein tyrosinases in the mushroom lysate seen in this study may be a way to prevent browning, because neither the EDTA-treated nor heat-treated lysate became pigmented.

The other enzyme type, laccase (EC 1.10.3.2), is a multicopper oxidase that is capable of oxidizing a wide range of aromatic compounds [30]. These enzymes also catalyze the oxidation of o- and p-diphenols into their corresponding quinones for the enzymatic browning reaction. Although the roles of plant laccase in the radical coupling of monolignols to form lignin and flavonoid polymerization in the cell wall [31] and in the protection of plants against environmental stresses [32,33] are well-understood, the in vivo biological functions of mushroom enzymes belonging to the PPO family are still unclear. Nevertheless, the broad substrate specificities of laccases make them enzymes with potential involvement in the pigment formation via oxidation of endogenous aromatic compounds in mushrooms [34].

According to the expression pattern in the fruiting body and mycelium growth phases, a transcription analysis was performed using RNA-seq sequencing, which showed that the number of laccase transcripts was more diverse than tyrosinase transcripts, and the expression levels of certain laccase genes, namely lac1, lac2a, and lac2b, were significantly increased in the fruiting body compared with levels in the mycelium (Figure 5). As PPOs, these laccases would be abundant in the tissue of the mushroom; furthermore, these enzymes could be the predominant enzymes that form the dark pigments in the fruiting-body lysate. Nevertheless, the color of the fresh mushroom fruiting body did not change in spite of these highly expressed proteins. The enzyme activities of PPOs, including tyrosinases and laccases, could be strongly inhibited by the endogenous antioxidant inhibitory compounds in the mushroom metabolites. However, in chromatography, PPO enzymes with high molecular weights in the mushroom lysate were separated into low-molecular-weight inhibitors of metabolites when spontaneously exposed to oxygen; thus, these enzymes could be activated for the dark browning reaction.

The profile analysis of the mushroom metabolites in the different development phases revealed putative antioxidant compounds (Table 1) as well as enzymatic browning substrates (Table 2). Interestingly, some phenolic compounds, such as caffeic acid and chlorogenic acid, could be potential antioxidants as well as substrates of PPOs in the browning reaction in the mushroom fruiting-body extract. Adenine, a purine nucleobase, was also found to be a bilateral substrate in the mushroom. However, there were few data about how these compounds function as antioxidants or browning-reaction substances under certain conditions. Interestingly, contents of the endogenous tyrosinase inhibitors ascorbic acid and quinic acid were high only in the fruiting body (Table 1). Of these substances, ascorbic acid is already a well-known compound and additive that effectively prevents browning due to tyrosinase in *A. bisporus* [35]. Quinic acid is known to be an inducer of the antioxidant mechanism that increases urinary excretion of tryptophan and nicotinamide in the mammalian system [36], but it is also unknown whether this compound might play a similar role in mammals. Other antioxidant substances, such as ergosterol and ergothioneine, with the ability to scavenge free radicals and reactive oxygen species [37] as well as to chelate divalent metal ions [38], could inhibit the enzymatic browning reaction of PPOs in the mushroom fruiting body [39,40].

Indeed, these antioxidant compounds from mushrooms are already known for their significant bioactivity [19,20,41,42]. Antioxidant metabolites, mainly ascorbic acid, ergosterol, and ergothioneine, in the *H. erinaceus* fruiting body could prevent the oxidation of polyphenolic compounds and serve as anti-browning substances. Moreover, these functional compounds are potentially useful nutrient components themselves, as a food source to help the organism decrease oxidative stress [5,43]. Other potential applications of these active mushroom compounds include acting as natural ingredients for cosmetics because of their antioxidant, anti-tyrosinase, and anti-collagenase activities, among others [42,44]. Mushroom-based ingredients for skincare or anti-aging to fight against UV radiation and free radicals are already commercially available in cosmetic markets [42]. Mushroom metabolites identified as tyrosinase inhibitors, antioxidants/free-radical-scavenging compounds, and polyamines could be potential nutrient substances as well as cosmetic ingredients for nutricosmetics, combining foods, cosmetics, and pharmaceuticals [42–44].

3. Material and Methods

3.1. Mushroom Strains and Culture Conditions

Hericium erinaceus mushrooms were obtained from a local mushroom farm (Seojong, Korea). *H. erinaceus* mycelia isolated from the mushroom fruiting body were molecularly identified as NEU-2L and cultivated in potato dextrose broth (Difco, Detroit, MI, USA) at 20 °C for 20 days. The fruiting bodies and mycelia were collected, frozen immediately in liquid nitrogen, and then stored at −80 °C for total RNA extraction and preparation of the mushroom lysate.

3.2. Preparation of the Mushroom Fruiting-Body Lysates

The fruiting bodies were ground into a fine powder with a mortar and pestle while being cooled by liquid nitrogen. The powder was then suspended in 10 mM Tris-HCl buffer (pH 8.0), which contained 1 mM phenylmethylsulfonyl fluoride (PMSF) and a cOmplete™ EDTA-free protease inhibitor cocktail (Roche, Basel, Switzerland), overnight for extraction. The homogenate was centrifuged at 10,000×g for 30 min, and the resultant supernatant was centrifuged again at 20,000× g for 60 min. The clarified solution was used for ion-exchange chromatography.

3.3. Fractionation of the Mushroom Fruiting-Body Lysate Using Ion-Exchange Chromatography

The clarified crude lysate was applied to the purification columns. The crude protein solution was applied to a diethylaminoethanol (DEAE)-sepharose column (2.6 × 20 cm) (GE Healthcare, Chocago, IL,

USA) equilibrated with 10 mM Tris-HCl buffer (pH 8.0) containing 1 mM PMSF and a protease-inhibitor cocktail. After the unbound proteins were washed with five column volumes of buffer, the protein fractions were eluted with three bed volumes of the buffer, followed by either a linear gradient of 0.0–1.0 M NaCl or a step gradient of 0.25 M, 0.5 M, 1.0 M, and 2.5 M NaCl. Fractions (7 mL) were collected at a flow rate of 1 mL/min. The fractions were stored at 4 °C.

3.4. Browning Pigment Analysis

For ultraviolet-visible (UV-vis) spectroscopy analysis, the protein purification fractions (100 μL) collected in the DEAE-sepharose column were transferred to a 96 well polystyrene microtiter plate (Greiner Bio-one, Monroe, LA, USA). The microtiter plate was scanned at room temperature by a Synergy^{TM} H1 Multi-Mode Reader (BioTek, Winooski, VT, USA) microplate reader, using wavelengths ranging from 210 to 750 nm.

For the fluorescence spectroscopy analysis, the protein purification fractions (100 μL) were transferred to a 96 well black microtiter plate (Greiner Bio-one, Monroe, LA, USA). The microtiter plate was also scanned by a Synergy^{TM} H1 Multi-Mode Reader microplate reader. For identification of the optimum excitation wavelength, the emission wavelength (Em) was fixed at 460 nm, and then excitation wavelengths (Ex) ranging from 250 to 420 nm were scanned by the microplate reader. For identification of the optimum Em wavelength, the Ex wavelength was fixed at 380 nm, and then Em wavelengths ranging from 450 to 650 nm were scanned by the microplate reader. Finally, all of the fractions with dark color pigment were measured at the optimum Ex and Em wavelength conditions ($Ex_{380nm} \sim Em_{444nm}$).

3.5. RNA Sequencing of the Mushroom Fruiting Body

RNA sequencing with total RNA extracted from the fruiting bodies and mycelia was performed as described previously [45]. The procedure is described here briefly. Total RNA were extracted from the mushroom mycelia and fruiting bodies using the RNeasy Plant Mini Kit (Qiagen, Hilden, Germany) according to the manufacturer's protocol, and then the extracted RNAs were further treated with DNase I. The DNase-I-treated RNA samples were purified using an RNeasy column (Qiagen, Hilden, Germany). The RNA quality and quantity were analyzed by UV spectrophotometry and gel electrophoresis. The cDNA library was constructed using a TruSeq RNA sample prep kit v2 (Illumina, San Diego, CA, USA), and RNA sequencing with an Illumine HiSeq 4000 sequencing system (Illumina, San Diego, CA, USA) was performed by MACROGEN (Seoul, Korea).

3.6. Tyrosinase Activity Assay

Among polyphenol oxidases, tyrosinase activity with 4-*tert*-butylphenol (Sigma-Aldrich, St. Louis, MI, USA) as a substrate was determined by measuring the amount of *t*-butyl-quinone oxidized by the enzyme. The enzyme activity was assayed in a total volume of 100 μL with 1 mM substrate in a 96 well microtiter plate at 30 °C for 30 min. After the enzymatic reaction, the oxidized monophenol product was measured on a Synergy^{TM} H1 Multi-Mode Reader microplate reader at 400 nm. One unit of tyrosinase was defined as the amount of the enzyme required to produce 1 μmol of *t*-butyl-quinone ($\varepsilon_{400\ nm} = 1150\ M^{-1}cm^{-1}$) per min under the reaction conditions. The background oxidation of the substrate was deduced by using a reference sample with an identical composition to the reaction mixture without the enzyme.

3.7. Deactivation or Inactivation of the Polyphenol Oxidases' Activities in the Mushroom Lysate

For deactivation of the tyrosinase activities, the mushroom lysate was treated by heating at 100 °C. After heat treatment, residual enzyme activities were assayed under standard conditions after incubating the mushroom lysate for the appropriate time.

For the inactivation of enzyme activities, the lysate was treated with a chelating agent, ethylenediaminetetraacetic acid (EDTA), at concentrations ranging from 1 to 25 mM. The appropriate concentration of EDTA was added to the lysate, and then the sample was incubated at room temperature for 1 h. After incubation, residual enzyme activities were measured using the standard assay described above. All enzyme activities were determined in triplicate.

3.8. Metabolite Extracts and Analysis in the Mushroom Fruiting Body and Mycelium

Metabolome measurements were carried out by Human Metabolome Technology (HMT) Inc. (Tsuruoka, Japan). For capillary electrophoresis time-of-flight mass spectrometry (CE-TOF-MS) analysis, the frozen hydrate sample (approximately 30 mg of each sample) was mixed with 500 μL of methanol containing internal standards (50 μM) and homogenized using a homogenizer (1500 rpm, 120 s × 2 times). Subsequently, 500 μL of ultrapure water was added to the homogenates, mixed thoroughly, and centrifuged at 2300× g for 5 min at 4 °C. The clarified solution was filtrated through a 5 kDa cut-off filter (ULTRAFREE-MC-PLHCC, HMT Inc., Yamagata, Japan) to remove macromolecules. The filtrate was centrifugally concentrated and resuspended in Milli-Q ultrapure water for CE-TOF-MS analysis at HMT. CE-TOF-MS was carried out using an Agilent CE Capillary Electrophoresis system (Agilent Technologies, Waldbronn, Germany) equipped with an pump an Agilent 6210 time of flight mass spectrometry, an Agilent 1100 isocratic HPLC pump, an Agilent G1603A CE-MS adapter kit, and an Agilent G1607A CE-ESI-MS spray kit. System were controlled using Agilent G2201AA ChemStation software version B.03.01 for CE. Metabolites were analyzed by using a fused silica capillary (50 μm × 80 cm), with commercial electrophoresis buffer (Solution ID: H3301-1001 for cation and I3302-1023 for anion analysis; HMT. Inc.) as the electrolyte. Samples were injected at 50 mbar for 10 s for cation analysis and 22 s for anion analysis. The spectra were scanned from m/z 50 to 1000.

For liquid chromatography time-of-flight/mass spectrometry (LC-TOF-MS) analysis, the frozen hydrate sample was mixed with 500 μL of 1% formic acid in acetonitrile (v/v) containing internal standards (10 μM) and homogenized using a homogenizer. The mixture was homogenized once more after adding 167 μL of Milli-Q ultrapure water, and then centrifuged at 2300× g for 5 min at 4 °C. After the supernatant was collected, an additional 500 μL of 1% (v/v) formic acid in acetonitrile and 167 μL of ultrapure water were added to the precipitation. The homogenization and centrifugation were performed as described above to extract the residual compounds in the precipitant. The supernatant was combined with the previously collected sample. The mixed supernatant was filtrated through a 3 kDa cut-off filter to remove proteins and filtrated through a solid-phase extraction column (Hybrid SPE phospholipid 55261-U Supeleo, Bellefonte, PA, USA) to remove phospholipids. The filtrate was desiccated and resuspended in 20 μL of 50% (v/v) isopropanol in ultrapure water for LC-TOF-MS at HMT, as described above. LC-TOF-MS was carried out using an Agilent 1200 series RRLc systems SL (Agilent Technologies, Waldbronn, Germany) coupled with an Agilent-6224 TOF mass spectrometer equipped with an electrospray interface. Chromatographic analysis was performed on an Agilent Zorbax ODS column (2 mm × 50 mm, 2 μm). The autosampler and column were maintained at 4 and 40 °C, respectively. The mobile phase was water with 0.1% acetic acid (A) and water/ isopropanol/ acetonitrile (5/65/30) with 0.1% acetic acid and 2 mM ammonium acetate (B). The elution gradient was optimized as follows: 0–0.5 min, 1% B; 0.5–13.5 min, 100% B; 13.5–20 min, 100%, which was delivered at 0.3 mL min^{-1}. Re-equilibration duration was 7.5 min between individual runs. The optimized conditions of TOF/MS were: capillary voltage, 4.0 kV/−3.5 kV; drying gas (N_2) temperature, 350 °C; drying gas (N_2) flow rate, 10.0 L min^{-1}; Oct RFV, 750 V; skimmer voltage, 65 V; fragmentor voltage, 175 V; nebulizer pressure, 40 psi; scan range, m/z 100–1700.

For analysis of scanned metabolites, peaks detected in CE-TOF-MS and LC-TOF-MS were extracted using the automatic integration software MasterHands software version 2.17.1.11 (Keio University, Tsuruoka, Japan). Signal peaks corresponding to isotopomers, adduct ions, and other product ions of known metabolites were excluded, and the remaining peaks were annotated with putative metabolites from the HMT metabolite database on the basis of their migration times (MTs) in CE-TOF-MS or retention times (RTs) in LC-TOF-MS and m/z values determined by TOF-MS. In addition, peak areas were normalized against those of the internal standards, and the resultant relative area values were further normalized by the sample amount.

4. Conclusions

The dark browning pigment in the extract of the edible mushroom *H. erinaceus* was identified as potential byproducts of oxidization by endogenous polyphenol oxidases, including tyrosinases and laccases, measurable by fluorescence spectroscopy with 380 nm and 444 nm excitation and emission wavelengths, respectively.

Moreover, chelation deactivation and heat inactivation clearly showed that a tyrosinase—a metalloprotein as well as a PPO-family enzyme—affected the enzymatic browning reaction in the mushroom lysate. Furthermore, the identification of the putative PPO-coding genes and the analysis of their transcripts in the different growth phases suggested that the highly expressed tyrosinase and laccase isoenzymes in the mushroom fruiting body could be potential PPOs involved in the enzymatic browning reaction. The metabolite analysis of these growth statuses of the mushroom presented a number of prospective compounds as enzymatic browning substances, and they were grouped into amino acids and their derivatives, phenolic compounds, and purine and pyrimidine nucleobases. Nevertheless, the mushroom contained relatively high amounts of natural antioxidant compounds for the inhibition of tyrosinase and the scavenging of free radicals. These endogenous antioxidant substances could effectively decrease the browning reaction via PPO-inhibitory mechanisms in the mushroom fruiting body. With further evaluation of their health-promoting effects, these antioxidant metabolites and related compounds in *H. erinaceus* could be used to develop nutrient substances for healthcare and cosmetic ingredients for nutricosmetics.

Funding: This work was supported by Cooperative Research Program for Agriculture Science & Technology Development (PJ009783) from Rural Development Administration (RDA), the bilateral collaboration program (NRF-2018K1A3A1A39071061) between Korea and Slovakia through the National Research Foundation of Korea (NRF) and partially by KRIBB Research Initiative Program grant.

Conflicts of Interest: The author declares that there is no conflict of interests regarding the publication of this paper.

References

1. Erjavec, J.; Kos, J.; Ravnikar, M.; Dreo, T.; Sabotič, J. Proteins of higher fungi—From forest to application. *Trends Biotechnol.* **2012**, *30*, 259–273. [CrossRef] [PubMed]
2. Friedman, M. Mushroom Polysaccharides: Chemistry and Antiobesity, Antidiabetes, Anticancer, and Antibiotic Properties in Cells, Rodents, and Humans. *Foods* **2016**, *5*, 80. [CrossRef]
3. Chang, S.T.; Wasser, S.P. Current and Future Research Trends in Agricultural and Biomedical Applications of Medicinal Mushrooms and Mushroom Products (Review). *Int. J. Med. Mushrooms* **2018**, *20*, 1121–1133. [CrossRef] [PubMed]
4. He, X.; Wang, X.; Fang, J.; Chang, Y.; Ning, N.; Guo, H.; Huang, L.; Huang, X.; Zhao, Z. Structures, biological activities, and industrial applications of the polysaccharides from *Hericium erinaceus* (Lion's Mane) mushroom: A review. *Int. J. Boil. Macromol.* **2017**, *97*, 228–237. [CrossRef]
5. Friedman, M. Chemistry, Nutrition, and Health-Promoting Properties of *Hericium erinaceus* (Lion's Mane) Mushroom Fruiting Bodies and Mycelia and Their Bioactive Compounds. *J. Agric. Food Chem.* **2015**, *63*, 7108–7123. [CrossRef]
6. Zhang, Y.; Liu, L.; Bao, L.; Yang, Y.; Ma, K.; Liu, H. Three new cyathane diterpenes with neurotrophic activity from the liquid cultures of *Hericium erinaceus*. *J. Antibiot.* **2018**, *71*, 818–821. [CrossRef] [PubMed]
7. Hetland, G.; Tangen, J.-M.; Mahmood, F.; Mirlashari, M.R.; Nissen-Meyer, L.S.H.; Nentwich, I.; Therkelsen, S.P.; Tjønnfjord, G.E.; Johnson, E. Antitumor, Anti-inflammatory and Antiallergic Effects of Agaricus blazei Mushroom Extract and the Related Medicinal Basidiomycetes Mushrooms, Hericium erinaceus and Grifola frondosa: A Review of Preclinical and Clinical Studies. *Nutrients* **2020**, *12*, 1339. [CrossRef]
8. Wong, J.H.; Ng, T.B.; Chan, H.H.L.; Liu, Q.; Man, G.C.W.; Zhang, C.Z.; Guan, S.; Ng, C.C.W.; Fang, E.F.; Wang, H.; et al. Mushroom extracts and compounds with suppressive action on breast cancer: Evidence from

studies using cultured cancer cells, tumor-bearing animals, and clinical trials. *Appl. Microbiol. Biotechnol.* **2020**, *104*, 4675–4703. [CrossRef]

9. Kim, S. A novel core 1 O-linked glycan-specific binding lectin from the fruiting body of *Hericium erinaceus*. *Int. J. Boil. Macromol.* **2018**, *107*, 1528–1537. [CrossRef]

10. Sullivan, M.L. Beyond brown: Polyphenol oxidases as enzymes of plant specialized metabolism. *Front. Plant Sci.* **2015**, *5*. [CrossRef] [PubMed]

11. Moon, K.M.; Kwon, E.-B.; Lee, B.; Kim, C.Y. Recent Trends in Controlling the Enzymatic Browning of Fruit and Vegetable Products. *Molecules* **2020**, *25*, 2754. [CrossRef] [PubMed]

12. Mayer, A.M. Polyphenol oxidases in plants and fungi: Going places? A review. *Phytochemistry* **2006**, *67*, 2318–2331. [CrossRef] [PubMed]

13. Soler-Rivas, C.; Arpin, N.; Olivier, J.M.; Wichers, H.J. Activation of tyrosinase in *Agaricus bisporus* strains following infection by *Pseudomonas talaasii* or treatment with a tolaasin-containing preparation. *Mycol. Res.* **1997**, *101*, 375–382. [CrossRef]

14. Van Duuren, B.L. The Fluorescence Spectra of Aromatic Hydrocarbons and Heterocyclic Aromatic Compounds. *Anal. Chem.* **1960**, *32*, 1436–1442. [CrossRef]

15. Halaouli, S.; Asther, M.; Sigoillot, J.-C.; Hamdi, M.; Lomascolo, A. Fungal tyrosinases: New prospects in molecular characteristics, bioengineering and biotechnological applications. *J. Appl. Microbiol.* **2006**, *100*, 219–232. [CrossRef]

16. Soler-Rivas, C.; Olivier, J.; Wichers, H.J.; Jolivet, S.; Arpin, N. Biochemical and physiological aspects of brown blotch disease of *Agaricus bisporus*. *FEMS Microbiol. Rev.* **1999**, *23*, 591–614. [CrossRef] [PubMed]

17. Choi, Y.; Lee, S.; Chun, J.; Lee, H.; Lee, J. Influence of heat treatment on the antioxidant activities and polyphenolic compounds of Shiitake (*Lentinus edodes*) mushroom. *Food Chem.* **2006**, *99*, 381–387. [CrossRef]

18. Kim, Y.-J.; Uyama, H. Tyrosinase inhibitors from natural and synthetic sources: Structure, inhibition mechanism and perspective for the future. *Cell. Mol. Life Sci.* **2005**, *62*, 1707–1723. [CrossRef] [PubMed]

19. Loizzo, M.R.; Tundis, R.; Menichini, F. Natural and Synthetic Tyrosinase Inhibitors as Antibrowning Agents: An Update. *Compr. Rev. Food Sci. Food Saf.* **2012**, *11*, 378–398. [CrossRef]

20. Zolghadri, S.; Bahrami, A.; Khan, M.T.H.; Muñoz-Muñoz, J.L.; Garcia-Molina, F.; Garcia-Canovas, F.; Saboury, A.A. A comprehensive review on tyrosinase inhibitors. *J. Enzym. Inhib. Med. Chem.* **2019**, *34*, 279–309. [CrossRef]

21. Cerny, A.C.; Davidek, T. Formation of Aroma Compounds from Ribose and Cysteine during the Maillard Reaction. *J. Agric. Food Chem.* **2003**, *51*, 2714–2721. [CrossRef] [PubMed]

22. Skoog, D.A.; Holler, F.J.; Nieman, T. Molecular luminescence spectrometry. In *Principle of Instrumental Analysis*, 5th ed.; Mrssina, F., Sherman, M., Bortel, J., Eds.; Harcourt Brace & Company: Orlando, CA, USA, 1998; pp. 355–379.

23. Zafar, K.S.; Siegel, D.; Ross, D. A Potential Role for Cyclized Quinones Derived from Dopamine, DOPA, and 3,4-Dihydroxyphenylacetic Acid in Proteasomal Inhibition. *Mol. Pharmacol.* **2006**, *70*, 1079–1086. [CrossRef] [PubMed]

24. Harrison, W.H.; Whisler, W.W.; Ko, S. Detection, and study by fluorescence spectrometry of stereospecificity in mushroom tyrosinase-catalyzed oxidations. Proposal of a copper-containing reaction rate control site. *J. Boil. Chem.* **1967**, *242*, 1660–1667.

25. Friedman, M. Food Browning, and Its Prevention: An Overview. *J. Agric. Food Chem.* **1996**, *44*, 631–653. [CrossRef]

26. Kuijpers, T.F.M.; van Herk, T.; Vincken, J.-P.; Janssen, R.H.; Narh, D.L.; van Berkel, W.J.H.; Gruppen, H. Potato and Mushroom Polyphenol Oxidase Activities Are Differently Modulated by Natural Plant Extracts. *J. Agric. Food Chem.* **2013**, *62*, 214–221. [CrossRef]

27. Kaur, K.; Sharma, A.; Capalash, N.; Sharma, P. Multicopper oxidases: Biocatalysts in microbial pathogenesis and stress management. *Microbiol. Res.* **2019**, *222*, 1–13. [CrossRef]

28. Smith, D.F.Q.; Casadevall, A. The Role of Melanin in Fungal Pathogenesis for Animal Hosts. *Curr. Top. Microbiol. Immunol.* **2019**, *422*, 1–30. [CrossRef]

29. Cordero, R.J.; Casadevall, A. Functions of fungal melanin beyond virulence. *Fungal Boil. Rev.* **2017**, *31*, 99–112. [CrossRef]

30. Pezzella, C.; Guarino, L.; Piscitelli, A. How to enjoy laccases. *Cell. Mol. Life Sci.* **2015**, *72*, 923–940. [CrossRef]

31. Tobimatsu, Y.; Schuetz, M. Lignin polymerization: How do plants manage the chemistry so well? *Curr. Opin. Biotechnol.* **2019**, *56*, 75–81. [CrossRef]

32. Pourcel, L.; Routaboul, J.-M.; Cheynier, V.; Lepiniec, L.; Debeaujon, I. Flavonoid oxidation in plants: From biochemical properties to physiological functions. *Trends Plant Sci.* **2007**, *12*, 29–36. [CrossRef] [PubMed]

33. Coman, C.; Moţ, A.C.; Gal, E.; Parvu, M.; Silaghi-Dumitrescu, R. Laccase is upregulated via stress pathways in the phytopathogenic fungus *Sclerotinia sclerotiorum*. *Fungal Boil.* **2013**, *117*, 528–539. [CrossRef]

34. Upadhyay, P.; Shrivastava, R.; Agrawal, P.K. Bioprospecting, and biotechnological applications of fungal laccase. *3 Biotech* **2016**, *6*, 1–12. [CrossRef]

35. Rescigno, A.; Sollai, F.; Sanjust, E.; Rinaldi, A.C.; Curreli, N.; Rinaldi, A. Diafiltration in the presence of ascorbate in the purification of mushroom tyrosinase. *Phytochemistry* **1997**, *46*, 21–22. [CrossRef]

36. Pero, R.W.; Lund, H.; Leanderson, T. Antioxidant metabolism induced by quinic acid. increased urinary excretion of tryptophan and nicotinamide. *Phytother. Res.* **2009**, *23*, 335–346. [CrossRef]

37. Stampfli, A.R.; Blankenfeldt, W.; Seebeck, F.P. Structural basis of ergothioneine biosynthesis. *Curr. Opin. Struct. Biol.* **2020**, *65*, 1–8. [CrossRef] [PubMed]

38. Hanlon, D.P. Interaction of ergothioneine with metal ions and metalloenzymes. *J. Med. Chem.* **1971**, *14*, 1084–1087. [CrossRef] [PubMed]

39. Halliwell, B.; Cheah, I.K.; Tang, R.M.Y. Ergothioneine—A diet-derived antioxidant with therapeutic potential. *FEBS Lett.* **2018**, *592*, 3357–3366. [CrossRef]

40. Kalač, P. A review of chemical composition and nutritional value of wild-growing and cultivated mushrooms. *J. Sci. Food Agric.* **2012**, *93*, 209–218. [CrossRef]

41. Heleno, S.A.; Barros, L.; Martins, A.; Queiroz, M.J.R.P.; Morales, P.; Fernández-Ruiz, V.; Ferreira, I.C.F.R.; Morales, P. Chemical composition, antioxidant activity and bioaccessibility studies in phenolic extracts of two Hericium wild edible species. *LWT* **2015**, *63*, 475–481. [CrossRef]

42. Taofiq, O.; González-Paramás, A.M.; Martins, A.; Barreiro, M.F.; Ferreira, I.C.F.R. Mushrooms extracts and compounds in cosmetics, cosmeceuticals and nutricosmetics—A review. *Ind. Crop. Prod.* **2016**, *90*, 38–48. [CrossRef]

43. Muszyńska, B.; Grzywacz-Kisielewska, A.; Kała, K.; Gdula-Argasińska, J. Anti-inflammatory properties of edible mushrooms: A review. *Food Chem.* **2018**, *243*, 373–381. [CrossRef] [PubMed]

44. Taofiq, O.; Heleno, S.A.; Calhelha, R.C.; Alves, M.J.; Barros, L.; Barreiro, M.F.; González-Paramás, A.; Ferreira, I.C. Development of Mushroom-Based Cosmeceutical Formulations with Anti-Inflammatory, Anti-Tyrosinase, Antioxidant, and Antibacterial Properties. *Molecules* **2016**, *21*, 1372. [CrossRef] [PubMed]

45. Kim, S. *Hericium erinaceus* isolectins recognize mucin-type O-glycans as tumor-associated carbohydrate antigens on the surface of K562 human leukemia cells. *Int. J. Boil. Macromol.* **2018**, *120*, 1093–1102. [CrossRef] [PubMed]

 *foods* 

Article

Impact of Stability of Enriched Oil with Phenolic Extract from Olive Mill Wastewaters

Rosa Romeo, Alessandra De Bruno, Valeria Imeneo, Amalia Piscopo * and Marco Poiana

Department of AGRARIA, University Mediterranea of Reggio Calabria, Vito, 89124 Reggio Calabria, Italy; rosa.romeo@unirc.it (R.R.); alessandra.debruno@unirc.it (A.D.B.); valeria.imeneo@unirc.it (V.I.); mpoiana@unirc.it (M.P.)
* Correspondence: amalia.piscopo@unirc.it

Received: 9 June 2020; Accepted: 26 June 2020; Published: 30 June 2020

Abstract: The disposal of olive mill wastewaters is a considerable subject for the development of a sustainable olive oil industry considering their high content of pollutants. Nevertheless, the selective extraction of phenolic compounds from olive mill wastewaters represents a promising approach to obtain phenolics suitable for food enrichment. This work aimed to evaluate the efficiency of phenolic extract addition (50 mg L^{-1}), used as natural antioxidant, in sunflower oil against oxidative deterioration; to this aim, XAD-7-HP resin was tested in the recovery of phenolic compounds from olive mill wastewaters. Ultra-high performance liquid chromatography was used to evaluate the single phenols contained in the extract; the most consistent amount was detected for hydroxytyrosol (834 mg 100 mL^{-1}). The change in the oxidation state of fortified sunflower oil was studied by measuring physicochemical (refractive index, peroxide value and oxidative resistance to degradation) and antioxidant parameters (DPPH, ABTS and ORAC assays) during 90 days of storage. Results showed an enhancement of oxidative stability of 50% in the fortified oil compared to control.

Keywords: macroporous resin XAD-7HP; olive mill wastewaters; oxidative stability; polyphenols; sunflower oil

1. Introduction

Olive oil production represents a very important sector for Mediterranean countries. Despite its economic importance, olive oil production is associated with some negative implications on the environment as contamination of soil, water body pollution, underground seepage and air emissions due to the large amount of waste generated [1,2]. Olive mass is composed of about 80% of olive pulp and stones, whereas the liquid and solid waste yield is greater than the oil after the production steps [3]. Specifically, the amount of produced olive mill wastewaters (OMWWs) varies from 0.3 to 1.1 m^3 for tons of processed olives, depending on the olive oil extraction system [4,5]. OMWW disposal is a serious problem for the development of a sustainable olive oil industry owing to the high content of pollutants as heavy metals, a considerable amount of suspended solid and organic compounds. As a result of their high water solubility, the polyphenol concentration in OMWW ranged from 5 to 25 g L^{-1} [6,7]. These compounds show strong antioxidant activity, principally based on their ability to transfer the hydrogen atom of phenolic hydroxyl group to the free radicals. Phenolic compounds have potential beneficial effects with their anti-inflammatory and antimicrobial properties [8–10].

However, the selective recovery of phenolic substances from industrial wastes, such as OMWW, represents a valid approach for the reduction of their environmental toxicity and an opportunity to obtain high added value molecules [11,12]. Several methods for the recovery of OMWW polyphenols have been investigated, such as solvent extraction, ultrasound treatment, supercritical fluid extractions, membrane, chromatographic separations and Amberlite XAD resins, with the final aim to maximize the phenolic yield without impurities and toxic substances, and after to use these in the food industry [8,13–15]. Literature is available on the application of solid extraction and purification methods with the use of polymeric resin. A variety of synthetic polymer adsorbents are nowadays available for OMWW treatment, such as polystyrenedivinylbenzene copolymers and divinylbenzeneethylvinylbenzene as acrylic ester-based Amberlite XAD7 and the styrenedivinylbenzene-based XAD16 resin copolymers [16–19]. The adsorption process is characterized by a simple design and operation, low operating costs, reduction of the amount of used solvent, reduction of the operation time and separation of large amounts of bioactive compounds [14,20,21]. Moreover, the relatively inexpensive resins are durable, chemically stable and safe. In this work, Amberlite-XAD7HP was used as mean for the obtainment of a natural phenolic extract (WE) suitable for oil enrichment. XAD7HP resin exhibits a suitable pore diameter, polarity and surface area to recovery low molecular weight phenols using solvents with low toxicity and safe for human use. In addition, the potential to be reused with a low loss of absorbing capacity over time makes it useful for industrial application.

Considering the OMWW dotation of phenols with recognised antioxidant activity and health benefits, in the last few years there was an increasing interest in their recovery to re-use these as natural antioxidants. The enrichment with natural phenolic extracts of foods, beverages, edible oils, etc. (that contain a low concentration of phenolic compounds), could be a viable alternative to obtain a healthy added-value product. Sunflower oil with higher amounts of unsaturated fatty acids, in particular polyunsaturated ones, is more exposed to oxidative rancidity or autoxidation. This process reduces the nutritional and sensorial qualities and also gives undesirable chemical products, such as organic acids, aldehydes and ketones, which are harmful for human health [22]. Chemical stability depends on the chemical constituents of the oil, like antioxidants and oxidizable components [23]. Sunflower oil is usually subjected to a refining process that involves the complete loss of valuable active components with interesting nutritional-, health- and antioxidant-related characteristics [24]. Although synthetic antioxidants initially have been used with the purpose of retarding the oxidative deterioration, their implications in some diseases, such as cancer and carcinogenesis, are still controversial [25]. Thus, the tendency to use natural antioxidants has been increasing in order to increase the food quality and satisfy the demand of consumers for healthy products [26].

The aim of this study was to evaluate the efficiency of the phenolic extract obtained from OMWW with a solid extraction method against sunflower oil oxidative deterioration.

2. Materials and Methods

2.1. Chemicals and Reagents

Standards of phenolic compounds gallic acid (99%), vanillic acid (97%), tyrosol (97%), ferulic acid (99%), *p*-coumaric acid (98%) and *o*-cumaric acid (98%) were purchased from Fluka (Steinheim, Germany). Caffeic acid (98%), Apigenin (99%), luteolin (99%), and oleuropein (99%) were purchased from Extrasynthèse (Z. I. Lyon Nord, France). Hydroxytyrosol ((3,4-dihydroxyphenyl) ethanol) was acquired from TCI (Saitama, Japan). Verbascoside (99%) was procured from Sigma-Aldrich (St. Louis, MO, USA). The solvents used for chromatographic analysis (methanol, water, and acetonitrile) were ultra-high performance liquid chromatography (UHPLC)-MS grade (Carlo Erba, Milan, Italy). 2,2′-Azino-bis (3-ethylbenzothiazoline-6-sulfonic acid) diammonium salt (ABTS), 2,2-diphenyl-1-picrylhydrazyl (DPPH), Folin–Ciocalteu's phenol reagent and Trolox were purchased from Sigma Chemical Co. (St. Louis, MO, USA). AAPH (2,2′-azobis (2-amidino-propane) dihydrochloride) and fluorescein sodium were purchased from Acros Organics (New Jersey, USA) and Panreac (Barcelona, Spain), respectively. The reagents used for chemical investigation (ethanol absolute anhydrous, chloroform, isooctane, acetic acid, Diethyl ether, n-Hexane, sodium hydroxide solution, sodium thiosulfate solution, acetic acid glacial) were purchased from Carlo Erba (Milan, Italy); potassium iodide was acquired from Honeywell Fluka (Steinheim, Germany), Amberlite® XAD-7HP 20-60 mesh from Sigma-Aldrich (St. Louis, MO, USA) and Lecithin from Carlo Erba (Milan, Italy).

2.2. Sample Collection

Olive Mill Wastewater (OMWW) is a secondary product of the olive oil extraction process containing soft tissues of the olive fruit and the water used in the various stages of the oil extraction treatment together with the water contained in the fruit. OMWW were obtained by mean of a three-phase centrifugation process from Ottobratica olive cultivar during the crop seasons 2019 and supplied by an olive oil mill located in the Calabrian region (Italy). Sunflower oil used for phenolic enrichment was purchased in a local market.

2.3. OMWW Extraction

In order to obtain an extract enriched of phenolic compounds, OMWW were processed with Amberlite XAD-7-HP resin following the literature [27] with some modifications. Amberlite XAD-7-HP resin have a large surface with a macroreticular structure that allows to recover a mixture of different sizes of polyphenols. Moreover its design and operation are simple, operating costs low and the resin regeneration is easy. Before performing the extraction procedure, the adsorbent was pre-conditioned with NaOH 0.1 N for 2 h, rinsed with distilled water, immersed in HCl 0.1 N for 2 h and finally washed with distilled water.

For the extraction, 50 g of OMWW sample were mixed with 10 g of resin under stirring for 20 min. The adsorbent was washed three times with water; successively, it was eluted by a mean of three volumes of 50 mL of EtOH. The combined ethanol extract (WE) was desolvented in a rotary vacuum at 25 °C, then the dried residue was recovered with 100 mL of water, filtered using a 0.45-µm PTFE (Ø 15 mm) syringe filter and stored at 4 °C for the successive analytical determinations.

2.4. Production of Enriched Sunflower Oil

Enriched sunflower oil (MBoil) was produced in a Food Technologies laboratory of the Mediterranean University of Reggio Calabria (Italy) following the literature [28] with some modifications. An aliquot of WE and lecithin were added to sunflower oil and mixed for five hours until complete homogenization, in order to obtain oil samples enriched with a final concentration of 50 mg L^{-1} of hydroxytyrosol. Sunflower oil samples were used as control. Oil samples were kept in dark glass bottles (150 mL) at 10 and at 25 °C (three independent replicates for each thesis and time) and periodically analysed at different times (0, 15, 45 and 90 days of storage).

Extraction of Antioxidant Compounds

Phenols of MBoil were obtained by liquid–liquid extraction using methanol and according to the method [29] opportunely modified. Five grams of oil were added with 2 mL of methanol:water (70:30) and 2 mL of hexane and mixed with a Vortex for 10 min. The hydro-alcoholic phase was separated from the oil phase in a refrigerated (NF 1200R) centrifuge apparatus (Nüve, Ankara, Turkey) at 5000 rpm, 4 °C for 10 min. Hydro-alcoholic extracts (WE) were recovered with a syringe, filtered through a 0.45-µm nylon filter, diameter 15 mm (Thermo Fischer Scientific, Waltham, MA, USA) and utilised for the phenolic compounds quantification and antioxidant activity.

2.5. Determination of Total Phenol Content and Evaluation of Antioxidant Activity

Total Phenol content (TPC) of WE was determined in accordance to referred method [12] with minor modifications. Briefly, an aliquot of diluted WE was placed in a volumetric flask and mixed with deionized water (20 mL) and Folin–Ciocalteau reagent (0.625 mL). Then, 2.5 mL of saturated solution of Na_2CO_3 (20%) were then added after 3 min and made up to the 25 mL with deionized water. Thereafter, the mixture was left to react for 12 h in the dark and at room temperature. Sample absorbance was measured at 725 nm using a double-beam ultraviolet-visible spectrophotometer (8453 UV–Vis, Agilent, Waldbronn, Germany).

For the determination of TPC of MBoil, 0.05 mL of WE were mixed with 0.300 mL of Folin reagent and 0.25 mL of deionised water and, after 4 min, with 2.4 mL of an aqueous solution of Na_2CO_3 (5%). The mixture was maintained in a 40 °C water bath for 20 min and TPC was determined at 750 nm. Quantification was performed by mean of a calibration curve obtained at gallic acid concentrations from 1 to 10 mg L^{-1}. The results were expressed as mg of gallic acid equivalent 100 mL^{-1} of WE and mg gallic acid 100 g^{-1} of oil.

The evaluation of antioxidant capacity of WE and MBoil was performed by DPPH and ABTS assays [30,31]. An aliquot of diluted WE (1:50) was mixed with DPPH solution (6×10^{-5} mM) to the final volume of 3 mL and left in the dark for 30 min. The absorbance decrement was measured against methanol at 515 nm using a spectrophotometer (8453 UV–Vis, Agilent, Waldbronn, Germany) at 20 °C. The radical scavenging activity was expressed as mmol Trolox 100 mL^{-1} of WE and µmol Trolox 100 g^{-1} sample of oil. Then, 7 mM ABTS and 2.4 mM potassium persulphate ($K_2S_2O_8$) solutions were mixed for the ABTS assay and placed at room temperature for 12 h in the dark for stabilization. The resulting ABTS·+ solution was diluted with ethanol to obtain a blue-green chromogen with an absorbance of 0.70 (±0.02) at 734 nm. Then, 10 µL of diluted sample were added to the radical solution up to 3 mL and after 6 min the absorbance was measured. The quenching of initial absorbance was plotted against Trolox concentration (from 1.5 to 24 µM) and the results were expressed as TEAC values (mmol Trolox 100 mL^{-1} of WE and µmol Trolox 100 g^{-1} of oil).

In addition, the antioxidant activity of MBoil samples was also analysed by ORAC assay according to our previous study [32]. The ORAC assay was carried out on VICTOR X2 2030 Multilabel Plate Readers (PerkinElmer, Boston, Massachusetts, USA) in 96-well black microplate (PerkinElmer, Boston, Massachusetts, USA) using a fluorescence filter with an excitation wavelength of 485 nm and emission wavelength of 520 nm. The mix for reaction consisted of 130 µL of fluorescein solution, 50 µL of AAPH solution and 20 µL of phenolic extract. The fluorescence was measured at 37 °C immediately after the addition of fluorescein (time 0) and measurements of fluorescence kinetic were taken every minute for 30 times until the relative fluorescence intensity was less than 5% of the initial value. Results were expressed as µmol Trolox 100g^{-1} of oil.

Identification and Quantification of Phenolic Compounds

Identification and determination of the principal bioactive phenolic compounds of WE and MBoil were performed by UHPLC in accordance with [32]. The UHPLC system consisted of an UHPLC PLATINblue (Knauer, Berlin, Germany) equipped with a binary pump system using a Knauer blue orchid column C18 (1.8 µm, 100 × 2 mm) coupled with a PDA–1 (Photo Diode Array Detector) PLATINblue (Knauer, Berlin, Germany). The Clarity 6.2 software was used. Before the injection, phenolic compounds of MBoil were extracted using a variation of method [33]. One millilitre of oil was extracted with 1 mL of a methanol:water (80:20, *v:v*) in 2 mL Eppendorf reaction tubes. The mixture was shaken vigorously for 1 min using a vortex and then centrifuged (Micro Centrifuge Model 1K15 SIGMA, Laborzentrifugen, Osterode am Harz, Germany) at 13,000 rpm for 10 min at 10 °C. The methanolic phase was filtered with 0.22-µm nylon syringe filters, diameter 13 mm (Thermo Fischer Scientific, Waltham, MA, USA). The flow rate was 0.4 mL min^{-1} and the injection volume 5 µL. Acidified water (pH 3.10) (A) and acetonitrile (B) were the mobile phases and the applied gradient was the following: 95% A and 5% B (0–3 min), 95–60% A and 5–40% B (3–15 min); 60–0% A and 40–100% B (15–15.5 min). Quantification was performed by external standards (1–100 mg L^{-1}) and results expressed as mg kg^{-1} of sample.

2.6. Measurement of Chemical and Physical Properties of Oils

Free acidity (% oleic acid), peroxide value (mEq O_2 kg^{-1}) analyses and extinctions parameters K232 and K270 were performed according to Official and standard methods [34–36].

The moisture of samples was tested in an Electronic Moisture Analyser MA37 (Sartorius, Goettingen, Germany). The analysis was performed using 5 g of sample at 105 °C. The results were expressed as percentages.

Sunflower Oil Oxidative Stability in Accelerated Storage Test

OXITEST Oxidation Test Reactor (VELP Scientifica, Usmate Velate, MB, Italy) was used in order to evaluate the opposition to fat oxidation. This method is recognized by AOCS International Standard Procedure (Cd 12c–16) for the determination of oxidation stability of food, fats and oils [37]. The analysis consists of monitoring the oxygen uptake of the reactive constituent of food samples to determine the oxidative stability under conditions of accelerated oxidation. Briefly, 5 g of oil sample were distributed homogenously in a hermetically sealed titanium chamber; oxygen was purged into chamber up to a pressure of 6 bar. The reactor temperature was set at 90 °C. These reaction working conditions allow obtaining the sample Induction Period (IP) within a short time. The OXITEST allows to measure the modification of absolute pressure inside the two chambers and, through the OXISoft™ Software (Version 10002948 Usmate Velate, MB, Italy), automatically generates the IP expressed as hours by the graphical method.

2.7. Statistical Analysis

Results of the present study were expressed as mean ± *SD* of three measurements (n = 3). Multivariate and One-way analysis of variance with Tukey's *post hoc* test at $p < 0.05$ were performed by SPSS Software (Version 15.0, SPSS Inc., Chicago, IL, USA).

3. Results and Discussion

3.1. Water Extract Characterization

As is well known, OMWW represent a complex medium in which more than 50% of the total phenolic components of the olive drupes are present [38]. UHPLC analysis provided identification of individual phenols in WE, as illustrated in Figure 1, but only the principal compounds were quantified (Table 1). The principal constituents of the Amberlite-desorption fraction were phenyl acids (caffeic, chlorogenic, *p*-coumaric and vanillic acids), phenyl alcohols (hydroxytyrosol and tyrosol), secoridoids (oleuropein), flavonoids (apigenin and luteolin), derivatives of hydroxycinnamic acid (verbascoside). WE showed a high content of hydroxytyrosol (834.51 mg 100 mL^{-1} of sample) and tyrosol (147.55 mg 100 mL^{-1} of sample) in agreement with other authors [39]. The TPC of WE was instead about 788.96 mg 100 mL^{-1} of sample. The amount of recovered phenolic compounds is related to interactions between the adsorbates and adsorbent as well as chemical structure of the compounds themselves. Non-polar resins, or weakly polar, such as XAD7HP, allow the recovery of low molecular weight phenols, especially when ethanol is used as desorbing solvent [15]. On the other hand, a high concentration of ethanol promotes the solubilisation of alcohol-soluble impurities resulting in a drop-in desorption capacity [40]. Considering the high initial phenolic concentration and the amount of soluble impurities in OMWW, an increased amount of adsorbed molecules could occur per unit mass of absorbent, leading to saturation and reduction of desorption yield. This could explain why the TPC value results lower than the sum of individual compounds quantified by UHPLC.

The antioxidant activity of WE was measured by mean of DPPH and ABTS$^{\bullet+}$ assays. Results obtained with the ABTS assay were higher as mmol TE 100 mL^{-1} than those obtained by DPPH, according to our previous study [32]. As reported in the literature, antioxidant activity is due to the synergism between the various phenolic compounds and the assay responses are affected by the functional group's reactivity and characteristics, reaction time and complexity of the reaction kinetics [41,42].

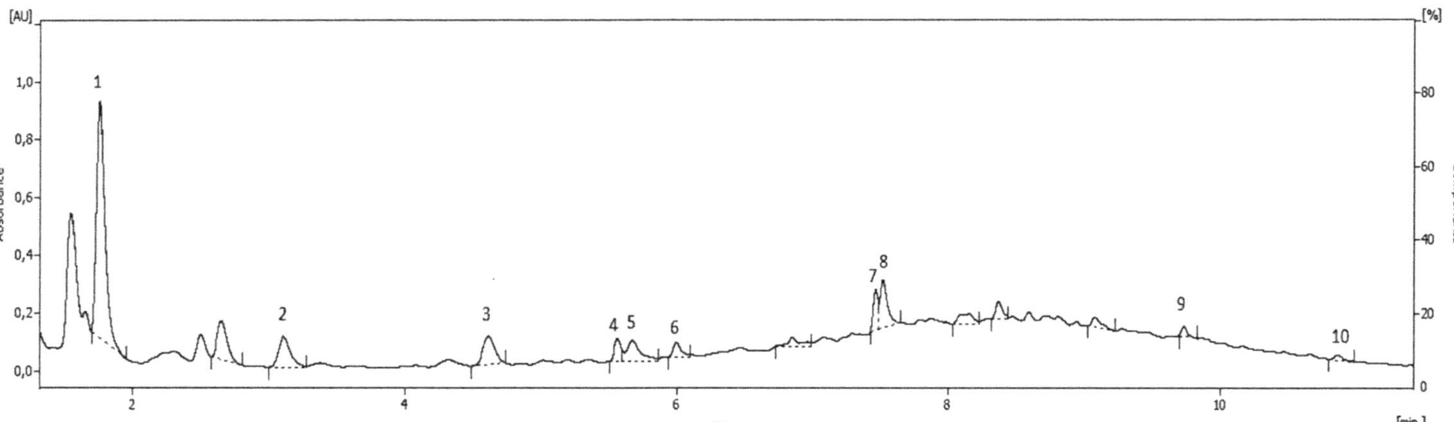

Figure 1. Chromatogram of phenolic compounds in wastewaters extract (WE). (1) hydroxytyrosol; (2) tyrosol; (3) chlorogenic acid; (4) vanillic acid; (5) caffeic acid; (6) *p*-coumaric acid; (7) verbascoside; (8) luteolin; (9) oleuropein; (10) apigenin.

Table 1. Phenolic characterisation and antioxidant activity of wastewaters extract (WE). Data are expressed as mg 100 mL^{-1} for phenols and mmol TE 100 mL^{-1} for ABTS and DPPH assays.

Hydroxytyrosol	834.51 ± 0.71
Tyrosol	147.55 ± 0.70
Chlorogenic Acid	16.06 ± 0.70
Vanillic Acid	40.25 ± 0.17
Caffeic Acid	20.53 ± 0.47
P-Cumaric Acid	61.06 ± 0.71
Oleuropein	65.09 ± 0.67
Apigenin	74.62 ± 0.71
Verbascoside	876.91 ± 0.91
Luteolin	14.11 ± 0.89
Total Polyphenol Content	788.96 ± 1.41
ABTS	2569.19 ± 399.90
DPPH	114.37 ± 151.87

3.2. Evaluation of Effect of WE on Sunflower Oil Stability

3.2.1. Qualitative Parameters

In sunflower oil enrichment, lecithin was used in order to promote the dispersion of WE into the lipid matrix. Lecithin stabilizes the added phenolic compounds in the oil matrix due to its amphiphilic behaviour, producing reverse micelles that include the extract [28]. Firstly, the effectiveness of enrichment was evaluated in terms of qualitative parameters and oxidative stability. The characteristics of sunflower oil used in this work are shown in Table 2. The oil showed a low free acidity (0.05%) while 6.1 mEq O_2 kg^{-1} of PV, according to the data about the commercial sunflower oil [43]. Spectrophotometric indices at 232 and 270 nm evidence the presence of dienes and trienes and the detected values (2.45 and 1.24) were characteristic of refined oils. Regarding enriched sunflower oil (Mboil), it is important to observe as the addition of extract involved the inclusion of water in the oil, as result of an increase in moisture content from 0.3% to 1% (Table 3). Free acidity of Mboil showed a rising trend during the storage: from 0.28 after enrichment to 0.32 at 10 °C and 0.35 at 25 °C after 90 days of storage, the formation of peroxides was reduced to about 49% in sunflower oil enriched with WE comparing with control sample at the beginning of storage. It could be linked to the phospholipids (the main constituents of soy lecithin) conferring to the oil oxidative stability [44]. According to the data in the literature [45–47], enriched samples, stored at different temperatures, showed a significant change ($p < 0.01$) of PV during storage. It can be noticed that the value of the PV was fluctuating progressively in time, reaching its maximum value on the 45th day (5.51 ± 0.21 at 10 °C and 5.47 ± 0.18 at 25 °C). The assessment of conjugated diene (K232) and conjugated triene (K270) is a reliable parameter for the measurement of oxidative deterioration of oils and, thus, the effectiveness of antioxidants in oils. The enriched samples had a little higher K270 values than the control; moreover, the conjugated triene content in the oil samples could be linked to the secondary oxidation compounds and conjugated trienes in the used commercial lecithin [48]. ANOVA data elaboration showed that no significant variations ($p > 0.05$) were observed during the storage at different temperatures. Likewise, other authors [46] have investigated the antioxidant efficacy of OMWW for the stabilization of lipid matrix obtaining the same our results.

Table 2. Qualitative parameters of sunflower oil.

Free acidity (Oleic acid %)	0.05 ± 0.00
Peroxide value (mEq O_2 kg−1)	6.10 ± 0.15
Moisture (%)	0.31 ± 0.01
Induction Period (minutes)	576 ± 0.01
K232	2.45 ± 0.09
K270	1.24 ± 0.08

Table 3. Qualitative parameters of Mboil during storage at 10 and 25 °C.

Temperature	Time (days)	Free Acidity (Oleic acid %)	Peroxide Value (mEq O_2 kg^{-1})	Moisture (%)	K232	K270
	0	0.28 ± 0.02 [b]	3.07 ± 0.03 [c]	1.08 ± 0.14	2.49 ± 0.13	1.40 ± 0.01
	15	0.28 ± 0.03 [b]	2.95 ± 0.13 [c]	0.95 ± 0.03	2.57 ± 0.35	1.33 ± 0.20
10 °C	45	0.23 ± 0.03 [c]	5.51 ± 0.21 [a]	1.0 ± 0.9	2.55 ± 0.21	1.50 ± 0.04
	90	0.32 ±0.02 [a]	3.85 ± 0.16 [b]	1.0 ± 0.09	2.55 ± 0.21	1.50 ± 0.04
	Significance	**	**	ns	ns	ns
	0	0.28 ± 0.02 [b]	3.07 ± 0.03 [d]	1.08 ± 0.14	2.49 ± 0.13	1.40 ± 0.01
	15	0.28 ± 0.00 [b]	4.32 ± 0.01 [b]	1.04 ± 0.16	2.84 ± 0.17	1.42 ± 0.03
25 °C	45	0.23 ± 0.00 [c]	5.47 ± 0.18 [a]	1.80 ± 0.53	2.86 ± 0.39	1.40 ± 0.03
	90	0.35 ± 0.03 [a]	3.78 ± 0.10 [c]	0.77 ± 0.54	2.55 ± 0.21	1.37 ± 0.05
	Significance	**	**	ns	ns	ns

Different letters show significant differences among mean values by Tukey's *post hoc* test. ** Significance at $p < 0.01$; ns: not significant.

3.2.2. Oxidative Stability

To evaluate the resistance of fat oxidation, the oil samples were subjected to a high-oxidative stress environment using OXITEST reactor that shows a curve of oxidation characterized by an Induction Period (IP). It is the time necessary to reach an end point of oxidation that corresponds to a detectable rancidity or a rapid change in the oxidation rate. Oil stability was measured on control and Mboil just after the addition of WE and during storage, to evaluate the effect in protection from oxidation. In Figure 2, two examples of oxidation curves of oils stored at different temperatures were reported. The addition of the WE significantly involved an increase of the oxidative stability of the oils: Mboil samples had an average rise of oxidative stability of 50% (IP of 1022 min) with respect to the control that showed an Induction Period of about 540 min. The resistance of oxidation did not show a significant variation over time regardless of the storage temperature, from 0 to 90 days as reported in the tables included in Figure 2. The higher value of oxidative stability observed in enriched oils can be linked to the incorporation of phenolic compounds that are able to donate a hydrogen atom to the radical formed during the propagation phase of lipid oxidation [49]. Moreover, sunflower oil added with only lecithin was also analysed in order to evaluate the effect of the addition of lecithin on the sample stability against oxidation. From the Rancimat analysis of oil plus lecithin, a high Induction Period was observed compared with oil without emulsifier [50]. In our investigation, sunflower oil added with lecithin showed a lower value than control (420 min, Figure S1, Supplementary Materials). This confirms that the oxidative stability of enriched sample was related to the added phenolic extract. Previous studies also showed that the antioxidant protection of lecithin, attributed to phospholipids, was not effective for sunflower oil [51].

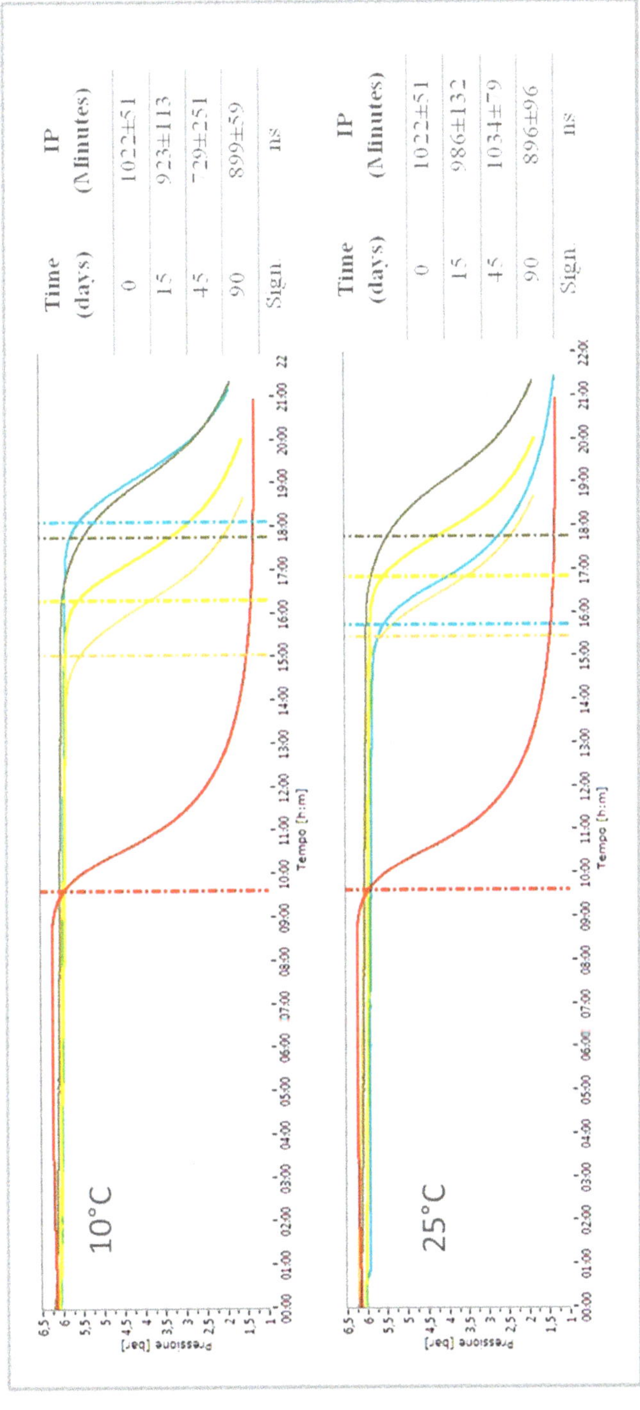

Time (days)	IP (Minutes)
0	1022±51
15	923±113
45	729±251
90	899±59
Sign.	ns

Time (days)	IP (Minutes)
0	1022±51
15	986±132
45	1034±79
90	896±96
Sign.	ns

Figure 2. Oxidation curves during storage at 10 and 25 °C: red (sunflower oil), green (Mboil at 0 days), yellow (Mboil after 15 days), blue (Mboil after 45 days) and orange (Mboil after 90 days). ns: not significant.

3.2.3. Phenolic Composition and Antioxidant Activity

The evolution of individual phenolic compounds added to oil was analysed during storage by UHPLC and the determination was repeated at different times (0, 45 and 90 days of storage). The analysis of the samples stored at different temperatures showed a similar phenolic composition (Table 4). According to the literature [52], samples were preferentially enriched with 50 mg L^{-1} of hydroxytyrosol. After the enrichment process, the chromatographic analysis of the samples showed (at time 0) a higher content of hydroxytyrosol and tyrosol, and a lower amount of caffeic acid, luteolin, oleuropein and verbascoside was also detected. In general, a significant decrease ($p < 0.01$) of phenolic compounds was observed during the storage, particularly after 45 days: about 59% of hydroxytyrosol and 32% of tyrosol were lost. Not significant variations were detected at the 90th day compared to the loss detected at 45th day, except for tyrosol, which showed another loss of 40% at the end of storage. In this regard, it is important to point out that authors ascribe to the amphiphilic character of lecithin the reduction of extraction yield of phenolic compounds due to the development of stable emulsion between lecithin and phenols [48]. This also could explain the low amount of phenolic compounds quantified compared to those added. In addition, we speculated that the formation of small drops of WE affected the reaction with Folin-Ciocalteau, underestimating results (Table 5).

Table 4. Phenolic compounds (mg kg^{-1} of sample) in Mboil during the storage (days) at 10 and 25 °C. letters and ** see Table 3.

10 °C	0	45	90	Significance
Hydroxytyrosol	39.65 ± 0.6 [a]	16.25 ± 0.47 [b]	15.06 ± 1.95 [b]	**
Tyrosol	36.15 ± 0.16 [a]	24.78 ± 0.5 [b]	14.44 ± 1.23 [c]	**
Vanillic acid	1.25 ± 0.01 [a]	0.70 ± 0.02 [c]	0.70 ± 0.01 [b]	**
Caffeic acid	20.79 ± 0.59 [a]	13.66 ± 0.02 [b]	13.66 ± 0.00 [b]	**
Verbascoside	9.11 ± 0.12 [a]	8.74 ± 0.04 [b]	8.92 ± 0.06 [b]	**
Luteolin	18.19 ± 0.17 [a]	12.19 ± 0.00 [b]	12.24 ± 0.04 [b]	**
Apigenin	11.05 ± 0.23 [a]	4.59 ± 0.03 [b]	4.69 ± 0.03 [b]	**
25 °C	**0**	**45**	**90**	**Significance**
Hydroxytyrosol	39.65 ± 0.6 [a]	15.77 ± 0.35 [b]	15.88 ± 1.24 [b]	**
Tyrosol	36.15 ± 0.16 [a]	25.54 ± 0.32 [b]	15.47 ± 1.29 [c]	**
Vanillic acid	1.25 ± 0.01 [a]	0.79 ± 0.04 [c]	0.98 ± 0.04 [b]	**
Caffeic acid	20.79 ± 0.59 [a]	13.72 ± 0.02 [b]	13.79 ± 0.00 [b]	**
Verbascoside	9.11 ± 0.12 [a]	8.73 ± 0.00 [b]	8.87 ± 0.02 [b]	**
Luteolin	18.19 ± 0.17 [a]	12.19 ± 0.00 [b]	12.18 ± 0.02 [b]	**
Apigenin	11.06 ± 0.23 [a]	4.61 ± 0.04 [b]	4.74 ± 0.03 [b]	**

The antioxidant activity of Mboil samples was analysed during storage using three different methods: ABTS, DPPH and ORAC assays. No single method is enough to determine the food antioxidant property, since different methods can give widely different results [53]. Moreover, the application of a single method can yield only a limited suggestion of the antioxidant activity of the samples under investigation [54]. The antioxidant activity of sample assays showed a significant variation ($p < 0.01$) over the time of storage of the enriched oils. A decrease of antioxidant activity was detected by ABTS assay according to the trend observed for TPC ($r > 0.9$). In contrast, a negative correlation was detected between TPC, DPPH and ORAC values. In addition, a non-linear trend was observed for DPPH and ORAC results with a minimum value detected at the 45th day. It should be considered that the added compounds have hydrophilic nature, thus their distribution in lipid phase is linked to their partition coefficient that may determinate also the distribution speed. It is conceivable that these compounds could be at first aggregated and only afterwards a suitable period distributed to the matrix. Multivariate analysis revealed that different temperatures did not significantly influence the total phenol content and the antioxidant activity measured by different assays while time seemed to affect them. Considering the oxidative stability of samples over time, the decrease of antioxidant activity is probably linked to the use of lecithin that led to the formation of lamellar structures in which hydrophilic antioxidants may be entrapped [48]. Despite this, from the start to the end of storage, at both temperatures, only about 3% of reduction was observed for the ORAC test: this is in line with the slow decrease of TPC in enriched oils and the observed oxidative stability of samples, so it confirms the robustness and the validity of ORAC method for the determination of the radical scavenging activity of the lipid matrix [50].

Table 5. TPC (mgGA 100 g^{-1}) and total antioxidant activity by ABTS, DPPH and ORAC assays (μmol TE 100 g^{-1}) of the samples stored at different temperatures.

Temperature	Time (days)	TPC	ABTS	DPPH	ORAC
	0	37 ± 1 [a]	1536.18 ± 1.55 [a]	74.67 ± 2.79 [b]	157.39 ± 0.86 [a]
	15	26 ± 5 [b]	1283.06 ± 4.93 [b]	80.82 ± 2.50 [a]	143.82 ± 0.88 [c]
10 °C	45	23 ± 2 [b]	1203.61 ± 7.69 [c]	56.63 ± 1.66 [d]	127.17 ± 0.91 [d]
	90	23 ± 3 [b]	1111.41 ± 9.78 [d]	68.29 ± 0.88 [c]	151.58 ± 1.22 [b]
	Significance	**	**	**	**
	0	37 ± 1 [a]	1536.18 ± 1.55 [a]	74.67 ± 2.79 [a]	157.39 ± 0.86 [a]
	15	25 ± 3 [b]	1285.05 ± 5.09 [b]	68.29 ± 3.45 [a]	153.21 ± 0.85 [b]
25 °C	45	25 ± 2 [b]	1240.37 ± 11.13 [b]	55.19 ± 2.76 [b]	128.90 ± 0.80 [c]
	90	24 ± 2 [b]	1204.21 ± 83.51 [b]	69.58 ± 1.27 [b]	152.15 ± 1.53 [b]
	Significance	**	**	**	**

Means within a row with different letters are significantly different by Tukey's post hoc test. ** Significance at $p < 0.01$.

The obtained results confirm that the use of a low concentration of WE improve the nutritional quality of refined oil, whereas in previous studies the effectiveness of retarding lipid oxidation of oil is correlated to addition of high concentration of phenolic extract [9]. It is possible to hypothesize the use of WE for the fortification of different kinds of food products.

4. Conclusions

In conclusion, the addition of a phenolic extract obtained by OMWW in sunflower oil permitted the production of an enriched oil with a higher content of polyphenols and antioxidant properties for up to 90 days of storage. The two tested temperatures did not affect these results, so it can be considered an initial step to the enriched sunflower oil production, with increased antioxidant characteristics for a more prolonged time. The successful results are also linked to the valorisation of olive industry by-product, such as OMWW, converted from waste to resource, through an efficient methodology of extraction.

Supplementary Materials: The following are available online at http://www.mdpi.com/2304-8158/9/7/856/s1. Figure S1. Oxidation curves: red (sunflower oil), blue (sunflower oil plus lecithin), yellow, violet, orange and green (Mboil during the storage).

Author Contributions: Sampling and laboratory analysis R.R. and V.I.; literature search R.R.; Conceptualization, R.R. and A.P.; writing original draft, data provision R.R. and A.D.B.; review, editing and approval of final text, A.P. and M.P.; supervision and approval M.P. All authors have read and agreed to the published version of the manuscript.

Funding: This research was funded by AGER 2 PROJECT, grant number 2016-0105".

Conflicts of Interest: The authors declare no conflict of interest.

References

1. Yay, A.; Oral, H.V.; Onay, T.T.; Yengun, O. A study on olive mill wastewater management in Turkey: A questionnaire and experimental approach. *Resour. Conserv. Recycl.* **2012**, *60*, 64–71.
2. Espadas-Aldana, G.; Vialle, C.; Belaud, J.P.; Vaca-Garcia, C.; Sablayrolles, C. Analysis and trends for Life Cycle Assessment of olive oil production. *Sust. Prod. Consum.* **2019**, *19*, 216–230. [CrossRef]

3.   Cinar, O.; Alma, M.H. Environmental assessment of olive oil production: Olive oil mill wastes and their disposal. *Acta Hortic.* **2008**, *791*, 645–649. [CrossRef]

4.   Markou, G.; Georgakakis, D.; Plagou, K.; Salakou, G.; Christopoulou, N. Balanced waste management of 2- and 3-phase olive oil mills in relation to the seed oil extraction plant. *Terr. Aquat. Environ. Toxicol.* **2010**, *4*, 109–112.

5.   El Hassani, F.Z.; Fadile, A.; Faouzi, M.; Zinedine, A.; Merzouki, M.; Benlemlih, M. The long term effect of Olive Mill Wastewater (OMW) on organic matter humification in a semi-arid soil. *Heliyon* **2020**, *6*, e03181. [CrossRef]

6.   Yangui, A.; Abderrabba, M. Towards a high yield recovery of polyphenols from olive mill wastewater on activated carbon coated with milk proteins: Experimental design and antioxidant activity. *Food Chem.* **2018**, *262*, 102–109. [CrossRef]

7.   Dutournié, P.; Jeguirim, M.; Khiari, B.; Goddard, M.L.; Jellali, S. Olive mill wastewater: From a pollutant to green fuel, agricultural water source and biofertilizer. Part 2: Water Recovery. *Water* **2019**, *11*, 768. [CrossRef]

8.   Soberón, L.F.; Carelli, A.A.; González, M.T.; Ceci, L.N. Method for phenol recovery from "alperujo": Numerical optimization and predictive model. *Eur. Food Res. Technol.* **2019**, *245*, 1641–1650. [CrossRef]

9.   Caporaso, N.; Formisano, D.; Genovese, A. Use of phenolic compounds from olive mill wastewater as valuable ingredients for functional foods. *Crit. Rev. Food Sci. Nutr.* **2018**, *58*, 2829–2841. [CrossRef]

10. Giuffrè, A.M.; Sicari, V.; Piscopo, A.; Louadj, L. Antioxidant activity of olive oil mill wastewater obtained from different thermal treatments. *Grasas y Aceites* **2012**, *63*, 209–213. [CrossRef]

11. Xynos, N.; Abatis, D.; Argyropoulou, A.; Polychronopoulos, P.; Aligiannis, N.; Skaltsounis, A.L. Development of a sustainable procedure for the recovery of hydroxytyrosol from table olive processing wastewater using adsorption resin technology and centrifugal partition chromatography. *Planta Med.* **2015**, *81*, 1621–1627. [CrossRef] [PubMed]

12. De Bruno, A.; Romeo, R.; Fedele, F.L.; Sicari, A.; Piscopo, A.; Poiana, M. Antioxidant activity shown by olive pomace extracts. *J. Environ. Sci. Health* **2018**, *53 Pt B*, 526–533. [CrossRef]

13. Galanakis, C.M. Emerging technologies for the production of nutraceuticals from agricultural by-products: A viewpoint of opportunities and challenges. *Food Bioprod. Process.* **2013**, *91*, 575–579. [CrossRef]

14. Frascari, D.; Bacca, A.E.M.; Wardenaar, T.; Oertléc, E.; Pinellia, D. Continuous flow adsorption of phenolic compounds fromolive mill wastewater with resin XAD16N: Life cycle assessment, cost–benefit analysis and process optimization. *J. Chem. Technol. Biotechnol.* **2019**, *94*, 1968–1981. [CrossRef]

15. Ferri, F.; Bertin, L.; Scoma, A.; Marchettia, L.; Fava, F. Recovery of low molecular weight phenols through solid-phase extraction. *Chem. Eng. J.* **2011**, *166*, 994–1001. [CrossRef]

16. Hellwig, V.; Gasser, J. Polyphenols from waste streams of food industry: Valorisation of blanch water from marzipan production. *Phytochem. Rev.* **2020**. [CrossRef]

17. Scoma, A.; Bertin, L.; Zanaroli, G.; Fraraccio, S.; Fava, F. A physicochemical-biotechnological approach for an integrated valorization of olive mill wastewater. *Bioresour. Technol.* **2011**, *102*, 10273–10279. [CrossRef]

18. Zagklis, D.P.; Vavouraki, A.I.; Kornaros, M.E.; Paraskeva, C.A. Purification of olive mill wastewater phenols through membrane filtration and resin adsorption/desorption. *J. Hazard. Mater.* **2015**, *285*, 69–76. [CrossRef]

19. Ochando-Pulido, J.M.; González-Hernández, R.; Martinez-Ferez, A. On the effect of the operating parameters for two-phase olive-oil washing wastewater combined phenolic compounds recovery and reclamation by novel ion exchange resins. *Sep. Purif. Technol.* **2018**, *195*, 50–59. [CrossRef]

20. Bertin, L.; Ferri, F.; Scoma, A.; Marchettia, L.; Fava, F. Recovery of high added value natural polyphenols from actual olive millwastewater through solid phase extraction. *Chem. Eng. J.* **2011**, *171*, 1287–1293. [CrossRef]

21. Papaoikonomou, L.; Labanaris, K.; Kaderides, K.; Goula, A.M. Adsorption–desorption of phenolic compounds from olive mill wastewater using a novel low-cost biosorbent. *Environ. Sci. Pollut. Res.* **2019**. [CrossRef] [PubMed]

22. Bai, Z.; Yu, R.; Li, J.; Wang, N.; Wang, N.; Niu, L.; Zhang, Y. Application of several novel natural antioxidants to inhibit oxidation of tree peony seed oil. *Cyta J. Food* **2018**, *16*, 1071–1078. [CrossRef]

23. Castelo-Branco, V.N.; Santana, I.; Di-Sarli, V.O.; Freitas, S.P.; Torres, A.G. Antioxidant capacity is a surrogate measure of the quality and stability of vegetable oils. *Eur. J. Lipid Sci. Technol.* **2016**, *118*, 224–235. [CrossRef]

24. Gotor, A.A.; Rhazi, L. Effects of refining process on sunflower oil minor components: A review. *OCL* **2016**, *23*, D207. [CrossRef]

25. Shahidi, F.; Ambigaipalan, P. Phenolics and polyphenolics in foods, beverages and spices: Antioxidant activity and health effects—A review. *J. Funct. Foods* **2015**, *18*, 820–897. [CrossRef]

26. Lourenço, S.C.; Moldão-Martins, M.D.; Alves, V. Antioxidants of natural plant origins: From sources to food industry applications. *Molecules* **2019**, *24*, 4132.

27. Jiménez-Aspee, F.; Quispe, C.; del Pilar, C.S.M.; Gonzalez, J.F.; Hüneke, E.; Theoduloz, C.; Schmeda-Hirschmann, G. Antioxidant activity and characterization of constituents in copao fruits (*Eulychnia acida* Phil., Cactaceae) by HPLC–DAD–MS/MSn. *Food Res. Int.* **2014**, *62*, 286–298. [CrossRef]

28. Suàrez, M.; Valls, R.M.; Motilva, M.P.; Màcia, A.; Fernàndez, S.; Giralt, M.; Solà, R.; Motilva, M.J. Study of stability during storage a phenol-enriched olive oil. *Eur. J. Lipid Sci. Technol.* **2011**, *113*, 894–903. [CrossRef]

29. Baiano, A.; Gambacorta, G.; Terracone, C.; Previtali, M.A.; Lamacchia, C.; La Notte, E. Changes in phenolic content and antioxidant activity of Italian Extra-Virgin Olive Oils during storage. *J. Food Sci.* **2009**, *4*, 177–183. [CrossRef]

30. Brand-Williams, W.; Cuvelier, M.E.; Berset, C. Use of free radical method to evaluate antioxidant activity. *Lebensm. Wiss. Technol.* **1995**, *28*, 25–30. [CrossRef]

31. Re, R.; Pellegrini, N.; Proteggente, A.; Pannala, A.; Yang, M.; Evans, C.R. Antioxidant activity applying an improved ABTS radical cation decolorazation assay. *Free Radic Biol. Med.* **1999**, *26*, 1231–1237. [CrossRef]

32. Romeo, R.; De Bruno, A.; Imeneo, V.; Piscopo, A.; Poiana, M. Evaluation of enrichment with antioxidants from olive oil mill wastes in hydrophilic model system. *J. Food Proc. Pres.* **2019**, *43*, e14211. [CrossRef]

33. Pizarro, M.L.; Becerra, M.; Sayago, A.; Beltrán, M.; Beltrán, R. Comparison of different extraction methods to determine phenolic compounds in virgin olive oil. *Food Anal. Methods* **2013**, *6*, 123–132. [CrossRef]

34. AOCS. Method Ca 5a 40. In *Official Methods and Recommended Practices of the American Oil Chemists' Society*, 6th ed.; AOCS Press: Champaign, IL, USA, 2017.

35. AOCS. Method Cd 8–53. In *Official Methods and Recommended Practices of the American Oil Chemists' Society*, 6th ed.; AOCS Press: Champaign, IL, USA, 2017.

36. AOCS. Method Ch 5-91. In *Official Methods and Recommended Practices of the American Oil Chemists' Society*; AOCS Press: Champaign, IL, USA, 1989.

37. AOCS. Method Cd 12c-16. In *Official Methods and Recommended Practices of the American Oil Chemists' Society*, 6th ed.; AOCS Press: Champaign, IL, USA, 2017.

38. Senol, A.; Hasdemir, İ.M.; Hasdemir, B.; Kurdaş, İ. Adsorptive removal of biophenols from olive mill wastewaters (OMW) by activated carbon: Mass transfer, equilibrium and kinetic studies. *Asia Pac. J. Chem. Eng.* **2017**, *12*, 128–146. [CrossRef]

39. Wang, Z.; Wang, C.; Yuan, J.; Zhang, C. Adsorption characteristics of adsorbent resins and antioxidant capacity for enrichment of phenolics from two-phase olive waste. *J. Chrom. B* **2017**, *1040*, 38–46. [CrossRef]

40. Zhang, Y.; Wang, B.; Jia, Z.; Scarlett, C.J.; Sheng, Z. Adsorption/desorption characteristics and enrichment of quercetin, luteolin and apigenin from Flos populi using macroporous resin. *Rev. Bras. Farmacogn.* **2019**, *29*, 69–76. [CrossRef]

41. Lins, P.G.; Pugine, S.M.P.; Scatolini, A.M.; de Melo, M.P. In vitro antioxidant activity of olive leaf extract (*Olea europaea* L.) and its protective effect on oxidative damage in human erythrocytes. *Heliyon* **2018**, *4*, e00805. [CrossRef]

42. Abramovič, H.; Grobin, B.; Ulrih, N.P.; Cigic, B. Relevance and standardization of in vitro antioxidant assays: ABTS, DPPH, and Folin–Ciocalteu. *J. Chem.* **2018**. [CrossRef]

43. Pal, U.S.; Patra, R.K.; Sahoo, N.R.; Bakhara, C.K.; Panda, M.K. Effect of refining on quality and composition of sunflower oil. *J. Food Sci. Technol.* **2015**, *52*, 4613–4618. [CrossRef]

44. Ramadan, M.F.; Kroh, L.W.; Mörsel, J.T. Radical scavenging activity of black cumin (*Nigella sativa* L.), coriander (*Coriandrum sativum* L.), and niger (*Guizotia abyssinica* Cass.) crude seed oils and oil fractions. *J. Agric. Food Chem.* **2003**, *51*, 6961–6969. [CrossRef]

45. Fki, I.; Allouche, N.; Sayadi, S. The use of polyphenolic extrcat, purified hydroxytyrosol and 3, 4-dihidroxyphenyl acetic acid from olive mill wastewater for the stabilization of refined oils: A potential alternative to synthetic antioxidants. *Food Chem.* **2005**, *93*, 197–204. [CrossRef]

46. De Leonardis, A.; Macciola, V.; Lembo, G.; Aretini, A.; Nag, A. Studies on oxidative stabilisation of lard by natural antioxidants recovered from olive-oil mill wastewater. *Food Chem.* **2007**, *100*, 998–1004. [CrossRef]

47. Sayyari, Z.; Farahmandfar, R. Stabilization of sunflower oil with pussy willow (Salix aegyptiaca) extract and essential oil. *Food Sci. Nutr.* **2017**, *5*, 266–272. [CrossRef] [PubMed]

48. Koprivnjak, O.; Škevin, D.; Valić, S.; Majetić, V.; Petričević, S.; Ljubenkov, I. The antioxidant capacity and oxidative stability of virgin olive oil enriched with phospholipids. *Food Chem.* **2008**, *111*, 121–126. [CrossRef]

49. Lafka, T.I.; Lazou, A.E.; Sinanoglou, V.J.; Lazos, E.S. Phenolic extracts from wild olive leaves and their potential as edible oils antioxidants. *Foods* **2013**, *2*, 18–31. [CrossRef]

50. Suarez, M.; Romero, M.P.; Motilva, M.J. Development of a phenol-enriched olive oil with phenolic compounds from olive cake. *J. Agric. Food Chem.* **2010**, *58*, 10396–10403. [CrossRef]

51. Judde, A.; Villeneuve, P.; Rossignol-Castera, A.; Le Guillou, A. Antioxidant effect of soy lecithins on vegetable oil stability and their synergism with tocopherols. *J. Am. Oil Chem. Soc.* **2003**, *80*, 1209–1215. [CrossRef]

52. de Medina, V.S.; Priego-Capote, F.; de Castro, M.D.L. Characterization of refined edible oils enriched with phenolic extracts from olive leaves and pomace. *J. Agric. Food Chem.* **2012**, *60*, 5866–5873. [CrossRef]

53. Huang, D.; Ou, B.; Prior, R.L. The chemistry behind antioxidant capacity assays. *J. Agric. Food Chem.* **2005**, *53*, 1841–1856. [CrossRef]

54. Sacchetti, G.; Maietti, S.; Muzzoli, M.; Scaglianti, M.; Manfredini, S.; Radice, M.; Bruni, R. Comparative evaluation of 11 essential oils of different origin as functional antioxidants, antiradicals and antimicrobials in foods. *Food Chem.* **2005**, *91*, 621–632. [CrossRef]

Article

A Novel and Simpler Alkaline Hydrolysis Methodology for Extraction of Ferulic Acid from Brewer's Spent Grain and its (Partial) Purification through Adsorption in a Synthetic Resin

Pedro Ideia [1], Ivo Sousa-Ferreira [2] and Paula C. Castilho [1,*

1   CQM—Centro de Química da Madeira, Universidade da Madeira, Campus da Penteada, 9020-105 Funchal, Portugal; pedro.ideia@staff.uma.pt
2   Centro de Estatística e Aplicações, Faculdade de Ciências, Universidade de Lisboa, 1749-016 Lisboa, Portugal; ivo.ferreira@staff.uma.pt
*   Correspondence: pcastilho@staff.uma.pt

Received: 25 March 2020; Accepted: 18 April 2020; Published: 8 May 2020

Abstract: This work aims to develop simpler methodologies of extracting ferulic acid (FA) from brewer's spent grain (BSG). BSG is produced by brewing companies at high amounts all over the year and does not possess a direct application. Thus, its use as raw material for extraction of bioactive compounds has gained attention in the last years. FA has different interesting applications in cosmetics, food industry, and pharmaceutics. Several studies aim for its extraction from BSG by various methods, namely alkaline hydrolysis. In the present work, we suggest the use of autoclave to process higher amounts of BSG in a lab scale. A simplification of the regular post-hydrolysis procedures is also proposed to decrease the number of experimental steps and energy costs and to simultaneously increase the extraction yield (up to 470 mg of FA per 100 g of BSG). The adsorption of extracted FA in a synthetic resin is suggested as a partial purification method.

Keywords: ferulic acid; brewer's spent grain; alkaline hydrolysis; adsorption; synthetic resin

---

1. Introduction

Due to the great political and social pressure in reducing pollution arising from industrial activities, large companies no longer consider residues as a waste but as valuable raw materials for other processes [1].

Brewer's spent grain (BSG) is the main solid by-product of brewing industry, produced during the wort elaboration step of beer production [1]. BSG is produced in a ratio of 20 kg per 100 L of beer [2], and the worldwide production is around 38.6×10^6 tons/year [3]. Despite being commonly used for animal and even human feed [4–7], excessive BSG availability is gaining attention for other applications, among which are the production or the extraction of high value added compounds, namely oligosaccharides [8], xylitol [9,10], and ferulic acid (FA) [11–13], a phenolic compound belonging to the hydroxycinnamic acids family.

Due to its physiological functions—anti-oxidant, anti-inflammatory, anti-thrombosis anti-microbial, and anti-cancer—as well as its protective effect against coronary disease, FA is considered one of the most important phenolic compounds [14]. Several applications are described, such as vanillin production [15], as preservatives [16,17], and as an ingredient for dermatologic lotions [14], among others. Because FA is covalently linked to the structure of lignocellulosic biomass by ether and ester bonds, conventional extraction techniques (e.g., solid–liquid extraction) are not effective in its separation from the matrix. Other techniques, such as alkaline hydrolysis, are needed in order to cleave these bonds and release FA. In fact, alkaline hydrolysis is able to cleave the lignin/phenolic-carbohydrate complexes structure, resulting in a phenolic portion, soluble sugars, insoluble lignin, and carbohydrates [18]. Other methods such as enzymatic hydrolysis were developed to recover ferulic acid from lignocellulosic biomass, namely wheat bran [19,20]. The main disadvantages of enzymatic hydrolysis are the cost of enzymes and/or the reaction time. Additionally, for the process to be efficient, control of reaction temperature and pH is required.

Different procedures for alkaline hydrolysis reactions are described in literature, namely those reported by Mussatto et al. (2007) [11] and McCarthy et al. (2013) [6]. The most common is to perform the reaction into auto-pressurized tubes or cylindrical stainless steel reactors at high temperatures and high pressures. Moreia and co-workers [13,21] suggested using microwave assisted extraction (MAE) to promote FA release from lignocellulosic materials, such as BSG. Despite the short time required for extraction, MAE's main limitation is the small amount of BSG the system is able to process in each batch. The present work aimed to optimize FA extraction from BSG by the simplification of the methods described in literature, which resulted in reducing the time and the resources required for extraction. Additionally, extraction in an autoclave allows one to scale-up the FA extraction, bridging the limitations of other processes already described.

Currently, adsorption technology is widely used for the removal of organic compounds from aqueous solutions and heterogeneous mixtures. The main disadvantage associated with the most used adsorbents is the high regeneration cost. This has stimulated the research on new adsorbents such as macrobead synthetic resins, which may provide a cheap and effective chemical regeneration process [22].

Several authors refer to the adsorption of ferulic acid in resins such as Amberlite XAD-16 [23] or Lewatit-type resins [24], aiming at its purification. These last authors studied three different polystirene-based macroporous resins, Lewatit S6328 A (an anionic, strongly alkaline exchange resin), Lewatit S2328 (food grade cationic exchange, strongly acidic), and Lewatit S7968, a resin without functional groups, which gave the best performance on the adsorption of chlorogenic acids from artichoke residues with little sugar co-adsorption. A similar resin Lewatit VPOC1064 was chosen for purification of the extracted FA in the present work.

Our goal in this work was to use the principles of methods already reported and simplify the processes for cleaning as well as partially purify the FA extracted from BSG.

2. Materials and Methods

2.1. Chemicals and Reagents

All reagents and standards were of analytical reagent (AR) grade. Folin–Ciocalteu reagent, gallic acid (99%), and sodium hydroxide (98%) were purchased from Panreac (Madrid, Spain). Ferulic acid (99%) and sodium carbonate (99.8%) were from Sigma-Aldrich (St. Louis, MO, USA). Absolute ethanol was obtained from Riedel-de Haën, acetonitrile (99.9%) from Fisher Scientific, and formic acid (≥98%) from Sigma-Aldrich. Synthetic resin (Lewatit VPOC1064 MD PH®) was purchased from LANXESS (Köln, Germany).

Dimethylsulfoxid-d_6 (DMSO-d_6) (≥99%) was purchased from Merck (Darmstadt, Germany).

2.2. Raw Material

Brewer's spent grain (BSG) was provided by a local brewing company (ECM, Empresa de Cervejas da Madeira). The samples were freeze-dried immediately after delivery to our laboratory (NatLab—CQM). Dry BSG was stored at −20 °C until use. These samples were used throughout the work with the simple designation of "BSG". A sample of fresh BSG was used for moisture and ash content determinations.

2.3. Physicochemical Characterization

The moisture content determination was carried out on a KERN DBS 60-3 moisture balance running a semi-automatic program, which heats the sample at 120 °C until the moisture content is stable for 30 s. Ash content (%) was determined after incineration of BSG samples in a muffle furnace at (500 °C for 24 h). Particle size distribution was studied by passing three freeze-dried BSG portions though a set of sieves with decreasing mesh size (1.0, 0.5, 0.25 and 0.125 mm).

2.4. Alkaline Hydrolysis—Basic Procedure

Alkaline hydrolysis was performed according to the procedure outlined in Figure 1A and described in literature. Reactions were performed in Ace pressure tubes using BSG and NaOH solution in a solid:liquid ratio of 1:20 (*w/v*). Experimental conditions were optimized and set as 120 °C for 1.5 h with 20 mL of NaOH (2%) for 1 g of BSG. Optimization was performed by individual variation of each of the extraction parameters.

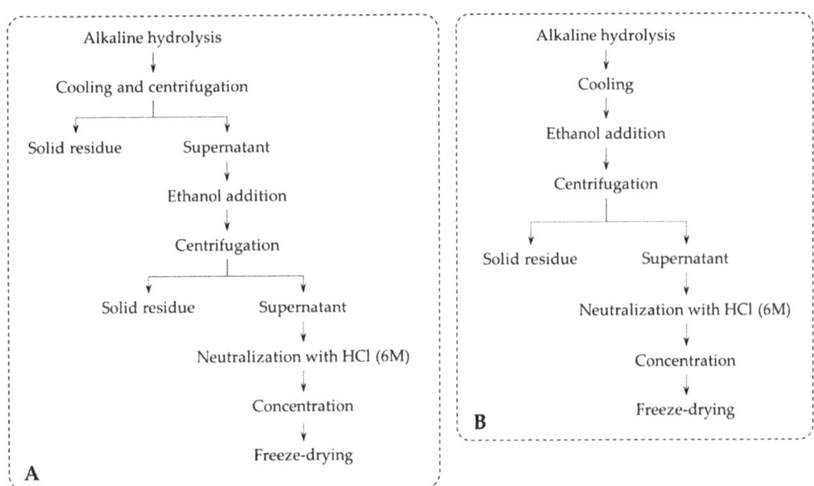

Figure 1. Proceeding to the alkaline hydrolysis reaction. (**A**) classic procedure; (**B**) simplified methodology.

The tubes were placed into an oil bath and heated. After alkaline hydrolysis reaction, the mixture was cooled to room temperature, and the solid residue, containing mainly cellulose and lignin, was separated by centrifugation. Precipitation of the hemicellulose fraction was triggered by addition of ethanol to a final concentration of 30% (*v/v*), and its separation was performed by centrifuging. The supernatant, which contained the ferulic acid, was neutralized with HCl (6M) and concentrated on a rotary evaporator. The aqueous phase was freeze-dried and stored at −20 °C until analysis.

2.5. Pretreatment by Solid–Liquid Extraction with Acetone

A portion of BSG was extracted with acetone (60%) in a solid/liquid ratio of 50 mL/g in an ultrasound bath for 1 h. After solid–liquid extraction, the mixture was allowed to cool to room temperature, and the solid residue was separated from the supernatant by centrifugation followed by filtration. The solid residue was washed with distilled water and freeze-dried. Alkaline hydrolysis was carried out in Ace tubes at 120 °C for 1.5 h with 20 mL of NaOH (2%) for 1 g of BSG. For comparison, a portion of BSG not subjected to pretreatment was extracted by alkaline hydrolysis in the same conditions.

2.6. Extraction by Alkaline Hydrolysis in Autoclave

An alkaline hydrolysis assay was carried under the same conditions (120 °C for 1.5 h with 20 mL of NaOH (2%) for 1 g of BSG) in autoclave. The subsequent procedure was similar to that described for Ace tubes. The resultant extracts were stored at −20 °C until analysis.

2.7. Simplification of the Procedure

A modification of the process described in the literature was introduced in order to diminish the number of separation steps and thus improve the yield in the desired product; after alkaline hydrolysis, the mixture was cooled, and ethanol was added (Figure 1B). Precipitated hemicellulose was separated by centrifugation together with lignin and cellulose fraction. The supernatant was neutralized, concentrated, and the aqueous phase was freeze-dried and stored at −20 °C until analysis.

2.8. Analysis of Extracts

The extracts obtained after the procedures described in the previous sections were analyzed in terms of their total soluble solids (TSS), total phenolic content (TPC), and quantification of ferulic acid by HPLC-diode-array detector (DAD).

2.8.1. Total Soluble Solids (TSS) Determination

For TSS determinations [25], extracts were resuspended in water (10 mg/mL) and filtered through membrane filters (0.45 μm). Using an ATAGO RX-1000 refractometer, the TSS was measured based on a calibration curve of sucrose (5–50 mg/L). The results are expressed in milligrams of sucrose equivalent (SE) per 100 g of dry BSG.

2.8.2. Total Phenolic Content (TPC) Determination

TPC was determined by Folin–Ciocalteu method [26]; fifty microliters of the sample in methanol (5 mg/mL) was mixed with 1.25 mL of Folin–Ciocalteu solution (1:10) and 1 mL of Na_2CO_3 (7.5%). The absorbance at 765 nm was measured after 30 min, and results were expressed as milligrams of gallic acid equivalent (GAE) per 100 g of BSG (dry weight).

2.8.3. Quantification of Ferulic Acid

Ferulic acid concentrations were determined by high performance liquid chromatography (HPLC) using a UV detector (at 320 nm) and a Phenomenex Gemini C18 (5 μm, 250 × 0.3 mm i.d.) column. The HPLC analysis, adapted from Gouveia and Castilho (2011) [26], was performed on a Dionex ultimate 3000 series instrument (Dionex, Sunnyvale, CA, USA) coupled to a binary pump, a diode-array detector (DAD), an autosampler and a column compartment. Samples (5 mg/mL) were prepared in the mobile phase and filtered through 0.45 μm membranes (Millipore, Burlington, MA, USA). Then, samples were injected into the equipment under the following conditions: column at 30 °C, acetonitrile/0.1% formic acid as mobile phase (isocratic elution, 75:25), a flow rate of 400 μL/min, and injection volume of 10 μL. Results are expressed as mg of FA per 100 g of BSG (dry weight).

Identification was performed comparing retention times with those obtained from commercial ferulic acid standard. Quantification was based on the UV signal response at 320 nm, and the resultant peak areas in the chromatograms were plotted against concentrations obtained from standard. Calibration curve (5–100 mg/L) was prepared by diluting the stock solutions (1000 mg/L in methanol) with the initial mobile phase. Quantification was carried out by plotting peak area versus concentration ($R^2 = 0.9994$).

2.8.4. Statistical Analysis

All samples were assayed in triplicate ($n = 3$), and results are given as means ± standard deviations. Differences between means were tested by ANOVA using SPSS Statistics 22 software.

2.9. Purification by Adsorption on a Synthetic Resin

Synthetic resin (Lewatit VPOC1064 MD PH®) was used to promote adsorption of FA from the extract obtained after alkaline hydrolysis using the simplified method previously described (Figure 1B).

2.9.1. Kinetic Studies with FA Standard

Before purification of alkaline hydrolysis extract, adsorption kinetic studies with FA standard were performed. During these studies, several conditions were tested in three different assays. In the first assay, a proportional variation of FA concentration and the amount of adsorbent were carried up within three different tests. A second assay was achieved in two tests, where different initial concentrations of FA were studied. Finally, in the third assay, the effect of temperature in the adsorption was evaluated. Two tests were performed at room temperature (22–25 °C) and 6 °C (controlled ice bath), respectively. Table 1 resumes the conditions of the different assays.

Table 1. Conditions of the adsorption mixtures during in adsorption kinetic studies. The volume of ferulic acid (FA) solution used in each teste was constant (50 mL).

Assay	Test	[FA] (g/L)	Resin wt. (g)	Temperature
1	*AdsA*	1	0.5	Room temp.
	AdsB	0.5	0.25	
	AdsC	0.25	0.125	
2	*AdsHC*	1	0.5	Room temp.
	AdsLC	0.25	0.5	
3	*AdsRT*	0.25	0.5	Room temp.
	AdsT6	0.25	0.5	6 °C

A supernatant sample was collected every 2 min in the first 10 min, every 10 min during the next 90 min, and every 20 min until 180 min of adsorption. After filtration of supernatants, FA concentration was determined. Adsorption percentage, the amount of FA adsorbed per gram of resin, and C/C_0 were used to evaluate the adsorption process.

The equation below was applied to the experimental data, and the parameters α, β, γ, δ and θ were determined using the Solver Microsoft Excel add-in program.

$$\frac{C}{C_0} = \alpha + \beta \cdot e^{-\frac{t}{\gamma}} + \delta \cdot e^{-\frac{t}{\theta}}$$

where C_0 is the initial concentration of the absorbate (g/L), and C is the concentration at the time t (g/L).

2.9.2. Purification of an Alkaline Hydrolysis Extract

As outlined in Figure 2, a portion of 2 g of alkaline hydrolysis freeze dried extract was dissolved in 40 mL of water at 40 °C with vigorous stirring. After cooling the mixture to room temperature, a portion of 10 g of pre-activated resin (according to the supplier, treatment with 6% HCl and 4% NaOH and washing with distilled water) was added to the flask. The mixture was magnetically stirred for 2 h and filtered under reduced pressure. The filtrate was used to determine the FA adsorption yield, and the loaded resin was further stirred with 50 mL of ethanol:water (70%) in order to promote desorption. A new filtration allowed us to separate the resin from the liquid phase containing the FA. The determination of desorbed FA was possible through analysis of the filtrate.

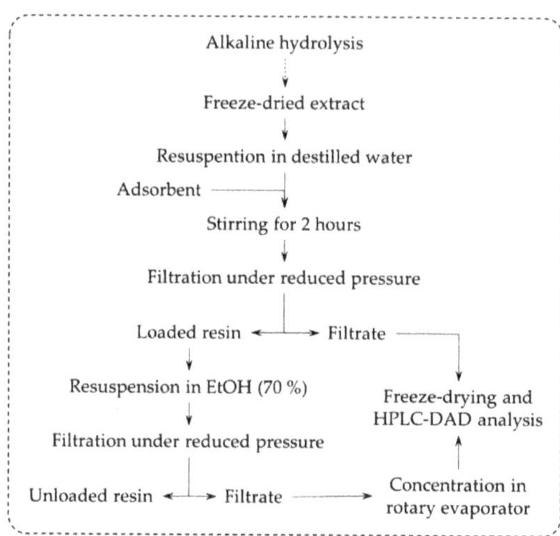

Figure 2. Procedure for the partial purification of FA by adsorption on a synthetic resin.

The extracts were analyzed in terms of FA quantification by HPLC-DAD according to the conditions mentioned before. ^{1}H NMR was performed to verify the partial purification of the extract obtained by alkaline hydrolysis of BSG and treated with the Lewatit resin. For NMR analysis, 10 mg of each extract were dissolved in 1 mL of DMSO-d$_6$ and transferred to 5 mm NMR tubes. ^{1}H spectra were recorded on a Bruker UltraShield 400 Plus NMR (Bruker, Billerica, MA, USA) at 10,061 MHz and 400 MHz. Acquisition parameters for ^{1}H were: size of fit 65 k; spectral width 4401 Hz; acquisition time 64 k; relaxation delay 1 s; number of scans 512.

3. Results and Discussion

3.1. Physicochemical Characterization of BSG

The moisture content of supplied BSG was 68.44 ± 0.93%, and the ash content was 4.18 ± 0.03%. Sifting of freeze-dried material revealed that BSG was provided as a fine powder with particle size between 1 and 0.25 mm (Table 2), and it was further used without any separation.

Table 2. Particle size distribution.

Particle Size (mm)	Percentage (%)		
>1	3.83	±	0.37
1–0.5	36.00	±	1.48
0.5–0.25	41.06	±	2.44
0.25–0.125	12.57	±	2.65
≤0.125	6.76	±	1.12

3.2. Optimization of Alkaline Hydrolysis Conditions

Table 3 summarizes the results for optimization of alkaline hydrolysis reaction in Ace pressure tubes. Different reaction temperatures (60, 80, 100, and 120 °C) were tested for 1 h with NaOH (2%). Reaction time (1 to 3 h) was tested at 100 °C using NaOH (2%). Optimization of alkali solution concentration was performed in reactions at 100 °C for 1 h. Optimal conditions were set as 120 °C for 1.5 h with 20 mL of NaOH (2%) for 1 g of BSG.

Table 3. Conditions and results obtained for optimization of the parameters for alkaline hydrolysis reaction in Ace pressure tubes. FA yield is expressed as mg of FA per 100 g of brewer's spent grain (BSG).

Temp. (°C)	Time (Hours)	NaOH (%)	FA Yield (mg FA/100 g)		
60	1	2	46.95	±	9.72
80	1	2	81.52	±	17.89
100	1	2	214.18	±	30.43
120	1	2	234.98	±	8.03
100	1	2	214.67	±	4.70
100	1.5	2	223.97	±	6.82
100	2	2	223.05	±	1.47
100	2.5	2	202.62	±	19.65
100	3	2	223.09	±	17.85
100	1	0.5	174.76	±	23.64
100	1	1	199.63	±	11.21
100	1	1.5	214.53	±	23.26
100	1	2	204.44	±	9.64
100	1	2.5	200.10	±	22.07

3.3. Pretreatment by Solid–Liquid Extraction with Acetone

The results compiled in Table 4 reveal that extraction with acetone is effective in removing free sugars from the matrix, resulting in a significant decrease in TSS for pretreated BSG compared with the untreated portion. FA concentration in the extracts obtained by alkaline hydrolysis of both untreated and pretreated BSG portions did not show a statistically significant difference. Results obtained in previous works [27] showed that solid–liquid extraction with 60% acetone is efficient in the extraction of free form compounds from BSG. The present data show that solid–liquid extraction is not efficient in the extraction of FA from lignocellulosic materials, since it is covalently bonded to their structure, but soluble solids such as mono and disaccharides are partially removed.

Table 4. Comparison of the results obtained for alkaline hydrolysis of not pretreated and treated BSG (1 h). Total soluble solids (TSS) is expressed as mg of sucrose equivalent per 100 g of BSG, total phenolic content (TPC) is expressed as gram of gallic acid equivalent per 100 g of BSG, and FA yield is expressed as mg of FA per 100 g of BSG.

	Not Pretreated			Pretreated		
TSS mg SE/100 g	94.50	±	8.39 [a]	69.39	±	3.28 [b]
TPC g GAE/100 g	1010.44	±	1.50 [a]	1323.24	+	143.30 [b]
FA yield mg FA/100 g	259.21	±	35.95 [a]	270.32	±	65.86 [a]

[a] indicates not significant differences and [b] indicates significant differences. SE: sucrose equivalent; GAE: gallic acid equivalent.

3.4. Extraction by Alkaline Hydrolysis in Autoclave

The characterization of the extracts (Table 5) showed that differences on TSS were not significant. However, there was a statistically significant increase of TPC and FA yield for those obtained by alkaline hydrolysis in autoclave compared to those obtained in Ace pressure tubes. This might have been due to a higher contact exchange between BSG and alkali solution in autoclave, resulting from the greater volume of the reaction vessels in which hydrolysis was performed. Taking into account the aspects mentioned above, alkaline hydrolysis of BSG in an autoclave might be an interesting process for a possible scale-up of the extraction process, since even everyday laboratory equipment is capable of processing large amounts of BSG.

Table 5. Comparison of the results obtained for alkaline hydrolysis on Ace pressure tubes and in an autoclave (1.5 h). TSS is expressed as mg of sucrose equivalent per 100 g of BSG, TPC is expressed as gram of gallic acid equivalent per 100 g of BSG, and FA yield is expressed as mg of FA per 100 g of BSG.

	Ace Pressure Tubes			Autoclave		
TSS mg SE/100 g	81.45	±	2.59 [a]	82.44	±	9.17 [a]
TPC g GAE/100 g	1194.20	±	21.34 [a]	1439.73	±	102.02 [b]
FA yield mg FA/100 g	203.41	±	5.37 [a]	280.61	±	5.77 [b]

[a] indicates not significant differences and [b] indicates significant differences.

3.5. Simplification of the Procedure

Table 6 shows the results for the simplification of the procedure after alkaline hydrolysis in comparison with the normal procedure (schematized in Figure 1). Both TSS and TPC showed a large increase (of about 30.9% for TSS and 122.97% for TPC). FA yield was increased in about 80% with the change of procedure, up to 476.99 ± 25.94 mg (FA)/100 g (BSG, dry weight). The increase was probably due to the solvent washing during the addition of the ethanol to the alkaline liquor, which resulted in reduction of losses associated with the process. Thus, in addition to reducing the experimental steps and the energetic resources required for extraction, simplification of the procedure also permits obtaining a greater amount of FA. TSS increase may be an indication that larger carbohydrates are degraded into smaller, soluble sugar molecules.

Table 6. Comparison of the results obtained for normal and simplified procedures applied after alkaline hydrolysis. TSS is expressed as mg of sucrose equivalent per 100 g of BSG, TPC is expressed as gram of gallic acid equivalent per 100 g of BSG, and FA yield is expressed as mg of FA per 100 g of BSG.

	Normal Procedure			Simplified Procedure		
TSS mg SE/100 g	94.50	±	8.39 [a]	123.70	±	1.47 [b]
TPC g GAE/100 g	1483.72	±	90.03 [a]	3342.86	±	71.21 [b]
FA yield mg FA/100 g	259.21	±	35.95 [a]	476.99	±	25.94 [b]

[a] indicates not significant differences and [b] indicates significant differences.

3.6. Purification by Adsorption on a Synthetic Resin

3.6.1. Kinetic Studies with FA Standard

Three assays were performed to study the adsorption kinetics under different experimental conditions, according to Table 1.

After determination of FA concentration in the various supernatant samples, C/C_0 was calculated for each time t, and parameters α, β, γ, δ and θ were determined.

Figure 3 resumes the evolution of C/C_0 along the 180 min of adsorption in the different assays. Figure 3a is related to assay 1, where dispersion of both FA and resin increased in proportion within the tests *AdsA*, *AdsB*, and *AdsC*. Adsorption isotherms show that the variation in the concentration of adsorbed FA during the equilibrium was proportional to the dispersion of both adsorbate and adsorbent in the mixture. Assay 2 (Figure 3b) was performed to study the effect of the initial concentration of FA in the adsorption process. The FA concentration in test *AdsHC* was four times higher than that in test *AdsLC*. Because the amount of resin was similar in both tests, isotherms indicated a more efficient adsorption process for *AdsLC*. Regarding the study of the effect of temperature in the adsorption process, Figure 3c resumes the results obtained under room temperature (*AdsRT*) and 6 °C (*AdsT6*). Results showed that, at room temperature, the equilibrium was achieved faster than at 6 °C. In all tests, it was possible to establish that 100–120 min is enough to reach equilibrium.

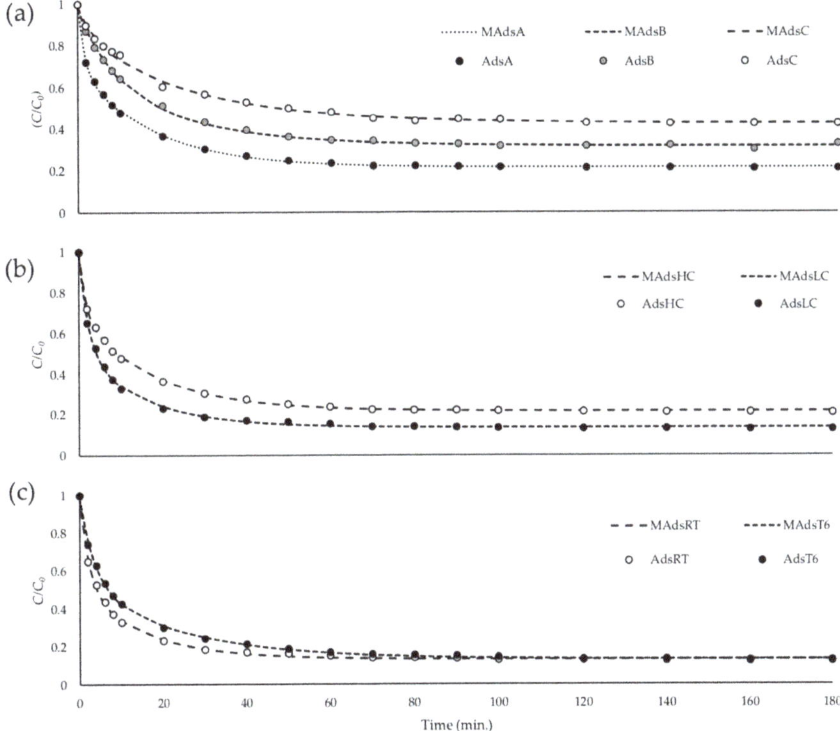

Figure 3. Adsorption kinetic isotherms for the assays performed to study the effect of (**a**) dispersion of FA and resin, (**b**) the initial concentration of FA, and (**c**) the temperature in the adsorption mixture.

Parameters for adsorption isotherms equations were determined and compiled in Table 7. Based on kinetic studies results, the best conditions for adsorption of FA standard may be an intermediate between the conditions of assays *AdsA*, *AdsLC*, and *AdsRT*. For real samples, where the goal is to purify FA from a variety of compounds of different chemical nature, the ideal conditions might be different, namely because of the competition for adsorption in the resin.

Table 7. Determination of parameter estimates and sum of squared differences (SSD) by MS Solver for adjustment of adsorption kinetic curves.

Assay	Test	Parameter					SSD
		α	β	γ	δ	θ	
1	AdsA	0.2144	0.4721	18.1123	0.3122	1.6490	0.0006
	AdsB	0.3169	0.3620	24.1647	0.3031	7.9226	0.0014
	AdsC	0.4262	0.4271	27.1620	0.1391	4.1121	0.0022
2	AdsHC	0.2144	0.4721	18.1123	0.3122	1.6490	0.0006
	AdsLC	0.1343	0.3759	15.7397	0.4861	2.2909	0.0014
3	AdsRT	0.1343	0.3759	15.7401	0.4861	2.2910	0.0014
	AdsT6	0.1335	0.3992	24.5310	0.4604	3.4751	0.0013

3.6.2. Purification of an Alkaline Hydrolysis Extract

Adsorption of FA from the alkaline hydrolysis extract showed to be effective (90.83%), where around 4.6 milligrams of compound were adsorbed into the resin. The desorption was achieved by adding two portions of 25 mL of 70% ethanol followed by continuous stirring for 30 min and filtration. The filtrates were combined, concentrated, and the concentration of FA was determined by HPLC-DAD. Results showed that 68.70% of the adsorbed FA was desorbed under these conditions, indicating that about 1.4 milligrams of FA remained adsorbed.

^{1}H NMR spectra (Figure 4) suggests a partial purification of the extract obtained by alkaline hydrolysis of BSG in an autoclave and a simplified procedure. Figure 4A corresponds to FA standard spectrum. FA peaks were identified in both Figures 4B and 4C for initial and partially purified extracts, respectively. Because FA concentration increased after purification, the peaks between 6.3 and 7.5 ppm and the peak at 3.8 ppm slightly increased in the final extract when compared to the original extract. On the contrary, peaks at 8.4 and 5.3 ppm as well as regions 3.5 to 3.7, 3.0 to 3.4, 2.6 to 2.8, 1.5 to 2.3, and 1.0 to 1.3 ppm decreased in the final extract compared with the initial extract. This suggests a decrease in the concentration of these compounds after partial purification with synthetic resin.

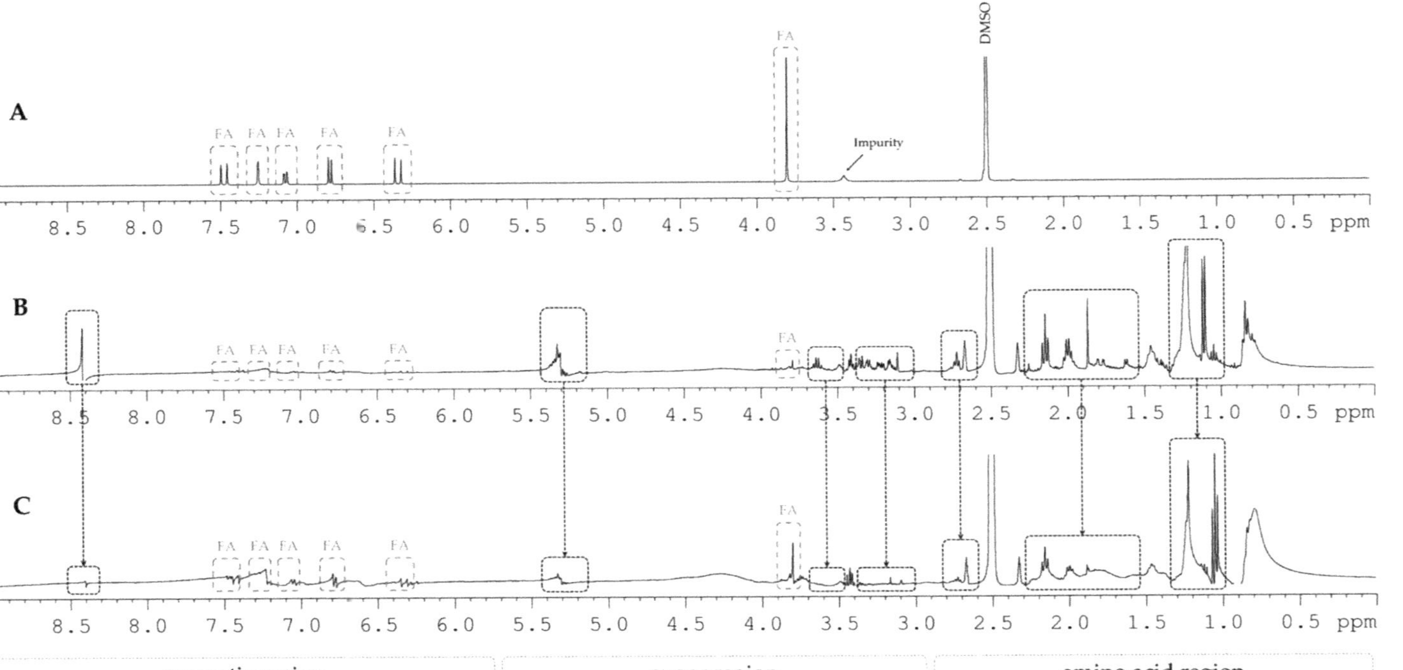

Figure 4. ^{1}H NMR spectra for (**A**) ferulic acid standard, (**B**) initial extract obtained by alkaline hydrolysis of BSG in autoclave with simplified post-extraction process and (**C**) final extract partially purified by adsorption/desorption in the Lewatit resin. Peaks corresponding to ferulic acid are marked in a grey box with the inscription "FA". The black boxes indicate the regions where the differences were more significant. The arrows make the correspondence between initial and final extracts.

4. Conclusions

Because of its high availability and potential, several works are being developed around BSG. One of the interesting compounds present in the lignocellulosic structure of BSG is ferulic acid. Since FA is not extracted from the BSG matrix by conventional solid–liquid extraction methods, alternative techniques such as alkaline hydrolysis are often applied. The methods described in literature usually comprise a pretreatment step to remove compounds resulting from the brewing process. Alkaline hydrolysis is commonly followed by centrifugation, precipitation of hemicellulose fraction in the alkaline liquor, and neutralization. In the present work, the use of autoclave to perform the alkaline hydrolysis and a simplification of the post-extraction process were suggested and were shown to increase FA yield. The extraction in an autoclave resulted in an increase of FA extraction yield of around 38% when compared with the extraction in pressure tubes. An increase of about 84% in the extraction yield was achieved in a small scale extraction in pressure tubes when a simplification of the post-extraction process was applied. A partial purification by adsorption on a synthetic resin was also suggested, constituting a potential approach to obtain ferulic acid in a higher degree of purity.

During the studies, p-coumaric acid was co-extracted in molar proportion (p-CA:FA) between 1:4 and 1:8, depending on the extraction conditions.

Author Contributions: Conceptualization: P.I. and P.C.C.; methodology; P.I. and P.C.C.; software; P.I. and I.S.-F.; validation; P.I. and I.S.-F: formal analysis: P.I. and I.S.-F; investigation: P.I. and P.C.C.; resources: P.C.C.; data curation: P.I., I.S.-F and P.C.C.: writing—original draft preparation, P.I. and P.C.C.; writing—review and editing: P.I., I.S.-F and P.C.C.; visualization: P.I.; supervision: P.C.C.; project administration: P.C.C.; funding acquisition: P.C.C. All authors have read and agreed to the published version of the manuscript.

Funding: This research was partially supported by FCT—Fundação para a Ciência e a Tecnologia with funds from the Portuguese Government, under de projects UIDB/00674/2020 (Centro de Química da Madeira) and UIDB/00006/2020 (Centro de Estatística e Aplicações). PI grant through the project M1420-01-0145-FEDER-000005—Centro de Química da Madeira—CQM+ (Madeira 14–20) is also acknowledged.

Acknowledgments: Authors acknowledge Empresa de Cervejas da Madeira for the supply of BSG samples.

Conflicts of Interest: The authors declare no conflict of interest.

References

1. Mussatto, S.I.; Dragone, G.; Roberto, I.C. Brewers' spent grain: Generation, characteristics and potential applications. *J. Cereal Sci.* **2006**, *43*, 1–14. [CrossRef]
2. Gupta, M.; Abu-Ghannam, N.; Gallaghar, E. Barley for brewing: Characteristic changes during malting, brewing and applications of its by-products. *Compr. Rev. Food Sci. Food Saf.* **2010**, *9*, 318–328. [CrossRef]
3. Mussatto, S.I. Brewer's spent grain: A valuable feedstock for industrial applications. *J. Sci. Food Agric.* **2014**, *94*, 1264–1275. [CrossRef] [PubMed]
4. Aliyu, S.; Bala, M. Brewer's spent grain: A review of its potentials and applications. *Afr. J. Biotechnol.* **2011**, *10*, 324–331. [CrossRef]
5. McCarthy, A.L.; O'Callaghan, Y.C.; Piggott, C.O.; FitzGerald, R.J.; O'Brien, N.M. Brewers' spent grain; bioactivity of phenolic component, its role in animal nutrition and potential for incorporation in functional foods: A review. *Proc. Nutr. Soc.* **2013**, *72*, 117–125. [CrossRef]
6. McCarthy, A.; O'Callaghan, Y.C.; Neugart, S.; Piggott, C.O.; Connolly, A.; Jansen, M.A.; Krumbein, A.; Schreiner, M.; Fitzgerald, R.J.; O'Brien, N.M. The hydroxycinnamic acid content of barley and brewers' spent grain (BSG) and the potential to incorporate phenolic extracts of BSG as antioxidants into fruit beverages. *Food Chem.* **2013**, *141*, 2567–2574. [CrossRef]
7. Kaur, V.I.; Saxena, P.K. Incorporation of brewery waste in supplementary feed and its impact on growth in some carps. *Bioresour. Technol.* **2004**, *91*, 101–104. [CrossRef]
8. Carvalheiro, F.; Esteves, M.P.; Parajó, J.C.; Pereira, H.; Gírio, F.M. Production of oligosaccharides by autohydrolysis of brewery's spent grain. *Bioresour. Technol.* **2004**, *91*, 93–100. [CrossRef]
9. Mussatto, S.I.; Roberto, I.C. Acid hydrolysis and fermentation of brewer's spent grain to produce xylitol. *J. Sci. Food Agric.* **2005**, *85*, 2453–2460. [CrossRef]

10. Carvalheiro, F.; Duarte, L.C.; Medeiros, R.; Gírio, F.M. Xylitol production by Debaryomyces hansenii in brewery spent grain dilute-acid hydrolysate: Effect of supplementation. *Biotechnol. Lett.* **2007**, *29*, 1887–1891. [CrossRef]

11. Mussatto, S.I.; Dragone, G.; Roberto, I.C. Ferulic and p-coumaric acids extraction by alkaline hydrolysis of brewer's spent grain. *Ind. Crop. Prod.* **2007**, *25*, 231–237. [CrossRef]

12. Bartolomé, B.; Gómez-Cordovés, C.; Sancho, A.; Diez, N.; Ferreira, P.; Soliveri, J.; Copa-Patiño, J.L. Growth and release of hydroxycinnamic acids from Brewer's spent grain by Streptomyces avermitilis CECT 3339. *Enzym. Microb. Technol.* **2003**, *32*, 140–144. [CrossRef]

13. Moreira, M.M.; Morais, S.; Barros, A.; Delerue-Matos, C.; Guido, L.F. A novel application of microwave-assisted extraction of polyphenols from brewer's spent grain with HPLC-DAD-MS analysis. *Anal. Bioanal. Chem.* **2012**, *403*, 1019–1029. [CrossRef] [PubMed]

14. Ou, S.; Kwok, K.C. Ferulic acid: Pharmaceutical functions, preparation and applications in foods. *J. Sci. Food Agric.* **2004**, *84*, 1261–1269. [CrossRef]

15. Priefert, H.; Rabenhorst, J.; Steinbüchel, A. Biotechnological production of vanillin. *Appl. Microbiol. Biotechnol.* **2001**, *56*, 296–314. [CrossRef] [PubMed]

16. Heinonen, M.; Rein, D.; Satue, M.T.; Huang, S.; German, J.B.; Frankel, E.N. Effect of Protein on the Antioxidant Activity of Phenolic Compounds in a Lecithin-Liposome Oxidation System. *J. Agric. Food Chem.* **1998**, *8561*, 917–922. [CrossRef]

17. Friedman, M.; Jürgens, H.S. Effect of pH on the stability of plant phenolic compounds. *J. Agric. Food Chem.* **2000**, *48*, 2101–2110. [CrossRef]

18. Buranov, A.U.; Mazza, G. Extraction and purification of ferulic acid from flax shives, wheat and corn bran by alkaline hydrolysis and pressurised solvents. *Food Chem.* **2009**, *115*, 1542–1548. [CrossRef]

19. Ferri, M.; Happel, A.; Zanaroli, G.; Bertolini, M.; Chiesa, S.; Commisso, M.; Tassoni, A. Advances in combined enzymatic extraction of ferulic acid from wheat bran. *New Biotechnol.* **2020**, *56*, 38–45. [CrossRef]

20. Long, L.; Ding, D.; Han, Z.; Zhao, H.; Lin, Q.; Ding, S. Thermotolerant hemicellulolytic and cellulolytic enzymes from *Eupenicillium parvum* 4-14 display high efficiency upon release of ferulic acid from wheat bran. *J. Appl. Microbiol.* **2016**, *121*, 422–434. [CrossRef]

21. Moreira, M.M.; Morais, S.; Carvalho, D.O.; Barros, A.A.; Delerue-Matos, C.; Guido Luís, F. Brewer's spent grain from different types of malt: Evaluation of the antioxidant activity and identification of the major phenolic compounds. *Food Res. Int.* **2013**, *54*, 382–388. [CrossRef]

22. Lin, S.H.; Juang, R.S. Adsorption of phenol and its derivatives from water using synthetic resins and low-cost natural adsorbents: A review. *J. Environ. Manag.* **2009**, *90*, 1336–1349. [CrossRef] [PubMed]

23. Dávila-Guzman, N.E.; Cerino-Córdova, F.J.; Diaz-Flores, P.E.; Rangel-Mendez, J.R.; Sánchez-González, M.N.; Soto-Regalado, E. Equilibrium and kinetic studies of ferulic acid adsorption by Amberlite XAD-16. *Chem. Eng. J.* **2012**, *183*, 112–116. [CrossRef]

24. Conidi, C.; Rodriguez-Lopez, A.; Garcia-Castello, E.M.; Cassano, A. Purification of artichoke polyphenols by using membrane filtration and polymeric resins. *Sep. Purif. Technol.* **2015**, *144*, 153–161. [CrossRef]

25. Javanmardi, J.; Kubota, C. Variation of lycopene, antioxidant activity, total soluble solids and weight loss of tomato during postharvest storage. *Postharvest Boil. Technol.* **2006**, *41*, 151–155. [CrossRef]

26. Gouveia, S.; Castilho, P.C. Antioxidant potential of Artemisia argentea L'Hér alcoholic extract and its relation with the phenolic composition. *Food Res. Int.* **2011**, *44*, 1620–1631. [CrossRef]

27. Meneses, N.G.T.; Martins, S.; Teixeira, J.; Mussatto, S.I. Influence of extraction solvents on the recovery of antioxidant phenolic compounds from brewer's spent grains. *Sep. Purif. Technol.* **2013**, *108*, 152–158. [CrossRef]

Article

Radical Scavenging and Antimicrobial Properties of Polyphenol Rich Waste Wood Extracts

Anita Smailagić [1], Petar Ristivojević [2], Ivica Dimkić [3], Tamara Pavlović [3],
Dragana Dabić Zagorac [1], Sonja Veljović [4], Milica Fotirić Akšić [5], Mekjell Meland [6]
and Maja Natić [2,*]

1   Innovation Center of the Faculty of Chemistry, University of Belgrade, P.O. Box 51, 11158 Belgrade, Serbia;
    anitasmailagic@yahoo.com (A.S.); naca10@gmail.com (D.D.Z.)
2   Faculty of Chemistry, University of Belgrade, P.O. Box 51, 11158 Belgrade, Serbia; ristivojevic@chem.bg.ac.rs
3   Faculty of Biology, University of Belgrade, Studentski trg 16, 11000 Belgrade, Serbia;
    ivicad@bio.bg.ac.rs (I.D.); tamara.pavlovic@bio.bg.ac.rs (T.P.)
4   Institute of General and Physical Chemistry, University of Belgrade P.O. Box 551, 11001 Belgrade, Serbia;
    pecic84@hotmail.com
5   Faculty of Agriculture, University of Belgrade, 11080 Belgrade, Serbia; fotiric@agrif.bg.ac.rs
6   Norwegian Institute of Bioeconomy Research-NIBIO Ullensvang, NO-5781 Lofthus, Norway;
    mekjell.meland@nibio.no
*   Correspondence: mnatic@gmail.com

Received: 12 February 2020; Accepted: 2 March 2020; Published: 10 March 2020

Abstract: The main focus of this study is to assess radical scavenging and antimicrobial activities of the 11 wood extracts: oak (*Quercus petraea* (Matt.) Liebl., *Q. robur* L., and *Q. cerris* L.), mulberry (*Morus alba* L.), myrobalan plum (*Prunus cerasifera* Ehrh.), black locust (*Robinia pseudoacacia* L.), and wild cherry (*Prunus avium* L.). High-performance thin-layer chromatography (HPTLC) provided initial phenolic screening and revealed different chemical patterns among investigated wood extracts. To identify individual compounds with radical scavenging activity DPPH-HPTLC, assay was applied. Gallic acid, ferulic and/or caffeic acids were identified as the compounds with the highest contribution of total radical scavenging activity. Principal component analysis was applied on the data set obtained from HPTLC chromatogram to classify samples based on chemical fingerprints: *Quercus* spp. formed separate clusters from the other wood samples. The wood extracts were evaluated for their antimicrobial activity against eight representative human and opportunistic pathogens. The lowest minimum inhibitory concentration (MIC) was recorded against *Staphylococcus aureus* for black locust, cherry and mulberry wood extracts. This work provided simple, low-cost and high-throughput screening of phenolic compounds and assessments of the radical scavenging properties of selected individual metabolites from natural matrix that contributed to scavenge free radicals.

Keywords: wood waste; phenolic profile; planar chromatography; DPPH-HPTLC assay; antimicrobial activity

---

1. Introduction

Ageing processes of some alcoholic beverages are one of the most important practices during their production. This contributes to improved sensory characteristics such as aroma, color, taste and astringency. The most commonly used material in cooperage is oak heartwood barrels. Alternative wood species such as chestnut, cherry and mulberry are also used in Balkan cooperages, in different forms such as wood chips and staves [1]. Nowadays, notable studies have showed that agri-food wastes and by-products, including waste from barrel production, represent an inexhaustible source of valuable biologically active compounds. Additionally, this waste represents a low-cost material, which can be used as material for the production of the extracts. Recently, various extraction techniques were reviewed and compared with classical extraction procedures used for recovery of the antioxidant compounds from wastes [2]. Using simple, fast and inexpensive eco-friendly extraction methods for phenolic compounds represents an efficient method and advantage for further implementation in the food, pharmaceutical and cosmetic industries [3–5].

From the production of wood barrels, it is estimated that more than 200 tons of wood waste is available annually in Serbia [6]. Forests and other wooded land occupy ~2.5 million hectares, which is about one third of the territory of the Republic of Serbia. These natural populations of Serbia contain a large number of economically important forest tree species (oak, beech, black locust, spruce, pine and fir) together with autochthonous and introduced wild fruit trees species (wild cherry, cherry plum, mulberry, wild pear, wild apple, cornelian cherry, hazel and walnut) which are used for timber production, afforestation and erosion prevention, for grafting, in human diet, in medicine, in industrial processing and in landscape architecture [7,8].

In Serbia, barrels are mostly made of oak (pedunculate oak, *Quercus robur*, or sessile oak, *Quercus petraea* (Matt.) Liebl. L.) and Turkey oak (*Quercus cerris* L.) but sometimes black locust (*Robinia pseudoacacia* L.), myrobalan plum (*Prunus cerasifera* Ehrh.), mulberry (*Morus alba* L.) and even wild cherry (*Prunus avium* L.) are used as a cheaper substitute. Oak is the most widespread deciduous tree in Serbia, a national tree with strong historical and religious importance. Myrobalan plum is a native tree in Southeast Europe, has great genetic importance for horticultural breeding, and has spread throughout the whole country in all kinds of micro-climatic and pedologic conditions. Mulberries are very common, since ex-Yugoslavia used to be the fifth largest silk producer in the world with more than 2.5 million white mulberry trees [9]. Wild cherry, a noble tree, is widely distributed by birds; its seeds are used for generative rootstock production and its fruits are suitable for table consumption and as a local medicine [10,11]. Black locust is mostly used in construction works, as technical or ornamental wood, and is most commonly used as firewood. It has special value as a honey species for beekeeping [12].

Wood waste has a potential to be reused in the food and pharmaceutical industry due to its richness in potentially bioactive phenolic compounds with high antioxidant and antimicrobial activity. In our previous research [6], ellagic acid was abundant in sessile and pedunculate oak wood. It was also found in Turkey oak, black locust and myrobalan plum, but in much lower quantities. Mulberry contained the largest concentration of *p*-hydroxybenzoic acid and stilbenoids in comparison with other wood species, while myrobalan plum showed the highest content of protocatechuic acid and 5-*O*-caffeoylquinic acid. Wild cherry was characterized by richness in flavonols, flavanones, flavones, isoflavones and flavanonols [6,13,14], with taxifolin as the most abundant phenolic compound [6]. Extracts from sessile and pedunculate oak, black locust, myrobalan plum, wild cherry and mulberry showed notable antioxidant capacity, with the highest radical scavenging activity in the latter extract. Turkey oak showed the lowest radical scavenging activity [6]. According to the literature, phenolic acids were identified as the major contributors to the antioxidant capacity in wood samples, including gallic, protocatechuic, *p*-coumaric and ellagic acid and all the ellagitannins, due to their characteristic structure [15]. The following phenolic acids: ferulic acid, caffeic acid, protocatechuic acid, gallic acid, *p*-coumaric acid and chlorogenic acid, also present in some wood species, exhibit strong free radical scavenging properties on silica plates [16].

It is proposed that phenolic compounds can damage the bacteria cell membrane by interacting with the proteins of the cell membrane, or can be involved in interaction with cellular enzymes [17], which may directly or indirectly cause metabolic dysfunction and finally bacterial death [18]. Phenolic compounds are able to inhibit bacterial quorum sensing signal receptors, enzymes and secretion of toxins [19]. The type, structure and concentration of phenolic compounds, as well as the microorganism used, will influence the bacterial growth. Large doses of phenolic compounds may be toxic for bacteria, but lower doses can be used as substrates [17].

Some phenolic compounds present in several wood species showed antimicrobial activity. Taxifolin exhibited antibacterial activity against six known clinical pathogens: *Escherichia coli*, *Listeria* sp., *Pseudomonas aeruginosa, Bacillus* sp., and *S. aureus* [20]. Oxyresveratrol, the most abundant stilbene in mulberry, was active against the methicillin-resistant *S. aureus* [21].

Among flavonoids present in wild cherry wood, flavonols were distinguished by effective antimicrobial activity against resistant bacteria [22]. Methanolic extract (80%, *v/v*) from oak bark (*Q. robur* L.) showed moderate bactericidal, fungicidal, bacteriostatic and fungistatic activity on *S. aureus*, *Enterobacter aerogenes* (today known as *Klebsiella aerogenes*) and *C. albicans* [23]. *Q. robur* bark showed strong antibacterial activities against *Pseudomonas aeruginosa, M. flavus* and *E. coli,* and moderate effects against other bacterial species [24]. Heartwood and resin of cherry wood exhibited cytochrome inhibition and antifungal activity [25], while cherry wood extracts possessed noticeable antimicrobial activity against 9 out of the 11 wine organisms tested [17]. Oak wood has abundant ellagitannins, which are toxic to microorganisms, and provides good resistance to fungal degradations [26].

Antimicrobial resistance presents a global problem since resistant pathogens can cause life-threatening conditions that become incurable with one or more known drugs. The mechanisms of the antibacterial activities of many plant-derived flavonoids are different than those of conventional drugs, which open new possibilities in enhancement of antibacterial therapy [27]. In addition, many synthetized drugs have side-effects, which are small in the case of plant-derived compounds [27]. Due to all these reasons, the development of alternative drugs derived from natural resources is an attractive option.

Radical scavenging activity using DPPH-HPTLC (high performance still layer chromatography) assay and antimicrobial activity on wood waste extracts are not investigated so far. Thus, the main aim of this research was to assess radical scavenging and antimicrobial activities of the wood waste extracts from mulberry (*M. alba* L.), myrobalan plum (*P. cerasifera* Ehrh.), black locust (*R. pseudoacacia* L.), wild cherry (*P. avium* L.), and different species of oaks (*Q. petraea* (Matt.) Liebl., *Q. robur* L. and *Q. cerris* L.) and consider their usage in the pharmaceutical and food industries. Phenolic compounds were separated by using HPTLC, while radical scavenging activity was determined using DPPH-HPTLC.

2. Materials and Methods

2.1. Chemicals

Ethyl acetate was purchased from Merck (KGaA, Darmstadt, Germany); formic acid, hexan 2,2-diphenyl-1-picrylhydrazyl (DPPH) and phenolic standards from Sigma-Aldrich (Steinheim, Germany); and 2-aminoethyl diphenylborinate (NTS) from Fluka (Steinheim, Germany). Gallic acid, ferulic acid and caffeic acid were supplied by Sigma Aldrich (Steinheim, Germany).

2.2. Samples and Preparation of Wood Extracts

Eleven different wood staves of different geographical origins were analyzed (Table 1). In total three samples of Pedunculate oaks (*Quercus robur* L.), three of sessile oaks (*Quercus petraea* (Matt.) Liebl), and one sample of Turkey oak (*Quercus cerris* L.), black locust (*Robinia pseudoacacia* L.), myrobalan plum (*Prunus cerasifera* Ehrh.), wild cherry (*Prunus avium* L.), and mulberry (*Morus alba* L.) were included. Nine staves were stored for the whole year in the open air at cooperage industry VBX-SRL. D.O.O. in Kraljevo, Central Serbia., while two samples (sessile oak from Kuršumlija and Turkey oak) were not seasoned [6]. The wood age of the oak wood staves was over 60 years, while the wood age of non-oak wood staves was more than 40 years.

Firstly, the staves were grinded in a mill for wood and sieved until granulation of 0.5–1.5 mm was obtained. The sawdust (2.5 g) was extracted with 25 mL of ethanol (60%, *v/v*), in Erlenmeyer flasks, with constant stirring in a magnetic stirrer for seven days in darkness and room temperature (20 ± 2 °C) [6]. The extracts were centrifuged twice (5 min at 8000 rpm). For investigation of antimicrobial activity, the extracts were evaporated with a rotary evaporator and diluted in methanol until the concentration of 50 mg mL^{-1} was reached. The extraction yield of each extract was calculated from the weight of the extract residue obtained after solvent removal and the weight of waste wood employed in the extraction procedure.

Table 1. Selected wood waste extracts of different forest trees for DPPH-HPTLC (high-performance thin-layer chromatography) and antimicrobial testing assay.

Sample No.	Tree	Geographical Origin	Extraction Yield (%)
1	Pedunculate oak—*Quercus robur* L.	Slavonija (Croatia)	4.44
2		Gornji Radan (Serbia)	4.40
3		Olovo (Bosnia and Herzegovina)	4.12
4	Sessile oak—*Quercus petraea* (Matt.) Liebl.	Kučaj (Serbia)	5.06
5		Kuršumlija (Serbia)	3.05
6		Ravna Gora (Serbia)	4.58
7	Turkey oak—*Quercus cerris* L.	Kuršumlija (Serbia)	1.63
8	Black locust—*Robinia pseudoacacia* L.	Kraljevo (Serbia)	6.37
9	Myrobalan plum—*Prunus cerasifera* Ehrh	Vrnjačka Banja (Serbia)	5.80
10	Wild cherry—*Prunus avium* L.	Ravna Gora (Serbia)	3.15
11	Mulberry—*Morus alba* L.	Vrnjačka Banja (Serbia)	7.29

2.3. High-Performance Thin-Layer Chromatography and Image Analysis

HPTLC Silica gel 60F$_{254}$ plates were used for both HPTLC fingerprint and DPPH-HPTLC assay (Merck, Germany). The oak and wild cherry samples (5 μL), black locust, myrobalan plum and mulberry (2 μL), and four standard compounds: gallic acid, ferulic acid, caffeic acid and *p*-coumaric acid (2 μL, c = 1000 ppm), were applied as bands (8 mm) using Linomat 5 system (Camag, Muttenz, Switzerland).

The mobile phase consisted of a mixture of ethyl acetate:hexan:formic acid:water (11:2:1:0.5 *v/v/v/v*). The plates were developed at room temperature (20 °C) in a twin-trough-chamber (CAMAG) saturated with the vapors of mobile phase for 15 min, at a developing distance of 70 mm. The obtained HPTLC chromatograms were derivatized with 2-aminoethyldiphenylborate solution (NTS - 0.2% in ethanol) in order to intensify the fluorescence of compounds.

For DPPH-HPTLC assay, a developed HPTLC chromatogram was immersed manually for 3 seconds (s) in DPPH·methanol solution (0.2%) and then photographed every 30 s for 15 min. Images of the plates were captured with mobile phone (Huawei P Smart) equipped with a 13-pixels camera. All developed plates were photographed both before and after derivatization and saved as TIF files.

Images of the HPTLC chromatograms were analyzed using free available Image J software. The obtained results for each sample were cropped and denoised by using median filter with three pixels width filter. Further, images were transformed and the tracks were outlined with a rectangular selection tool. The line profile plots were generated with Plot Profile option for each sample. Profile plot displays a 2-D graph of the intensities of pixels along a line.

2.4. Principal Component Analysis

The line profiles were obtained using ImageJ software [28]. Principal Component Analysis (PCA) was applied using PLS ToolBox, v.6.2.1 (Eigenvector Research, Inc. 196 Hyacinth Road Manson, WA 98831, USA), for MATLAB (7.12.0(R2011a) (http://www.eigenvector.com/software/pls_toolbox.htm, Eigenvector Research, Inc., Wenatchee, WA). The data were pre-processed using correlated optimized warping (COW), standard normal variate (SNV) and mean centering to improve multivariate models.

2.5. Bacterial Strains and Growth Conditions

Antibacterial activity was tested using eight indicator strains in line with their growth requirements (Table 2). Suspensions were adjusted to McFarland standard turbidity (0.5) (BioMérieux, Marcy-l'Étoile, France), which corresponds approximately to 1×10^8 CFU mL^{-1}.

Table 2. Indicator strains used in testing antimicrobial activity of selected extracts from forest trees.

Indicator Strains	Isolate Code	Growth Medium	Growth Temperature	The Origin of The Isolates
Streptococcus mutans	IBR S0001	LA		Oral cavity *
Streptococcus pyogenes	IBR S0004	ǁ		
Methicillin-resistant *Staphylococcus aureus* (MRSA)	ATCC33591	ǁ	37 °C	
Staphylococcus aureus	ATCC25923	ǁ		Reference strains
Escherichia coli	ATCC25922	ǁ		
Enterococcus faecalis	ATCC29212	ǁ		
Listeria monocytogenes	ATCC19111	BHA		
Candida albicans	ATCC10231	TSA		

* Strains isolated from the human oral cavity [29]. All reference strains belong to Department of Microbiology, Faculty of Biology, University of Belgrade.

2.6. Well-Diffusion Method

A modified well-diffusion method [30] was performed for initial screening of the antimicrobial potential of the selected 11 wood waste extracts. Wells were made of sterile bottom parts of pipette tips (200 µL) and placed on the LA/BHA/TSA solid medium (Table 2). According to growth requirements of used strains, 6 mL of LA/BHA/TSA soft agar was inoculated with 60 µL of the appropriate strain and poured into Petri dishes over the solid medium. Molds (5 mm in diameter) were removed after soft agars solidification and 20 µL of each extract (1 mg/well) was added. Vancomycin and nystatin were used as a positive control (antibiotic/mycotic, viz., 0.2 mg/well), for bacterial strains and *C. albicans*, respectively. As a negative control, 20 µL of methanol was used. The Petri dishes were incubated at 37 °C, for 24 h. After the incubation, bacterial susceptibility and zones of inhibition were measured and expressed in mm.

2.7. MIC Assay

A broth microdilution method was used to determine the minimum inhibitory concentration (MIC) and minimum bactericidal concentrations (MBC) of the selected 11 wood waste extracts. Extracts were tested in the concentration range from 0.02 to 2 mg mL^{-1} by performing two-fold serial dilutions with the appropriate medium in 96-well microtiter plates. Negative control (control of bacterial and yeast growth) and sterility control (blank, only appropriate medium) were also tested. The final concentration of the solvent control (methanol) in the first wells was 10%. Vancomycin, streptomycin and nystatin were tested as positive controls in concentration range from 0.001 to 0.4 mg mL^{-1}. Beside negative and sterility controls, each well was inoculated with 20 µL of bacterial/yeast culture (approx. 1×10^6 CFU mL^{-1}), reaching a final volume of 200 µL. In addition, 22 µL of resazurin indicator was added to each well. Microtiter plates were incubated for 24 h at 37 °C. In the presence of living bacterial cells, blue colored resazurin was being irreversibly reduced to pink colored and highly red fluorescent resorufin [31]. The lowest concentration of each extract which showed no change in color of resazurin was defined as MIC value. MBC/MFC values were determined by sub-culturing the dilutions from wells without color changes on agar plates. Plates were incubated 24 h at 37 °C and bacterial/yeast growth was monitored. The lowest concentration without growth was defined as MBC/MFC value. The results were expressed in mg mL^{-1}.

3. Results and Discussion

3.1. Line Profiles of Investigated Extracts

Investigated wood samples contained several characteristic phenolic compounds at R$_F$ values of: 0.28, 0.35, 0.43, 0.74, 0.86 (gallic acid), 0.91 (ferulic acid), 0.88 (caffeic acid) and 0.91 (*p*-coumaric acid)

(Figure 1). Based on HPTLC profiles, wood extracts contained bands with R_F values from 0.28 to 0.91. There are five different patterns in the investigated samples: *Quercus* samples showed one band of weak intensity with R_F value at 0.85, while sample 1 (Pedunculate oak—*Q. robur* L.) had the highest intensity peak of this compound. Further, black locust (sample 8) showed greenish bands with R_F at 0.75, 0.82 and 0.87, clearly different from standard phenolic acids (Figure 2a).

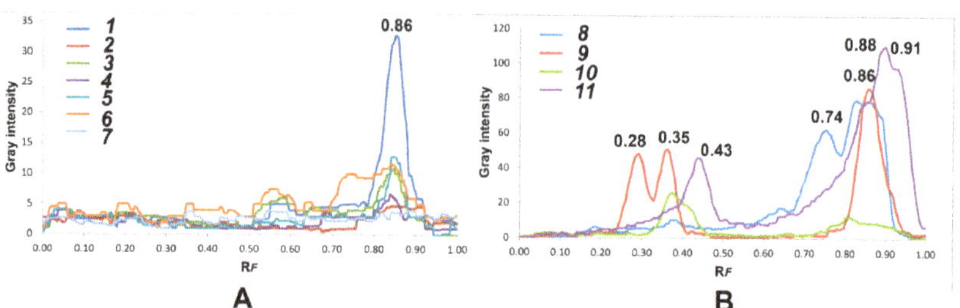

Figure 1. Line profiles of investigated wood extracts based on HPTLC analysis: (**A**) oak samples (no. 1–7); (**B**) non-oak samples (no. 8–11): black locust (*Robinia pseudoacacia* L.) (no. 8), myrobalan plum (*Prunus cerasifera* Ehrh.) (no. 9), wild cherry (*Prunus avium* L.) (no. 10) and mulberry (*Morus alba* L.) (no 11).

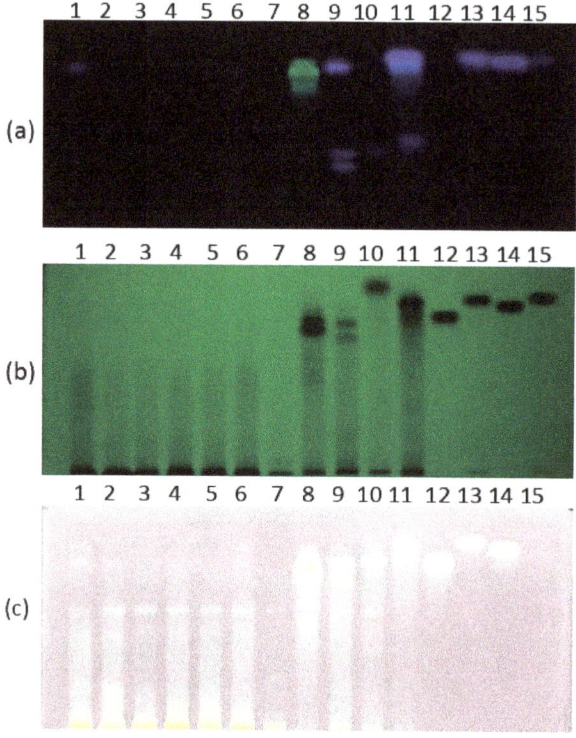

Figure 2. HPTLC chromatograms of samples: *Q. robur* (no. 1–3), *Q. petraea* (no. 4–6), *Q. cerris* (no. 7), *Robinia pseudoacacia* (no. 8), *Prunus cerasifera* (no. 9), *Prunus avium* (no. 10), mulberry (no. 11) and four standard compounds (gallic acid (no. 12), ferulic acid (no. 13), caffeic acid (no. 14) and *p*-coumaric acid(no. 15)); (**a**) under UV light at 366 nm; (**b**) under UV light at 254 nm; (**c**) DPPH-HPTLC chromatogram.

Wild cherry contained one characteristic peak at 0.36, whereas myrobalan plum had a different pattern from other wood samples with three characteristic peaks at R_F values of 0.28, 0.35 and caffeic acid (Figure 1). A different profile of myrobalan plum in comparison with other wood samples could be seen also by HPLC [6], where it contained significantly larger amounts of protocatechuic acid and 5-O-caffeoylquinic acid than other wood samples. The peak profiles for mulberry showed it contained peaks at 0.43, 0.84, gallic, ferulic and/or *p*-coumaric acids.

Mulberry sample showed hardly visible blue band with R_F at 0.86, recognized as gallic acid, while black locust and oak wood samples contained gallic acid in higher amounts. In addition, wild cherry contained *p*-coumaric and ferulic acid in greater quantities than mulberry, and caffeic acid in greater quantities than myrobalan plum, which was observed neither on HPTLC plates nor line profiles.

3.2. DPPH-HPTLC Assay

It was previously shown that the total antioxidant activity of each extract through the DPPH assay, mulberry and myrobalan plum wood extracts had significantly higher DPPH values in comparison to the other samples [6]. The single compounds with radical scavenging activity and their contribution to the total radical scavenging activity were investigated by DPPH˙-HPTLC assay. Substances exhibiting radical scavenging properties (yellow bands against a purple background) were located between R_F values at 75 and 92 (Figure 2c). The most dominant zones in the HPTLC-DPPH˙ fingerprints were compounds with R_F values at 0.87, 0.91 and 0.92, which could be recognized as gallic, ferulic and/or caffeic acids (zone 4). Extracts no. 8–11 showed strong radical scavenging activities, mainly due to the previously detected phenolic compounds, while *Quercus* samples revealed one weak band with R_F at 75 against the purple background. *P. cerasifera* contained two bands at 0.84 and caffeic acid, and were recognized as radical scavengers. The, *M. alba* sample showed radical scavengers with R_F values at 0.84, gallic, ferulic and/or *p*-coumaric acids. These compounds have been recognized before as strong radical scavengers on silica plates [16].

3.3. Principal Component Analysis

Visual inspection of HPTLC chromatograms is a subjective method and mainly depends on the analyst's perception. On the other hand, multivariate chemometrics analysis applied on the HPTLC chromatogram provides an objective classification of the investigated samples for identification of phenols most responsible for classification, as well as identification of outliers. The HPTLC system was optimized to separate and identify all phenols from different wood extracts.

Principal component analysis (PCA) is a commonly used multivariate technique. It accounts for most of the variation of total variability, visualizes the structure of data by grouping objects into two or three dimensions, and identifies important variables responsible for discrimination between wood samples. PCA as an initial multivariate technique was applied on the data matrix (11 samples × 389 variables) obtained from HPTLC chromatograms, where variables represent the intensities of pixels along the 389 length lines. The first two Principal Components (PCs) accounted for 33.35% and 20.09% of the total variability, respectively. The first five principal components describe 87.85% of total variability. From the PC score plot (Figure 3a), six *Quercus* samples were positioned on left side of PC score plot, while other four wood samples were misclassified and positioned on right side of PCs score plot. The loading plots (Figure 3b,c) demonstrated the significant contribution of polyphenolic compounds to the total variability. The most influential phenolic compounds discriminating between *Quercus* and the other wood samples were compounds with R_F values at 0.35, 0.43, 0.86 and 0.91. In contrast to other types of wood samples *Quercus* samples contained low amounts of phenolic compounds with R_F values at 0.35, 0.43, 0.86 and 0.91. Polar compounds with low R_F values could be some phenolic acids and/or glycosides. These phenolic compounds may be identified as characteristic taxonomical markers between wood species.

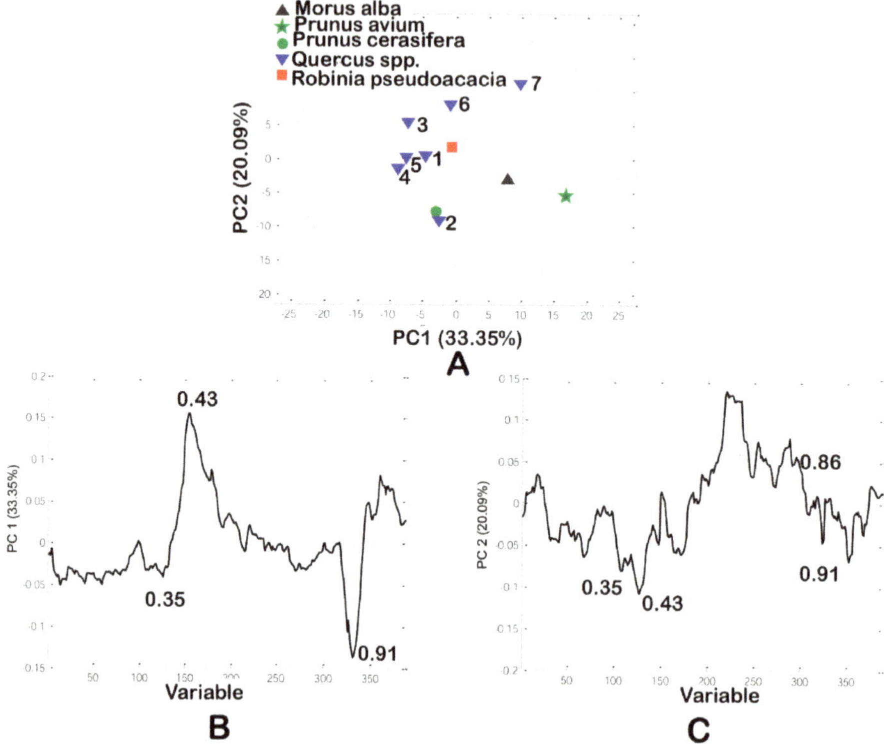

Figure 3. Principal component analysis (PCA) of HPTLC chromatogram: (**A**) The PC score plot; (**B**) and (**C**) The loading plots. 1–3 Pedunculate oaks (*Quercus robur* L.), 4–6—sessile oaks (*Quercus petraea* (Matt.) Liebl), 7—Turkey oak (*Quercus cerris* L.).

3.4. Well-Diffusion Method

Antimicrobial potential of the extracts was tested against eight representative human and opportunistic pathogens. Besides clear zones of inhibition, bacteriostatic/fungistatic effect of tested extracts was also observed. Wood waste extracts in general showed the highest antimicrobial potential against *S. mutans*, *S. pyogenes* and *L. monocytogenes* strains in tested concentration of 1 mg/well (Figure 4). The wild cherry extract (10) inhibited the growth of *S. mutans* and *S. aureus* yielding the largest zones of inhibition (21.7 and 19.8, respectively), compared to other extracts, towards to the mentioned pathogens. Additionally, only wild cherry and mulberry extracts (10,11) showed moderate bactericidal effect against *E. faecalis*. This indicator strain due to its higher resistance was excluded for further MIC testing. Mentioned extracts also showed high bacteriostatic effect against MRSA. Additionally, the wild cherry extract showed clear bactericidal effect only against *C. albicans* and *L. monocytogenes*, while other extracts acted more bacteriostatically. On the other hand, other wood extracts showed overwhelmingly bacteriostatic/fungistatic effect against almost all pathogens, including *E. coli*. All pathogens were susceptible to tested vancomycin and nystatin mycotic.

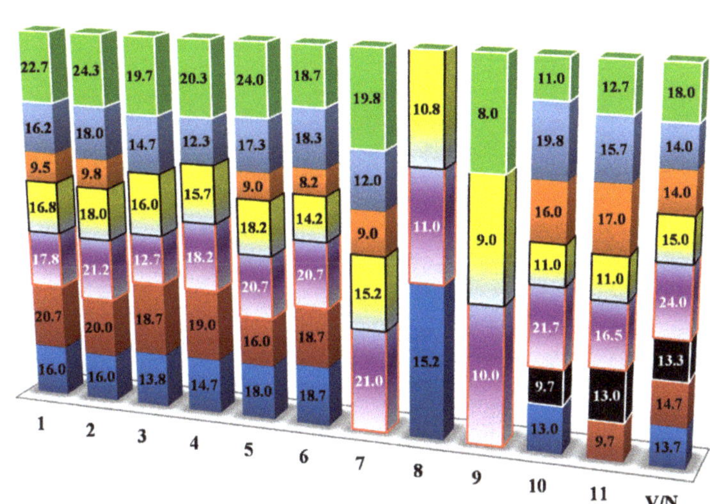

Figure 4. Antimicrobial potential of wood waste extracts in well-diffusion method. *V/N—Vancomycin/ Nystatin. Values within columns represent a mean of inhibition zones and expressed in mm. *Q. robur* (no. 1–3), *Q. petraea* (no. 4–6), *Q. cerris* (no. 7), *Robinia pseudoacacia* (no. 8), *Prunus cerasifera* (no. 9), *Prunus avium* (no. 10), mulberry (no. 11).

3.5. MIC Assay

For evaluation of new antimicrobials, the assessment of minimum inhibitory concentration (MIC) is usually the first step [27]. The MIC is the minimum concentration that causes visible inhibition of bacterial growth. Plant extracts with MIC < 100 µg mL^{-1} and purified compounds with MIC < 10 µg mL^{-1} are considered promising [27]. However, bactericidal activity, determined by MBC value in time-kill assays, is also an important parameter in assessing the antimicrobial activity. MBC and MIC parameters complement each other, and MBC below four times MIC value suggests the bactericidal action of a tested compound [27].

The obtained MIC values were in range from 0.02 mg mL^{-1} of extract 11, to MRSA to 2 mg mL^{-1} in the case of activity of extract 9 (myrobalan plum) against *C. albicans* (Table 3).

Table 3. Minimum inhibitory, minimum bactericidal and minimum fungicidal concentrations (MIC/MBC/MFC) of 11 wood extracts towards selected human pathogens (mg mL^{-1}).

Indicator Strains/MIC (mg mL^{-1})	1	2	3	4	5	6	7	8	9	10	11	Str	Van	Nys
S. mutans	0.25	0.13	0.25	0.25	0.25	0.25	0.13	0.25	1.00	0.05	0.13	0.020	0.006	NT
S. pyogenes	0.03	0.13	0.03	0.05	0.03	0.13	0.08	0.05	0.03	0.03	0.03	0.002	0.001	NT
S. aureus	0.08	0.05	0.05	0.03	0.03	0.03	0.03	0.09	0.06	0.05	0.09	0.009	0.002	NT
MRSA	0.06	0.06	0.13	0.06	0.03	0.05	0.03	0.03	0.13	0.13	0.02	-	-	NT
L. monocytogenes	0.50	0.50	0.75	0.75	0.19	0.63	0.50	0.13	0.06	0.06	0.03	0.019	0.002	NT
E. coli	0.75	0.75	1.50	1.50	0.75	1.50	-	0.75	-	0.75	0.75	0.009	0.200	NT
C. albicans	-	-	-	-	-	-	-	-	2.00	0.25	-	NT	NT	0.006

Indicator strains/MBC (mg mL^{-1})	1	2	3	4	5	6	7	8	9	10	11	Str	Van	Nys
S. mutans	2.00	2.00	2.00	2.00	2.00	2.00	0.50	2.00	2.00	0.06	0.50	0.050	0.150	NT
S. pyogenes	0.50	0.50	1.00	1.00	0.50	0.50	1.00	0.25	2.00	0.50	1.00	0.050	0.050	NT
S. aureus	0.25	0.13	0.25	0.25	0.13	0.25	0.50	0.13	0.25	0.13	0.13	0.025	0.003	NT
MRSA	0.50	0.50	0.50	0.50	0.25	0.50	0.63	0.25	1.00	0.25	0.03	-	-	NT
L. monocytogenes	1.00	1.00	1.00	1.00	1.00	1.00	1.00	0.50	1.00	0.13	0.25	0.025	0.006	NT
E. coli	1.00	1.00	2.00	2.00	1.00	2.00	-	1.00	-	1.00	1.00	0.013	0.400	NT
C. albicans	-	-	-	-	-	-	-	-	-	0.50	-	NT	NT	0.025

Str—Streptomycin; Van—Vancomycin; Nys—Nystatin; NT—not tested; (-)—not determined. Values are marked with lowest value (blue), middle value (yellow), and the highest value (red). *Q. robur* (no. 1–3), *Q. petraea* (no. 4–6), *Q. cerris* (no. 7), *Robinia pseudoacacia* (no. 8), *Prunus cerasifera* (no. 9), *Prunus avium* (no. 10), mulberry (no. 11).

The lowest MIC values (viz., 0.03 mg mL^{-1}) were recorded against MRSA (extracts 5, 7 and 8), *S. aureus* (extracts 4–7) and *S. pyogenes* (extracts 1, 3, 5, 9–11). *S. mutans* also showed high sensitivity to some of the tested extracts with MICs below 0.2 mg mL^{-1}. MIC values for *L. monocytogenes* were in range from 0.03–0.75 mg mL^{-1}, while extracts 9–11 significantly inhibited the growth rate of this pathogen. Compared to Gram-positive isolates, *E. coli* was less sensitive to the tested extracts. *Candida albicans* showed poor sensitivity to the action of all extracts, with the exception of extract 10 with obtained MIC value of 0.25 mg mL^{-1}. Alañón et al. [17] also concluded that yeasts had a stronger resistance to wood extracts than bacteria, since only toasted American oak wood and wild cherry wood extracts inhibited their growth. MICs for vancomycin, streptomycin and nystatin were lower compared to the tested extracts (0.001–0.4 mg mL^{-1}). Additionally, MRSA showed resistance to all antibiotics on the highest concentration tested (0.4 mg mL^{-1}). Interestingly, non-seasoned sessile oak (sample 5) showed lower MIC against MRSA and *L. monocytogenes* than seasoned oaks (samples 1, 2, 3, 4, 6).Comparing the results for *Q. robur* with the results for oak bark (*Q. robur*) [24], higher values for MIC were found against *L. monocytogenes* and *E. coli*, but lower values against *S. aureus*. In addition, the values of MIC for streptomycin were significantly lower than Elansary et al. [24] obtained. MBC and MFC values of tested extracts varied from 0.03–2 mg mL^{-1}. The lowest MBC was recorded against *S. aureus* for extracts 2, 5, 8, 10 and 11.

There was a strong simultaneous activity against all pathogens tested of extract 10 from the wild cherry wood. This could be explained by its richness of phenolic compounds which was observed in previous research [6]. For example, kaempferol is a potential candidate against different pathogenic microbes, and effective against fluconazole-resistant *Candida albicans* and Methicillin-resistant *S. aureus* (MRSA) [22]. In addition, galangin exhibited selective anti-cytochrome and antifungal activity [25], and showed antimicrobial activity against *S. aureus* [25,32], and methicillin-sensitive and methicillin-resistant *S. aureus*, *Enterococcus* spp., and *P. aeruginosa* [33]. Flavone apigenin showed strong activity against Gram-negative bacteria [34], while quercetin and apigenin derivatives showed strong antibacterial properties against Gram-negative and Gram-positive bacteria [35]. Some phenolic acids (gallic, caffeic and ferulic acids) showed antibacterial activity against Gram-positive (*S. aureus* and *Listeria monocytogenes*) and Gram-negative bacteria (*E. coli* and *P. aeruginosa*) with a greater efficiency than conventional antibiotics such as gentamicin and streptomycin [36]. Contrarily, chlorogenic acid, which was not abundant in wood species, showed no activity against Gram-positive bacteria [36]. Interestingly, noticeable antimicrobial activity of cherry wood against wine organisms was observed before [17], but, to our knowledge, its antimicrobial activity against human and opportunistic pathogens has not been investigated so far.

However, MIC values for extracts 1–7 against *S. aureus*, *L. monocytogenes* and *E. coli* were similar to MICs obtained from some other *Quercus* spp. bark extracts (Table 4), but MICs recorded towards *C. albicans* were lower compared to the results of this study.

Table 4. Summarized MICs values for other waste extracts obtained from literature data.

Bark Extracts Origin MIC (mg mL^{-1})	*Sa*	*Mr*	*Lm*	*Sm*	*Sp*	*Ec*	*Ca*	References
Picea abies	0.13	-	0.16	-	-	0.08	0.97	[37]
Larix decidua	0.21	-	0.15	-	-	0.33	0.60	
Quercus acutissima	0.23	-	0.27	-	-	0.17	0.40	
Quercus macrocarpa	0.22	-	0.29	-	-	0.13	0.34	[24]
Quercus robur	0.23	-	0.25	-	-	0.10	0.31	
Quercus robur	0.08	-	-	-	-	0.08	-	[38]
Quercus ilex	0.13	-	-	-	0.51	0.26	-	[39]
Quercus infectoria	-	1.25	-	-	-	-	-	[40]
Maclura tinctoria	-	-	-	0.08	-	-	-	[41]
Prunus africana	0.07	0.16	-	-	-	-	-	[42]
Prunus avium	6.25	-	-	-	-	12.50	-	[43]
Prunus cerasoides	5.00	1.00	-	-	-	-	1.00	[44]
Morus mesozygia	0.16	-	-	-	-	0.04	0.16	[45]

Sa—*S. aureus*; *Mr*—MRSA; *Lm*—*L. monocytogenes*; *Sm*—*S. mutans*; *Sp*—*S. pyogenes*; *Ec*—*E. coli*; *Ca*—*C. albicans*; (-)—not tested.

On the other hand, *P. avium* stem bark extracts from Nigeria showed lower antimicrobial activity against *S. aureus* and *E coli*, with MICs 6.25 mg mL^{-1} and 12.5 mg mL^{-1}, respectively [43]. Compared to extracts 9 and 10, *Prunus cerasoides* showed similar antibacterial activity towards MRSA [44]. Unlike extract 11, originating from *M. alba*, bark extracts originating from *Morus mesozygia* showed significant antimicrobial activity against *C. albicans*, with obtained MIC of 0.16 mg mL^{-1} (Table 4). Interestingly, higher susceptibility of *C. albicans* was also observed for *Picea abies* and *Larix decidua* bark extracts [37]. In the literature, no significant correlation was found between antimicrobial activity and total phenolic content [17,46], as well as between antimicrobial activities and antioxidant capacity. However, structure-function of the phenolic extracts have more influence on the antimicrobial activity than the total phenol content [17]. Finally, according to Cowan [47], a wide variety of specialized metabolites show antimicrobial activity *in vitro*, such as tannins, terpenoids and alkaloids, also found in wood.

4. Conclusions

Wood waste from forest trees is a source of different bioactive metabolites which could find application in the food and pharmaceutical industries. Radical scavenging and antimicrobial activities of the wood waste extracts appeared to be a valuable bio-functional source. In general, HPTLC fingerprint identify the main phenolic acids present in investigated samples and revealed chemical patterns among investigated wood extracts. DPPH-HPTLC assay identified gallic, ferulic and/or caffeic acids as compounds with the highest contribution to total radical scavenging activity. Based on PCA plot, six *Quercus* samples were separated from other extracts showing strong radical scavenging activities.

Wood samples were the most active against MRSA, *S. aureus* and *S. pyogenes*. The lowest MIC and MBC values were detected in mulberry extract against MRSA. Activities were also distinguished against MRSA (extracts of non-seasoned sessile oak (5), Turkey oak, black locust and mulberry) and *S. aureus* (Turkey oak and all sessile oak extracts). The largest zones of inhibition of the growth of *S. mutans* and *S. aureus* were observed for wild cherry extract. Among sessile and pedunculate oak extracts, non-seasoned sessile oak extract (5) was distinguished by lower MIC against MRSA and *L. monocytogenes*. Extracts of myrobalan plum, wild cherry and mulberry significantly inhibited the growth rate of *L. monocytogenes*. *E. coli* was less sensitive to the tested extracts. *C. albicans* showed poor sensitivity to the action of all extracts, with the exception of the wild cherry extract.

Wild cherry wood extract can be commercially important due to good simultaneous activity against all pathogens, and is a valuable source for various formulations: Wild cherry and mulberry wood extracts with given antimicrobial activities can be especially useful in preserving perishable foods with short shelf life.

Author Contributions: Conceptualization, M.N. and P.R.; methodology, P.R.; software, D.D.Z.; investigation, P.R., A.S., I.D. and T.P.; resources, S.V.; writing—original draft preparation, A.S., P.R. and T.P.; writing—review and editing, M.N., M.F.A. and M.M.; supervision, M.F.A. and M.M. All authors have read and agreed to the published version of the manuscript.

Funding: This research was funded by the Ministry of Education, Science and Technological Development, Republic of Serbia, grant number 172017, and Research Council of Norway (project No 11060-"NORWEGIAN FRUIT GENETIC RESOURCES—HEALTHY, TASTE & NO WASTE").

Acknowledgments: Authors would like to thank Ivanka Ćirić for technical assistance during the course of analysis.

Conflicts of Interest: The authors declare no conflict of interest. The funders had no role in the design of the study; in the collection, analyses, or interpretation of data; in the writing of the manuscript, or in the decision to publish the results.

References

1.   Fernández de Simón, B.; Sanz, M.; Cadahía, E.; Martínez, J.; Esteruelas, E.; Muñoz, A.M. Polyphenolic compounds as chemical markers of wine ageing in contact with cherry, chestnut, false acacia, ash and oak wood. *Food Chem.* **2014**, *143*, 66–76. [CrossRef]

2. Fierascu, R.C.; Fierascu, I.; Avramescu, S.M.; Sieniawska, E. Recovery of Natural Antioxidants from Agro-Industrial Side Streams through Advanced Extraction Techniques. *Molecules* **2019**, *24*, 4212. [CrossRef]

3. Squillaci, G.; Apone, F.; Sena, L.M.; Carola, A.; Tito, A.; Bimonte, M.; De Lucia, A.; Colucci, G.; La Cara, f.; Morana, A. Chestnut (*Castanea sativa* Mill.) industrial wastes as a valued bioresource for the production of active ingredients. *Process Biochem.* **2018**, *64*, 228–236. [CrossRef]

4. Matos, M.S.; Romero-Díez, R.; Álvarez, A.; Bronze, R.; Rodríguez-Rojo, S.; Mato, R.B.; Cocero, R.M.; Matias, A.A. Polyphenol-rich extracts obtained from winemaking waste streams as natural ingredients with cosmeceutical potential. *Antioxidants* **2019**, *8*, 355. [CrossRef]

5. Licursi, D.; Antonetti, C.; Fulignati, S.; Corsini, A.; Boschi, N.; Rasplolli Galletti, A.M. Smart valorization of waste biomass: Exhausted lemon peels, coffee silverskins and paper wastes for the production of levulinic acid. *Chem. Eng. Trans.* **2018**, *65*, 637–642. [CrossRef]

6. Smailagić, A.; Veljović, S.; Gašić, U.; Dabić Zagorac, D.; Stanković, M.; Radotić, K.; Natić, M. Phenolic profile, chromatic parameters and fluorescence of different woods used in Balkan cooperage. *Ind. Crop Prod.* **2019**, *132*, 156–167. [CrossRef]

7. Mratinić, E.; Fotirić-Akšić, M. Indigenous fruit species as a significant resource for sustainable development. *Bull Fac Forest.* **2014**, 181–194. [CrossRef]

8. Stojanović, D.B.; Matović, B.; Orlović, S.; Kržič, A.; Trudić, B.; Galić, Z.; Stojnić, S.; Pekeč, S. Future of the Main Important Forest Tree Species in Serbia from the Climate Change Perspective. *South East Eur For.* **2014**, *5*, 117–124. [CrossRef]

9. Natić, M.M.; Dabić, D.Č.; Papetti, A.; Fotirić Akšić, M.M.; Ognjanov, V.; Ljubojević, M.; Tešić, Ž.L. Analysis and characterisation of phytochemicals in mulberry (*Morus alba* L.) fruits grown in Vojvodina, North Serbia. *Food Chem.* **2015**, *171*, 128–136. [CrossRef]

10. Rakonjac, V.; Mratinić, E.; Jovković, R.; Fotirić Akšić, M. Analysis of morphological variability in wild cherry (*Prunus avium* L.) genetic resources from Central Serbia. *J. Agr. Sci. Tech.* **2014**, *16*, 151–162.

11. Mratinić, E.; Fotirić-Akšić, M.; Jovković, R. Analysis of wild sweet cherry (*Prunus avium* L.) germplasm diversity in South-East Serbia. *Genetika* **2012**, *44*, 259–268.

12. Mihajlović, L.; Glavendekić, M.; Jakovljević, I.; Marjanović, S. *Obolodiplosis robiniae* (haldeman) (*diptera*: Cecidomyiidae) a new invasive insect pest on black locust in Serbia. *Bullet. Faculty For.* **2008**, *97*, 197–208. [CrossRef]

13. Sanz, M.; Fernandez de Simón, B.; Cadahía, E.; Esteruelas, E.; Muñoz, A.M.; Hernández, M.T.; Estrella, I. Polyphenolic profile as a useful tool to identify the wood used in wine aging. *Anal. Chim. Acta.* **2012**, *732*, 33–45. [CrossRef]

14. Chinnici, F.; Natali, N.; Bellachioma, A.; Versari, A.; Riponi, C. Changes in phenolic composition of red wines aged in cherry wood. *LWT Food Sci. Technol.* **2015**, *60*, 977–984. [CrossRef]

15. Alañón, M.E.; Castro-Vázquez, L.; Díaz-Maroto, M.C.; Hermosín Gutiérrez, I.; Gordon, M.H.; Pérez–Coello, M.S. Antioxidant capacity and phenolic composition of different woods used in cooperage. *Food Chem.* **2011**, *129*, 1584–1590. [CrossRef]

16. Cieśla, Ł.; Kryszeń, J.; Stochmal, A.; Oleszek, W.; Waksmundzka-Hajnos, M. Approach to develop a standardized TLC-DPPH• test for assessing free radical scavenging properties of selected phenolic compounds. *J. Pharm. Biomed. Anal.* **2012**, *70*, 126–135. [CrossRef]

17. Alañón, M.E.; García-Ruíz, A.; Díaz-Maroto, M.C.; Pérez-Coello, M.S.; Moreno-Arribas, M.V. Antimicrobial and antioxidant activity of pressurized liquid extracts from oenological woods. *Food Control.* **2015**, *50*, 581–588. [CrossRef]

18. Hartmann, M.; Berditsch, M.; Hawecker, J.; Ardakani, M.F.; Gerthsen, D.; Ulrich, A.S. Damage of the bacterial cell envelope by antimicrobial peptides gramicidin S and PGLa as revealed by transmission and scanning electron microscopy. *Antimicrob Agents Chemother.* **2010**, *54*, 3132–3142. [CrossRef]

19. Cushnie, T.P.T.; Lamb, A.J. Recent advances in understanding the antibacterial properties of flavonoids. *Int. J. Antimicrob. Agents* **2010**, *38*, 99–107. [CrossRef]

20. Asmi, K.S.; Lakshmi, T.S.; Balusamy, R.; Parameswari, R. Therapeutic aspects of taxifolin—An update. *J. Adv. Pharmacy Educ. Res.* **2017**, 187–189.

21. Joung, D.K.; Mun, S.H.; Choi, S.H.; Kang, O.H.; Kim, S.B.; Lee, Y.S.; Zhou, T.; Kong, R.; Choi, J.G.; Shin, D.W.; et al. Antibacterial activity of oxyresveratrol against methicillin-resistant Staphylococcus aureus and its mechanism. *Exp. Ther. Med.* **2016**, *12*, 1579–1584. [CrossRef] [PubMed]

22. Shao, J.; Zhang, M.X.; Wang, T.M.; Li, Y.; Wang, C.Z. The roles of CDR1, CDR2, and MDR1 in kaempferol-induced suppression with fluconazole-resistant *Candida albicans*. *Pharm. Biol.* **2016**, *54*, 984–992. [CrossRef] [PubMed]

23. Andrenšek, S.; Simonovska, B.; Vovk, I.; Fyhrquist, P.; Vuorela, H.; Vuorela, P. Antimicrobial and antioxidative enrichment of oak (*Quercus robur*) bark by rotation planar extraction using ExtraChrom R. *Int. J. Food Microbiol.* **2004**, *92*, 181–187. [CrossRef] [PubMed]

24. Elansary, H.O.; Szopa, A.; Kubica, P.; Ekiert, H.; Mattar, M.A.; Al-Yafrasi, M.A.; El-Ansary, D.O.; El-Abedin, T.K.Z.; Yessoufou, K. Polyphenol Profile and Pharmaceutical Potential of Quercus spp. Bark Extracts. *Plants* **2019**, *8*, 486. [CrossRef]

25. McNulty, J.; Nair, J.J.; Bollareddy, E.; Keskar, K.; Thorat, A.; Crankshaw, D.J.; Holloway, A.C.; Khan, G.; Wright, G.D.; Ejim, L. Isolation of flavonoids from the heartwood and resin of *Prunus avium* and some preliminary biological investigations. *Phytochemistry* **2009**, *70*, 2040–2046. [CrossRef]

26. Zhang, B.; Cai, J.; Duan, C.Q.; Reeves, M.J.; He, F. A review of polyphenolics in oak woods. *Int. J. Mol. Sci.* **2015**, *16*, 6978–7014. [CrossRef]

27. Górniak, I.; Bartoszewski, R.; Króliczewski, J. Comprehensive review of antimicrobial activities of plant flavonoids. *Phytochem. Rev.* **2019**, *18*, 241–272. [CrossRef]

28. Ristivojević, P.; Andrić, F.L.; Trifković, J.Đ.; Vovk, I.; Stanisavljević, L.Ž.; Tešić, Ž.L.; Milojković-Opsenica, D.M. Pattern recognition methods and multivariate image analysis in HPTLC fingerprinting of propolis extracts. *J. Chemom.* **2014**, *28*, 302–310. [CrossRef]

29. Nikolić, M.; Marković, T.; Mojović, M.; Pejin, B.; Savić, A.; Perić, T.; Marković, D.; Stević, T.; Soković, M. Chemical composition and biological activity of *Gaultheria procumbens* L. essential oil. *Ind. Crop Prod.* **2013**, *49*, 561–567. [CrossRef]

30. Dimkić, I.; Ristivojević, P.; Janakiev, T.; Berić, T.; Trifković, J.; Milojković-Opsenica, D.; Stanković, S. Phenolic profiles and antimicrobial activity of various plant resins as potential botanical sources of Serbian propolis. *Ind. Crop Prod.* **2016**, *94*, 856–871. [CrossRef]

31. Elshikh, M.; Ahmed, S.; Funston, S.; Dunlop, P.; McGaw, M.; Marchant, R.; Banat, I.M. Resazurin-based 96-well plate microdilution method for the determination of minimum inhibitory concentration of biosurfactants. *Biotechnol. Let.* **2016**, *38*, 1015–1019. [CrossRef] [PubMed]

32. Cushnie, T.P.; Lamb, A.J. Detection of galangin-induced cytoplasmic membrane damage in *Staphylococcus aureus* by measuring potassium loss. *J. Ethnopharmacol.* **2005**, *101*, 243–248. [CrossRef] [PubMed]

33. Pepeljnjak, S.; Kosalec, I. Galangin expresses bactericidal activity against multiple-resistant bacteria: MRSA, *Enterococcus spp.* and Pseudomonas aeruginosa. *FEMS Microbiol. Lett.* **2004**, *240*, 111–116. [CrossRef] [PubMed]

34. Metsämuuronen, S.; Sirén, H. Bioactive phenolic compounds, metabolism and properties: A review on valuable chemical compounds in Scots pine and Norway spruce. *Phytochem. Rev.* **2019**, *18*, 623–664. [CrossRef]

35. Osonga, F.J.; Akgul, A.; Miller, R.M.; Eshun, G.B.; Yazgan, I.; Akgul, A.; Sadik, O.A. Antimicrobial Activity of a New Class of Phosphorylated and Modified Flavonoids. *ACS Omega* **2019**, *4*, 12865–12871. [CrossRef] [PubMed]

36. Daglia, M. Polyphenols as antimicrobial agents. *Curr. Opin. Biotechnol.* **2012**, *23*, 174–181. [CrossRef] [PubMed]

37. Salem, M.Z.M.; Elansary, H.O.; Elkelish, A.A.; Zeidler, A.; Ali, H.M.; Mervat, E.H.; Yessoufou, K. In vitro bioactivity and antimicrobial activity of *Picea abies* and *Larix decidua* wood and bark extracts. *Bioresources.* **2016**, *11*, 9421–9437. [CrossRef]

38. Brantner, A.; Grein, E. Antibacterial activity of plant extracts used externally in traditional medicine. *J. Ethnopharmacol.* **1994**, *44*, 35–40. [CrossRef]

39. Berahou, A.; Auhmani, A.; Fdil, N.; Benharref, A.; Jana, M.; Gadhi, C.A. Antibacterial activity of *Quercus ilex* bark's extracts. *J. Ethnopharmacol.* **2007**, *112*, 426–429. [CrossRef]

40. Khouzami, L.; Mroueh, M.; Daher, C.F. The role of methanolic extract of *Quercus infectoria* bark in lipemia, glycemia, gastric ulcer and bacterial growth. *J. Med. Plants Res.* **2009**, *2*, 224–230.

41. Lamounier, K.C.; Cunha, L.C.S.; de Morais, S.A.L.; de Aquino, F.J.T.; Chang, R.; do Nascimento, E.A.; de Souza, M.G.M.; Martins, C.H.G.; Cunha, W.R. Chemical analysis and study of phenolics, antioxidant activity, and antibacterial effect of the wood and bark of *Maclura tinctoria* (L.) D. Don ex Steud. *Evid. Based Complement. Alternat. Med.* **2012**, 451039. [CrossRef]

42. Bii, C.; Korir, K.R.; Rugutt, J.; Mutai, C. The potential use of *Prunus africana* for the control, treatment and management of common fungal and bacterial infections. *J. Med. Plants Res.* **2010**, *4*, 995–998. [CrossRef]

43. Oyetayo, A.M.; Bada, S.O. Phytochemical screening and antibacterial activity of *Prunus avium* extracts against selected human pathogens. *J. Complement. Altern. Med. Res.* **2017**, *4*, 1–8. [CrossRef]

44. Arora, D.S.; Mahajan, H. In vitro evaluation and statistical optimization of antimicrobial activity of *Prunus cerasoides* stem bark. *Appl. Biochem. Biotechnol.* **2018**, *184*, 821–837. [CrossRef] [PubMed]

45. Kuete, V.; Fozing, D.C.; Kapche, W.F.G.D.; Mbaveng, A.T.; Kuiate, J.R.; Ngadjui, B.T.; Abegaz, B.M. Antimicrobial activity of the methanolic extract and compounds from *Morus mesozygia* stem bark. *J. Ethnopharmacol.* **2009**, *124*, 551–555. [CrossRef]

46. Fattouch, S.; Caboni, P.; Coroneo, V.; Tuberoso, C.I.G.; Angioni, A.; Dessi, S.; Marzouki, N.; Cabras, P. Antimicrobial Activity of Tunisian Quince (*Cydonia oblonga* Miller) Pulp and Peel Polyphenolic Extracts. *J. Agric. Food Chem.* **2007**, *55*, 963–969. [CrossRef] [PubMed]

47. Cowan, M.M. Plant Products as Antimicrobial Agents. *Clin. Microbiol. Rev.* **1999**, *12*, 564–582. [CrossRef]

Article

Extract from Broccoli Byproducts to Increase Fresh Filled Pasta Shelf Life

Luisa Angiolillo, Sara Spinelli, Amalia Conte * and Matteo Alessandro Del Nobile

Department of Agricultural Sciences, Food and Environment, University of Foggia, Via Napoli, 2571121 Foggia, Italy; luisa.angiolillo@unifg.it (L.A.); sara.spinelli@unifg.it (S.S.); matteo.delnobile@unifg.it (M.A.D.N.)
* Correspondence: amalia.conte@unifg.it; Tel.: +39-088-158-9240

Received: 23 October 2019; Accepted: 22 November 2019; Published: 27 November 2019

Abstract: The aim of the study was to evaluate the efficacy of extract from broccoli byproducts, as a green alternative to chemical preservation strategies for fresh filled pasta. In order to prove its effectiveness, three different percentages (10%, 15%, and 20% *v/w*) of extract were added to the filling of pasta. A shelf life test was carried out by monitoring microbiological and sensory quality. The content of phenolic compounds before and after in vitro digestion of pasta samples was also recorded. Results underlined that the addition of the natural extract helped to record a final shelf life of about 24 days, that was 18 days longer in respect to the control sample. Furthermore, results highlighted that the addition of byproducts extract to pasta also increased its phenolic content after in vitro digestion. Therefore, broccoli byproducts could be valorized for recording extracts that are able to prolong shelf life and increase the nutritional content of fresh filled pasta.

Keywords: shelf life; byproducts; fresh pasta; vegetable extracts; antimicrobial activity

1. Introduction

Pasta is one of the main constituents of the Mediterranean diet as it contains significant amounts of complex carbohydrates, proteins, B-vitamins, and iron [1]. Pasta can be made with different kinds of flours (semolina, farina, wheat flour, etc.) mixed with water. Fresh pasta has more than 24% moisture and its water activity ranges from 0.92 to 0.99, thus it requires refrigeration [2]. It can be prepared with eggs in the dough or by filling a sheeted dough with a spiced mixture of ground meat, cheese or vegetables as for tortellini and ravioli. In the last decade, fresh filled pasta gained great national and international popularity, even though distribution beyond the Italian borders still represents a real problem due to the rapid microbial proliferation. In fact, this product is very susceptible to spoilage microorganisms and therefore, addition of preservatives or reduced oxygen packaging are necessary to prolong the shelf life that, even under refrigerated temperatures, lasts only two or three days [3]. Specifically, the pH of pasta without any preservatives may drop, thus indicating spoilage and increased coliforms [4]. Italian law [5] prescribes a pasteurization treatment before final packaging in order to reduce the growth of vegetative microbial forms and also to improve cooking behaviors. Generally, the thermal treatment is carried out in an injected steam belt pasteurizer and in addition to reducing water activity, it helps to increase starch gelatinization with consequently less water absorption during cooking [6]. For shelf life prolongation different methods have been applied to fresh pasta; the most common approach is based on chemical preservatives, as organic acids, and modified atmospheres (MAP) with low O_2 concentrations (below atmospheric levels) and high CO_2 concentrations (20% or higher), proper coupled with N_2 as an inert gas filler [7–10]. Castelvetri et al. [11] demonstrated that packaging atmospheres with more than 30% CO_2 were capable of extending fresh filled pasta shelf life, even if low residual O_2 in the package headspace may cause mold growth. The search for more natural technologies of food preservation greatly promoted the exploration of antimicrobial compounds of vegetable or animal origin, as essential oils, enzymes or chitosan [12]. Among natural antimicrobial compounds, substances recordable from leaves, flowers, seeds, and peels are becoming very interesting [13]. Many components with useful properties can be found in food byproducts [14]. The possibility to recycle food byproducts may also represent a mean to face their environmental and economic impact. Among plants, *Brassica oleracea,* that belongs to the Brassicaceae family, represents one of the most abundant byproducts producers. It comprises 3500 species such as cauliflower, broccoli, kale, cabbage, and Brussels sprouts. Byproducts from *Brassica oleracea* are rich in phenols, as flavonoids, phenolic acids, and tannins, usually extracted by solvent extraction [15]. This extraction method has low selectivity and utilizes high energy cost, elevated solvents, and high temperatures [16]. For this reason, alternative methods of extraction have been investigated. Among them, the supercritical fluid extraction (SFE) with improved selectivity, automation, and environmental safety, represents a valid alternative [17].

According to the above-reported considerations, the aim of the study was to evaluate the efficacy of broccoli byproducts extract, obtained by SFE, to improve fresh filled pasta shelf life. To this aim, bioactive substances were first extracted from broccoli byproducts and subsequently added to fresh filled pasta to verify their effects on microbial and sensory quality. The evaluation of the consequent polyphenols content in pasta samples was also assessed.

2. Material and Methods

2.1. Raw Materials

Broccoli stems and leaves (*Brassica oleracea*) were provided by a local company in Foggia, Southern Italy. The samples were dried at 30–35 °C in a dryer (SG600, Namad, Rome, Italy) for 48 h. The dried samples were reduced to fine powder (≤250 μm) by a hammer mill (16/BV-Beccaria s.r.l., Cuneo, Italy) and then stored at 4 °C until further utilization.

2.2. SFE

Supercritical fluid extraction was carried out to collect active compounds from byproducts, above all polyphenolic compounds. It was performed using process conditions (150 bars, 35 °C, 20% ethanol and 10 min of dynamic extraction time) previously described by Arnáiz et al. [18], by the supercritical fluid extractor Speed SFE-2 (Applied Separation, Allentown, USA). The extract was placed overnight in vacuum oven at 30 °C to remove ethanol. The solid residue was collected in 25 mL of water.

2.3. Fresh Filled Pasta Production

Fresh pasta samples were produced with durum semolina (provided by Agostini mill Montefiore dell'Aso, Ascoli Piceno, Italy). Semolina and distilled water (30% *v/w*) were mixed for about 20 min to prepare the pasta dough. The samples were prepared using a pilot scale extruder (60VR; Namad, Rome, Italy) equipped with a roller sheeter (Raff, Minipan, Massa Lombarda, Italy) and a compressor (mod. Rondostar, Rondo Doge, Burgdorf, Switzerland) in order to obtain a 6 mm sheeted dough. The filling was prepared by mixing 65% (*w/w*) ricotta cheese, 19% (*w/w*) grated cheese, 16% (*w/w*) fresh spinach, 0.30% (*w/w*) potato flour, and 0.01% (*w/w*) salt. The sheeted dough and filling were combined in a modified double sheet ravioli machine (mod. PRP 300, Genoa, Italy) to prepare 12 cm diameter fresh pasta samples in the form of ravioli. Each sample consisted of two square dough sheets containing the filling. Four different formulations were prepared: fresh filled pasta without any addition (CNT), and the other three samples with increasing concentrations of broccoli extract in the filling: 10% (*v/w*) (10-BE); 15% (*v/w*) (15-BE), and 20% (*v/w*) (20-BE). The product was conveyed through a 3-m chamber equipped with a perforated steel conveyor belt (Custom, Italgi, Genoa, Italy). By steam injection at 91 ± 1 °C for 9 min, the pasteurization was carried out. After pasteurization the product passed through two fans to eliminate the condensed vapor on the surface and then was cooled to 4 °C and packaged in bags with anti-fog high-barrier multilayer film made up of polyethylene terephthalate, ethylene-vinyl alcohol, and polyethylene. The film oxygen transmission rate (OTR) was 6.19 cc/m^2/day, the water vapor transmission rate (WVTR) was 1.208 g/m^2/day, and the thickness was 50 μm (Di Mauro Officine Grafiche spa, Salerno, Italy). All the samples were stored for about 2 months at 4 °C, without light.

2.4. Microbiological Analyses

For microbiological analyses, about 10 g of sample were aseptically removed from each package, placed in a stomacher bag, diluted with 90 mL of sterile NaCl solution, and homogenized with a stomacher LAB Blender 400 (Pbi International, Milan, Italy). Serial dilutions in sterile saline solution were plated onto appropriate media. The media and conditions were the following: plate count agar (PCA) incubated at 30 °C for 48 h for aerobic mesophilic bacteria and at 7 °C for 10 days for psychrotrophic bacteria; Violet Red Bile Glucose Agar (VRBGA) incubated at 37 °C for 24 h for *Enterobacteriaceae*; Baird-Parker Agar, supplemented with egg yolk tellurite emulsion, incubated at 37 °C for 48 h for *Staphylococcus* spp.; Sabouraud Dextrose Agar, added with 0.1 g/L chloramphenicol (C. Erba, Milan, Italy), incubated at 25 °C for 48 h for yeasts and 25 °C for 5 days for molds. Reinforced Clostridial Medium (Oxoid, Milan, Italy) was used for the sulfite-reducing clostridia; after heat treatment of samples at 80 °C for 10 min to destroy the vegetative cells, the plates were incubated at 37 °C for 48 h in anaerobic conditions, thus avoiding contact with air. The count was carried out with the most probable number (MPN) method. Aerobic spore-forming bacteria were detected and counted on Nutrient Agar (Oxoid, Milan, Italy) after 48 h at 30 °C; all vegetative forms were previously destroyed by heat treatment of samples at 80 °C for 10 min. All media and supplements were from Oxoid (Milan, Italy). All microbiological analyses were performed twice on two different samples (one sample from two different trays). In order to quantitatively determine the microbial acceptability limit (MAL), a modified version of the Gompertz equation was fitted to the experimental data, as reported in previous studies [19,20]. The Italian law [5] fixes the threshold for total microbial count (TMC), staphylococci, and clostridia at maximum values of 10^6, 5×10^3, and 10^3 Colony Forming Unit (CFU)/g, respectively.

2.5. Sensory Analysis

During the entire storage period, at selected times, both uncooked and cooked fresh pasta samples were subjected to a time intensity evaluation. Towards the aim, eight trained tasters were involved in the panel test. The panelists were asked to evaluate color, odor, and overall quality of uncooked samples and color, odor, taste, consistency, and overall quality of pasta cooked in food grade tap water at 100 °C. A nine-point rating scale, where 1 corresponded to 'extremely unpleasant' and 9 to 'extremely pleasant', was used to perform the panel test. [21]. The panelists were selected on the basis of their sensory skills (ability to accurately determine and communicate the sensory attributes, the appearance, odor, flavor, and texture). Prior to testing pasta, the panelists were trained in the sensory vocabulary and identification of particular attributes, by using commercial pasta. The analyses were performed in isolated booths, located in a standard taste panel kitchen. In order to determine the sensory acceptability limit (SAL), intended as the storage time to reach the sensory threshold, a modified version of the Gompertz equation was fitted to the sensory data [19,20]. The sensory threshold was set equal to 5.

2.6. Chemical Analyses

2.6.1. Extraction of Polyphenols from Cooked Pasta Samples

The extraction of polyphenols from both control and enriched pasta samples was based on the method also described by Rashidinejad et al. [22]. Briefly, 1 g of each cooked sample was homogenized and extracted in a water bath with 50 mL of 95% methanol containing 1% HCl at 50 °C and 200 rpm. The mixture was cooled, filtered, and washed with 2 mL of the same solvent.

2.6.2. In Vitro Digestion of Cooked Pasta Samples

Simulated gastric and intestinal digestions were carried out on both control and enriched pasta samples using the method of Rashidinejad et al. [22]. In brief, 1 g of each cooked sample was added with 10 mL of simulated filtered gastric fluid (SGF) at 37 °C and incubated in an orbital shaker at 37 °C at 235 rpm for 10 min. After adjusting the pH of the solution to 2.0, the treatment continued for a further 2 h at 95 rpm. Then, 36 mL at 37 °C of simulated intestinal fluid (SIF) were added to each gastric digestion sample and stirred at 37 °C at 95 rpm for 4 h. After 30 min from the beginning, the pH was adjusted to 6.8. In order to prepare the SGF sample, 2 g of NaCl, 7 mL of HCl (36%), and 3.2 g of purified porcine pepsin (in 1 L of deionized water, pH 1.2) were used, while to prepare the SIF sample a monobasic potassium phosphate solution (6.8 g in 250 mL of deionized water) with 77 mL of sodium hydroxide (0.2 M) and 500 mL of deionized water was mixed. Finally, 10 g of pancreatin and 0.05 g of porcine bile extract were added to the mixture and the pH was adjusted again to 6.8. For each sample, the digestion was carried out in triplicate.

2.6.3. Total Phenolic Content

To measure the total phenolic content (TPC) in all the undigested and digested samples, the Folin–Ciocalteu assay was used. TPC was determined as described by da Silva et al. [23] with slight modifications. Briefly, 0.5 mL of sample (that obtained in Section 2.6.1 and that recorded in Section 2.6.2) and 2.5 mL of Folin–Ciocalteu reagent diluted in water (1:10 ratio) were left to rest for 5 min. An amount of 2 mL of Na_2CO_3 (4 g/100 mL) was then added. The mixture was allowed to rest again for 2 h in darkness. The absorbance was read by a spectrophotometer (UV1800, Shimadzu Italia s.r.l.) at 740 nm. Total phenols content was expressed as mg of gallic acid equivalents (GAEs) per g of pasta, according to a previously recorded calibration curve. For each sample, the analyses were carried out in triplicate.

2.7. Statistical Analysis

Experimental data were compared by one-way ANOVA analysis. A Duncan's multiple range test, with the option of homogeneous groups ($p < 0.05$), was used to determine significance among differences. To this aim, Statistica 7.1 for Windows 152 (StatSoft Inc., Tulsa, OK, USA) was used.

3. Results and Discussion

3.1. Total Phenolic Content

TPC of both undigested and digested pasta samples is shown in Table 1. As can be seen, the TPC in the indigested control samples (0.63 mg GAEs/g) was significantly ($p < 0.05$) lower than that found in pasta enriched with broccoli extract (1.84–1.86 mg GAEs/g), even if it did not increase linearly with the quantity of added extract. This finding may have been due to the interaction of polyphenols with ricotta proteins and the concomitant formation of less active complexes [24]. Gallo et al. [25] also stated that milk protein fractions caused a decrease of the in vitro antioxidant activity of polyphenols, as a consequence of the weaker non-covalent bonds between proteins and polyphenols.

The other information that can be deduced looking at Table 1 is that the phenolic content of the digested samples appear to be higher in respect to the indigested ones, with the highest value found in the 20-BE sample (2.61 mg GAEs/g).

The explanation for this trend could be linked to the hydrolyzation exerted by digestive enzymes towards chemical bonds in the phenolic-protein complexes, promoting in this way a greater release and extractability of phenolics at the in vitro level [26,27]. In particular, Gumienna et al. [28] claimed that the action of gastric-digestive enzymes may lead to the development of aglycones phenolic compounds, more reactive than the corresponding glycoside forms. These findings positively encourage the consumption of pasta enriched with broccoli byproducts extract, as a way to promote intake of valuable food for human health, since it is widely demonstrated that phenolic compounds provide extraordinary anticancer, antiviral, antibacterial, cardio-protective, and anti-mutagenic activities [29].

Table 1. Total phenolic content (TPC) of undigested and digested cooked ravioli samples.

	TPC (mg GAEs/g Ravioli)	
	Undigested	**Digested**
CNT	0.63 ± 0.03 [a]	0.81 ± 0.01 [a]
10-BE	1.84 ± 0.08 [b]	2.13 ± 0.10 [b]
15-BE	1.84 ± 0.18 [b]	2.23 ± 0.10 [b]
20-BE	1.86 ± 0.12 [b]	2.61 ± 0.16 [c]

[a–c] Data in columns with different letters are significantly different ($p < 0.05$). GAEs: gallic acid equivalents; CNT = fresh filled pasta without any addition; 10-BE = fresh filled pasta with 10% broccoli extract; 15-BE = fresh filled pasta with 15% broccoli extract; 20-BE = fresh filled pasta with 20% of broccoli extract.

3.2. Quality of Fresh Filled Pasta

Figure 1 describes the growth of total mesophilic bacteria in all the experimental samples. Table 2 reports values of fitting parameters. As can be inferred from the data, in the CNT pasta an immediate growth with an ascendant trend was found, with values from 4.88 log CFU/g to 8.77 log CFU/g until the 22nd day.

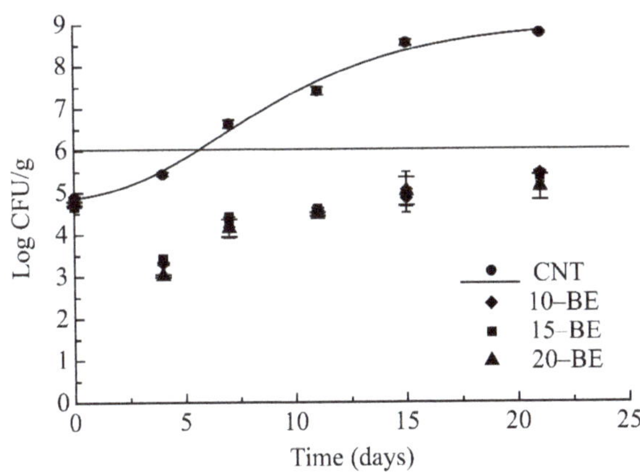

Figure 1. Evolution of mesophilic bacteria in pasta samples during storage at 4 °C. CFU: Colony Forming Unit.

In the control pasta the microbial acceptability limit was reached after six days. On the contrary, the three samples with increasing concentration of broccoli extract, revealed a positive microbial quality during the entire observation period, without significant difference among them. These data were similar to the microbial trend of psychrotrophic bacteria.

Regarding *Staphylococcus* spp., counts were found below the microbial limit for the entire 25 days of observation in all the active pasta samples, while in the CNT microbial growth started from the 15th day and reached the limit after 20 days of storage (Figure 2). Data of MAL are also reported in Table 2.

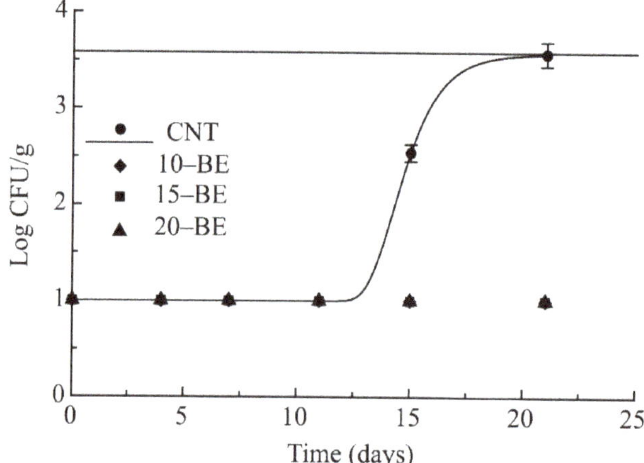

Figure 2. Evolution of *Staphilococcus* spp. in pasta samples during storage at 4 °C.

The addition of broccoli extract appeared to be also effective against *Enterobacteriacae*, molds, and yeasts as these microbial groups in the enriched pasta did not grow until the 25th day (data not shown). On the contrary, the CNT sample revealed molds growth, both visible and on plates, at the 13th day (Table 2), *Enterobacteriacae* and yeasts around 5.75 log CFU/g and 5.59 log CFU/g after 20 days. The trend of control samples is not surprising because it is in accordance with other studies on fresh filled pasta [10–12]. Clostridia and aerobic spore forming bacteria were never found in any samples (data not shown).

The results of the microbial quality confirm findings of other authors about the antimicrobial properties of vegetal extracts [30–32], even though the studies were all carried out under in vitro conditions. The application carried out in the current study, not only assessed the potential effects of extract from broccoli byproducts, but also demonstrated that it is possible to significantly extend microbial stability (Table 2). These data appeared to be of particular importance since the literature highlighted that fresh filled pasta packaged in ordinary atmosphere and stored at 4 °C generally reached very short shelf life values, accounting for hours to one week, depending on the hygienic production conditions [12]. Another import consideration is that the antimicrobial effect is not determined by the quantity of extract added to the experimental samples. As a fact, looking at Table 1, polyphenols appeared to be similar among samples with broccoli extract. It has been suggested by different authors that the bioactivity of polyphenols is rather related to their structure than to their quantity, with differences from one polyphenol to another [33–35]. Their structure is also dependent on temperature and, in particular, heat treatment [36].

According to the above-mentioned results, the microbiological acceptance of pasta was limited to about six days of storage for the CNT sample and lasted about 24 days for the active samples (Table 2).

Sensory evaluation was carried out for more than one month to know when the product became unacceptable for undesired sensory changes. Specifically, color evaluation allowed recording different results among samples, depending on the extract concentration added to the filling, because extract addition modified the filling color from a whitish green of the CNT to an intense green of the 10-BE and 15-BE ravioli samples, to a too dark green of the 20-BE pasta sample. Therefore, while the first two active samples were accepted for a long period, the 20-BE sample was refused within about two weeks.

Regarding pasta consistency, a gradual hardening of all the samples was found, an inevitable consequence of both pasteurization and storage. As described by [11] pasteurization may influence the texture of fresh filled pasta. In fact, during the thermal treatment hardening takes place due to a different distribution of water within the matrix. The increase in water–starch bonding results in a decrease of available water for the other components of the matrix, thus influencing protein denaturation and starch gelatinization. The addition of the extract helps to obtain a more hydrated structure, in fact, the CNT sample revealed a harder consistence in respect to the other three samples. The consistence of the 10-BE sample was found better than that of 15-BE and 20-BE samples because these two types of ravioli revealed an excessive liquid consistence of the filling. The taste of enriched pasta was perceived acceptable by the panelists, even if they underlined increasing the quantity of broccoli extract, the bitterness of the product increased.

Table 2. Shelf life (day) of pasta samples during storage at 4 °C, calculated as the lowest value between microbial acceptability limit (MAL) and sensory acceptability limit (SAL) (mean ± SD).

Sample	Microbial Quality (Day)			Sensory Quality Uncooked Pasta (Day)	Sensory Quality Cooked Pasta (Day)	Shelf Life (Day)
	MAL [Mesoph]	MAL [Staph]	VMT	SAL	SAL	
CNT	5.57 ± 0.33 [a]	22.4 ± 0.10	13	12.95 ± 0.26 [a]	12.45 ± 0.10 [a]	5.57 ± 0.33 [a]
10-BE	24.21 ± 0.11 [b]	>25	>25	40.91 ± 0.23 [d]	43.25 ± 0.20 [d]	24.21 ± 0.11 [c]
15-BE	23.72 ± 0.12 [b]	>25	>25	27.76 ± 0.15 [c]	39.16 ± 0.15 [c]	23.72 ± 0.12 [c]
20-BE	24.33 ± 0.10 [b]	>25	>25	14.54 ± 0.21 [b]	19.10 ± 0.17 [b]	14.54 ± 0.21 [b]

[a–d] Data in columns with different letters are significantly different ($p < 0.05$). VMT = visible molds time (day); CNT = fresh filled pasta without any addition; 10-BE = fresh filled pasta with 10% broccoli extract; 15-BE = fresh filled pasta with 15% broccoli extract; 20-BE = fresh filled pasta with 20% of broccoli extract.

The overall quality of both uncooked and cooked samples reflected the above discussed trends of specific sensory parameters without great differences between uncooked and cooked products. As an example, Figure 3 reports the trend of overall quality of uncooked pasta. All SAL values are reported in Table 2. It is evident looking at Figure 3 and at the sensory data of Table 2 that the CNT sample appeared to be the less appreciated, being rejected after about 13 days, due to undesired changes in color and consistence. The 20-BE samples were refused within 15 days of storage above all for the undesired color, 15-BE samples remained acceptable for less than one month when uncooked and more than one month when considered after cooking, whereas the 10-BE samples recorded the highest score before and after cooking, with sensory acceptability accounting for more than 40 days.

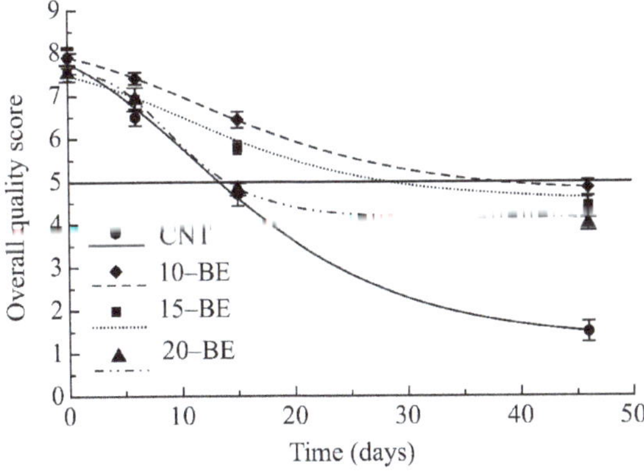

Figure 3. Sensory quality of uncooked fresh filled pasta during storage at 4 °C.

Taking into account both MAL and SAL values of Table 2, the shelf life was reported as the lowest value among them. From the results it is possible to highlight that excessive proliferation of mesophilic bacteria provoked in most cases the end of the product shelf life, whereas in the case of the pasta with the highest concentration of extract (20-BE) the main problem was the presence of unacceptable sensory defects. The effects of the extract were very tangible because while the CNT sample was refused after a few days, the two active samples 10-Be and 15-BE remained acceptable for more than 20 days because the extract controlled microbial proliferation and delayed undesired sensory changes.

4. Conclusions

Reuse of food byproducts in a sustainable way may be a possible way to reduce environmental impact and face costs related to their disposal. In this study extract from broccoli byproducts was added for the first time to the filling of fresh pasta to improve quality and prolong shelf life. Results underlined that the addition of broccoli extract helped to record a final shelf life of about 24 days, that was 18 days longer in respect to the control sample. In addition, pasta with broccoli extract showed a higher phenolic content in respect to the free samples, particularly after digestion. The most appropriate extract amount was 10%, with the pasta samples appreciated for color, taste, and consistency.

Author Contributions: Data collection: L.A. and S.S.; planning of the work and supervision: A.C. and M.A.D.N.; writing—review and editing: A.C.

Funding: This research received no external funding.

Conflicts of Interest: The authors declare no conflicts of interest.

References

1. Capozzi, V.; Russo, P.; Fragasso, M.; De Vita, P.; Fiocco, D.; Spano, G. Biotechnology and Pasta-Making: Lactic Acid Bacteria as a New Driver of Innovation. *Front. Microbiol.* **2012**, *3*, 94. [CrossRef] [PubMed]
2. ICMSF Members. *Microbial Ecology of Food Commodities*, 2nd ed.; Roberts, T.A., Cordier, J.L., Gram, L., Tompkin, R.B., Pitt, J.I., Gorris, L.G.M., Swanson, K.M.J., Eds.; Kluwer Academic/Plenum Publishers: New York, NY, USA, 2005.
3. Angiolillo, L.; Conte, A.; Del Nobile, M.A. Biotechnological approach to preserve fresh pasta quality. *J. Food Prot.* **2017**, *80*, 2006–2013. [CrossRef] [PubMed]
4. Costa, C.; Lucera, A.; Mastromatteo, M.; Conte, A.; Del Nobile, M.A. Shelf life extension of durum semolina-based fresh pasta. *Int. J. Food Sci. Technol.* **2010**, *45*, 1545–1551. [CrossRef]
5. Presidenza della Repubblica. Regolamento per la revisione della normativa sulla produzione e commercializzazione di sfarinati e paste alimentari, 2001 a norma dell'articolo 50 della legge 22 febbraio 1994. *Gazz. Uff. della Repubb. Ital.* **2001**, *187*, Art.9.
6. Zardetto, S.; Di Fresco, S.; Dalla Rosa, M. Effetto di trattamenti termici su alcune caratteristiche chimico-fisiche della pasta fresca farcita. *Tec. Molit.* **2002**, *53*, 113–130.
7. Lee, M.H.; No, H.K. Effect of chitosan on shelf life and quality of wet noodle. *J. Chitin Chitosan* **2010**, *7*, 14–17.
8. Sanguinetti, A.M.; Del Caro, A.; Scanu, A.; Fadda, C.; Milella, G.; Catzeddu, P.; Piga, A. Extending the shelf life of gluten-free fresh filled pasta by modified atmosphere packaging. *LWT-Food Sci. Technol.* **2016**, *71*, 96–101. [CrossRef]
9. Sanguinetti, A.M.; Del Caro, A.; Mangia, N.P.; Secchi, N.; Catzeddu, P.; Piga, A. Quality Changes of Fresh Filled Pasta during Storage: Influence of Modified Atmosphere Packaging on Microbial Growth and Sensory Properties. *Food Sci. Technol.* **2011**, *17*, 23–27. [CrossRef]
10. Zardetto, S. Effect of modified atmosphere packaging at abuse temperature on the growth of *Penicillium Aurantiogriseum* isolated from fresh filled pasta. *Food Microbiol.* **2005**, *22*, 367–371. [CrossRef]
11. Castelvetri, F. Il confezionamento in atmosfera modificata dei prodotti di pasta fresca. *Tec. Molit.* **1991**, *30*, 875–879.
12. Del Nobile, M.A.; Di Benedetto, N.; Suriano, N.; Conte, A.; Lamacchia, C.; Corbo, M.R.; Sinigaglia, M. Use of natural compounds to improve the microbial stability of amaranth-based homemade fresh pasta. *Food Microbiol.* **2009**, *26*, 151–156. [CrossRef] [PubMed]
13. Baydar, N.G.; Ozkan, G.; Sagdic, O. Total phenolic content and antibacterial activities of grape extracts. *Food Control* **2004**, *15*, 335–339. [CrossRef]
14. Martin, J.G.P.; Porto, E.; Correa, C.B.; Alencar, S.M.; Gloria, E.M.; Cabral, I.S.R.; Arquino, L.M. Antimicrobial potential and chemical composition of agro-industrial wastes. *J. Nat. Prod.* **2012**, *5*, 27–36.

15. Wijngaard, H.H.; Rößle, C.; Brunton, N. A survey of Irish fruit and vegetable waste and by product as a source of polyphenolic antioxidants. *Food Chem.* **2009**, *116*, 202–207. [CrossRef]
16. Pormortazavi, S.M.; Hajimirsadegui, S.H. Supercritical fluid extraction in plant essential and volatile oil analysis. *J. Chromatogr. A* **2007**, *1163*, 2–24. [CrossRef]
17. Brunner, G. Supercritical fluids: Technology and application to food processing. *J. Food Eng.* **2005**, *67*, 21–33. [CrossRef]
18. Arnaiz, E.; Bernal, J.; Martin, M.T.; Diego, J.C.; Bernal, J.L.; Toribio, L. Optimization of the supercritical fluid extraction of antioxidants from broccoli leaves. *Food Anal. Methods.* **2016**, *9*, 2174–2181. [CrossRef]
19. Conte, A.; Gammariello, D.; Di Giulio, S.; Attanasio, M.; Del Nobile, M.A. Active coating and modified-atmosphere packaging to extend the shelf life of Fior di Latte cheese. *J. Dairy Sci.* **2009**, *92*, 887–894. [CrossRef]
20. Del Nobile, M.A.; Corbo, M.R.; Speranza, B.; Sinigaglia, M.; Conte, A.; Caroprese, M. Combined effect of MAP and active compounds on fresh blue fish burger. *Int. J. Food Microbiol.* **2009**, *135*, 281–287. [CrossRef]
21. Del Nobile, M.A.; Conte, A.; Cannarsi, M.; Sinigaglia, M. Strategies for prolonging the shelf life of minced beef patties. *J. Food Saf.* **2009**, *29*, 14–25. [CrossRef]
22. Rashidinejad, A.; Birch, E.J.; Sun-Waterhouse, D.; Everett, D.W. Effects of catechin on the phenolic content and antioxidant properties of low fat cheese. *Int. J. Food Sci. Technol.* **2013**, *48*, 2448–2455. [CrossRef]
23. da Silva, F.C.; da Fonseca, C.R.; de Alencar, S.M.; Thomazini, M.; Baliero, J.C.; Pittia, P.; Favaro-Trindade, C.S. Assessment of production efficiency, physicochemical properties and storage stability of spry-dried propolis, a natural food additive, using gum arabic and osa starch-based carrier systems. *Food Bioprod. Process.* **2013**, *91*, 28–36. [CrossRef]
24. Mehanna, N.S.; Hassan, Z.M.R.; El-Din, H.M.F.; Ali, A.A.E.; Amarowicz, R.; EL-Messery, T.M. Effect of interaction phenolic compounds with milk proteins on cell line. *Food Nutr. Sci.* **2014**, *5*, 2130–2146. [CrossRef]
25. Gallo, M.; Vinci, G.; Graziani, G.; De Simone, C.; Ferranti, P. The interaction of cocoa polyphenols with milk proteins studied by proteomic techniques. *Food Res. Int.* **2013**, *54*, 406–415. [CrossRef]
26. Prajapati, M.R.; Patel, V.; Parekh, T.; Subhash, R. Effect of in bio-processing on antioxidant activity of selected cereals. *Asian J. Plant Sci. Res.* **2013**, *3*, 66–72.
27. Ti, H.; Zhang, R.; Li, Q.; Wei, Z.; Zhang, M. Effects of cooking and in vitro digestion of rice on phenolic profiles and antioxidant activity. *Food Res. Int.* **2015**, *76*, 813–820. [CrossRef]
28. Gumienna, M.; Lasik, M.; Czarnecki, Z. Influence of plant extracts addition on the antioxidative properties of products obtained from green lentil seeds during in vitro digestion process. *Pol. J. Food Nut. Sci.* **2009**, *59*, 295–298.
29. Cartea, M.E.; Francisco, M.; Soengas, P.; Velasco, P. Phenolic compounds in Brassica vegetables. *Molecules* **2010**, *16*, 251–280. [CrossRef]
30. Jaiswal, A.K.; Nissreen, A.G.; Gupta, S. A comparative study on the polyphenolic content, antibacterial activity and antioxidant capacity of different solvent extracts of Brassica oleracea vegetables. *Int. J. Food Sci. Technol.* **2012**, *47*, 223–231. [CrossRef]
31. Correa, C.B.; Martin, J.G.P.; Alencar, S.M.; Porto, E. Antilisterial activity of broccoli stems (*Brassica Oleracea*) by flow cytometry. *Int. Food Res. J.* **2014**, *21*, 395.
32. Sibi, G.; Abhilasha, S.; Dhananjava, K.; Rovikumar, K.R.; Mallesha, H. In vitro antimicrobial activities of Broccoli against food borne bacteria. *J. Appl. Pharm. Sci.* **2013**, *3*, 100–103.
33. Seberian, H.; Hamidi, E.Z.; Abbassi, S. Effect of conventional and ohmic pasteurization on some bioactive components of aloe Vera gel juice. *Iran. J. Chem. Chem. Eng.* **2015**, *34*, 99–108.
34. Ligor, M.; Trziszka, T.; Buszewski, B. Study of antioxidant activity of biologically active compounds isolated from green vegetables by coupled analytical techniques. *Food Anal. Methods.* **2013**, *6*, 630–636. [CrossRef]
35. Pandey, K.B.; Rizvi, S.I. Plant polyphenols as dietary antioxidants in human health and disease. *Oxid. Med. Cell. Longev.* **2009**, *2*, 270–278. [CrossRef]
36. Bhattacherjee, A.K.; Tandon, D.K.; Dikshit, A.; Kumar, S. Effect of pasteurization temperature on quality of aonla juice during storage. *J. Food Sci. Technol.* **2011**, *48*, 269–273. [CrossRef]

Article

Bioactive Compounds from Norway Spruce Bark: Comparison Among Sustainable Extraction Techniques for Potential Food Applications

Sara Spinelli [1], Cristina Costa [1], Amalia Conte [1,*], Nicola La Porta [2,3], Lucia Padalino [1] and Matteo Alessandro Del Nobile [1]

[1] Department of Agricultural Sciences, Food and Environment, University of Foggia, Via Napoli, 25–71121 Foggia, Italy; sara.spinelli@unifg.it (S.S.); cristina.costa@unifg.it (C.C.); lucia.padalino@unifg.it (L.P.); matteo.delnobile@unifg.it (M.A.D.N.)

[2] Department of Sustainable Agro-ecosystems and Bioresources, Research and Innovation Centre, Fondazione Edmund Mach, Via E. Mach 1, 38010 San Michele all'Adige, Italy; nicola.laporta@fmach.it

[3] The EFI Project Centre on Mountain Forests (MOUNTFOR), Via E. Mach 1, 38010 San Michele all'Adige, Italy

* Correspondence: amalia.conte@unifg.it; Tel.: +39-088-158-9240

Received: 25 September 2019; Accepted: 21 October 2019; Published: 23 October 2019

Abstract: *Picea abies* (L.) Karst, (Norway spruce) bark, generally considered as wood industry waste, could potentially be used as a valuable source of antioxidants for food applications. In this study, supercritical fluid extraction (SFE), pressurized liquid extraction (PLE), and ultrasound-assisted extraction (UAE) were carried out in order to recover bioactive compounds from bark of Norway spruce. Obtained results show that PLE with ethanol as solvent was the most effective method for extracting total flavonoid compounds (21.14 ± 1.42 mg quercetin g^{-1} sample) and consequently exerted the highest antioxidant activity measured by 2,2'-azino-bis (3-ethylbenzothiazoline-6-sulfonic acid) (257.11 ± 13.31 mg Trolox g^{-1} sample). On the other hand, UAE extract contained the maximum phenolic concentration (54.97 ± 2.00 mg gallic acid g^{-1} sample) and the most interesting antioxidant activity measured by the ferric-reducing antioxidant power (580.25 ± 25.18 µmol $FeSO_4$ g^{-1} sample). Additionally, PLE and UAE have demonstrated great efficiency in the extraction of *trans*-resveratrol, quantified by HPLC (0.19 and 0.29 mg *trans*-RSV g^{-1} sample, respectively).

Keywords: supercritical fluid extraction; pressurized liquid extraction; ultrasound-assisted extraction; *trans*-resveratrol; Norway spruce bark

---

1. Introduction

Annually, a considerable amount of bark waste is generated as by-product from the industrial wood transformation. This waste is usually discarded or used for energy, biogas production, or animal feed [1]. Nevertheless, it is known that tree barks contain a wide variety of bioactive compounds [2]. In particular, several authors have found a large amount of phenolic antioxidants in bark of *Picea abies* (Norway spruce), one of the most distributed conifer species in Eurasian forests [3–5]. Spruce bark is especially rich in glycosylated monomeric stilbenes (astringin, piceid, and isorhapontin) and their corresponding aglycone forms (piceatannol, resveratrol, and isorhapontigenin) [6]. Among these different kinds of stilbenes, *trans*-resveratrol (trans-3, 5, 4'-trihydroxystilbene; *trans*-RSV) has attracted great attention. It is a natural polyphenolic compound found in a variety of food and also in bark tree [7]. Actually, it is considered a powerful compound capable to improve health and prevent chronic disease in human [8,9]. Recently published studies have shown that resveratrol can also protect against some neurodegenerative diseases, obesity and diabetes [10], high blood pressure [11], as well as cancer [12] and osteoporosis [9]. In addition, it is widely used in cosmetics and dermatology [13].

Even though basic extractions from spruce bark do not represent a novelty in the literature [1,3,14,15], specific comparisons among available techniques are still very few.

Nowadays, there is a great attention to green extraction technologies able to reduce or eliminate the use of hazardous substances and limit the cost of solvent waste disposal [16]. Among them, a prominent position can be occupied by supercritical fluid extraction (SFE), pressurized liquid extraction (PLE), and ultrasound-assisted extraction (UAE), all considered sustainable techniques [17–19]. SFE is considered a fast, efficient, and clean method [20]. Carbon dioxide is the most common gas used as supercritical fluid due to its moderate critical temperature and pressure (31.3 °C and 72.9 atm, respectively) [17]. Applications for the extraction of essential oils, tocotrienols, alkaloids, phenolic compounds, carotenoids, and tocopherols from different food matrices were carried out [17,21,22]. PLE is a technique that uses liquid solvents at elevated pressure and temperature to enhance the extraction performance [19]. The PLE system provides protection to oxygen- and light-sensitive compounds and improves the extraction yield, thus also decreasing time and solvent consumption [19,23,24]. UAE is another efficient extraction method, with high reproducibility, which requires low energy and minimum consumption of solvent. Aromas, phenols, antioxidant pigments, and low-molecular-weight compounds have been extracted by this technique [18]. The extraction of antioxidant compounds and in particular of resveratrol from spruce bark could be an efficient way to reuse and enhance this voluminous biomass waste.

The aim of this work was to compare the extracts of Norway spruce bark obtained by SFE, PLE, and UAE. In particular, total extract yield (TEY), total phenolic content (TPC), total flavonoid content (TFC), and antioxidant capacity were measured to compare the efficacy of each technique. Chromatographic identification of *trans*-RSV was also performed for the extracts with the highest polyphenols content.

2. Materials and Methods

2.1. Wood Materials and Chemicals

Norway spruce bark was supplied by the timber sawmill company Vender Legnami s.r.l. (Trento, Italy) and dried at room temperature for two weeks. The bark was collected in July 2018, obtained by a stock of timber logs processed and debarked by the company. The spruce logs were coming from Trentino forests (Italy) at their final cutting phase. The trees were growing at an elevation ranging from 1000 to 1600 m a.s.l. The tree age was ranging from 90 to 110 years old. The bark thickness was ranging from 4 to 10 mm. The barks were ground in a knife mill at room temperature, and the powdered bark was sieved to select particles smaller than 1 mm.

Standard *trans*-resveratrol (3,5,4′-trihydroxystilbene) and all analytical grade reagents were purchased from Sigma-Aldrich (Milano, Italy). CO_2 with purity degree of 4.5 was supplied by Sapio (Monza, Italy), while N_2 with purity degree of 99.9% was provided by Air Liquide (Milan, Italy).

2.2. Supercritical Fluid Extraction

Supercritical fluid extractions from bark were carried out in triplicate with a Speed SFE-2 extractor (Applied Separation, Allentown, PA, USA). In particular, different concentrations of ethanol (10, 20, 40, and 70%; *v/v*) were tested (SFE_10, SFE_20, SFE_40, and SFE_70, respectively). According to Talmaciu et al. [14], for aqueous ethanol as co-solvent, a pressure of 100 bar and a temperature of 40 °C for a static time of 150 min and a dynamic time of 105 min were used. An amount of 2 g of spruce bark was used for all the experiments. Extractions were performed with 6 mL/min flow rate in the static phase and with CO_2 and ethanol as co-solvent, with 10:1 mL/min flow rate ratio for the dynamic phase.

2.3. Pressurized Liquid Extraction

Pressurized liquid extractions were performed in triplicate on a PLE-1 system (LabService Analytica srl, Anzola Emilia, Italy) using distillate water at 160 °C (PLE_H$_2$O) and absolute ethanol at

180 °C (PLE_EtOH) as solvent, according to Co et al. [3]. In particular, the extraction method included different steps: sample load into cell (30 g); cell preparation (3 min); pressurization and heating (5 min, 50 bar, and 160 °C or 180 °C for ethanol and water, respectively); depressurization (0.1 min); flush volume (60%), and finally, N_2 purge (2 min). To remove any process carryover, a washing cycle was made among the extractions.

2.4. Ultrasound-Assisted Extraction

Ultrasound-assisted extractions were performed in triplicate using an ultrasonic bath CP104 (C.E.I.A., Viciomaggio, Arezzo, Italy; bath frequency 39 kHz, power 200 W). The extraction conditions have been set according to the results obtained from Ghitescu et al. [15] for polyphenol recovery in spruce wood bark. In particular, a process time of 60 min, an extraction temperature of 54 °C, a concentration of ethanol of 70% (*v/v*), and a material/solvent ratio of 1:10 were taken into account.

2.5. Chemical Characterization

2.5.1. Total Extraction Yield (TEY)

The extracts were evaporated overnight in a vacuum oven (OPTO-LAB, Concordia, Modena, Italy) at 30 °C, and the obtained final mass was weighted to calculate the TEY. Prior to analysis, each extract recovered with ethanol (20 mL) was stored in the dark at 4 °C. Results were expressed as mg of dry extract per gram of samples.

2.5.2. Total Phenolic Content (TPC)

TPC was spectrophotometrically measured using Folin-Ciocalteu reagent, according to conditions previously described by Spinelli et al. [22]. The total phenol contents were evaluated using a standard curve with different gallic acid concentrations (3.125–100 mg L^{-1}; $R^2 = 0.99$). Results were expressed as mg gallic acid equivalents per gram of dry weight (dw).

2.5.3. Total Flavonoid Content (TFC)

The aluminum trichloride method was carried out to determine TFC, as described by Spinelli et al. [22]. The calibration curve was made with standard solutions of quercetin (6.25–400 mg L^{-1}; $R^2 = 0.99$) in order to express the total flavonoid content as mg quercetin equivalent per gram of dry weight (dw).

2.5.4. Antioxidant Activity

The ABTS [2,2′-azino-bis(3-ethylbenzothiazoline-6-sulfonic acid)] assay was measured as described by Marinelli et al. [25]. The ABTS values were calculated from a standard curve of different concentrations of Trolox (3.125–600 mg L^{-1}; $R^2 = 0.99$). The radical scavenging capacity of extracts was quantified as mg Trolox equivalent per gram of dry weight (dw).

The antioxidant capacity of extracts was also estimated in another assay, according to the FRAP (ferric reducing antioxidant power) procedure described by Lucera et al. [26]. For determination, a calibration curve of ferrous sulfate heptahydrate ($FeSO_4 \cdot 7H_2O$) was prepared, with dilutions from 600 μmol to 12.5 μmol ($R^2 = 0.99$).

2.6. HPLC Analysis of Trans-Resveratrol

Chromatographic identification of *trans*-RSV was performed using an Agilent 1100-Series HPLC system (Agilent Technologies Inc, Santa Clara, CA, USA), equipped with a degasser, binary pump solvent delivery, auto sampler, column oven, and DAD detector. An Agilent Zorbax Eclipse C18 (4.6 × 150 mm; 5 μm particles) and a guard column of the same stationary phase were used for trans-RSV separation. HPLC analysis was performed using the conditions described by Sun et al. [27] with slight modification, consisting of an isocratic elution by methanol–water (40:60 by volume). The flow rate was 1 mL min^{-1}, UV detection wavelength 300 nm, injection volume 10 μL, and column temperature 25 °C. The working standard solutions of *trans*-RSV (0.01–100 mg L^{-1}) were prepared by diluting the stock solution (250 mg L^{-1}) in mobile phase and stored at 4 °C in darkness to avoid oxidative degradation and isomerization of *trans*-RSV to *cis*-form. The method linearity was up to 100 mg L^{-1}. The identification of *trans*-RSV in the extracts was performed by comparison of the retention time (~7.5 min) and UV spectra with *trans*-RSV standard.

2.7. Statistical Analysis

A one-way ANOVA and a post-hoc Fisher's test were used to evaluate statistically significant differences among samples. The software was Statistica 7.1 for Windows (StatSoft Inc., Tulsa, OK, USA). All tests were carried out in triplicate.

3. Results and Discussion

In this study, three different extraction techniques, that is, supercritical fluid extraction, pressurized liquid extraction, and ultrasound-assisted extraction, were compared in order to obtain a valuable spruce bark extract rich in bioactive compounds with high antioxidant activity.

3.1. Supercritical Fluid Extraction

Table 1 summarizes the experimental results of the SFE, in terms of total extraction yield (TEY), total phenolic content (TFC), total flavonoid content (TFC), and antioxidant activity.

Table 1. Total extraction yield (TEY), total phenolic content (TPC), total flavonoid content (TFC), and antioxidant activity (ABTS and FRAP) of SFE spruce extracts with different ethanol concentrations: SFE_10 (10%; *v/v*); SFE_20 (20%; *v/v*); SFE_40 (40%; *v/v*).

	TEY	TPC	TFC	ABTS	FRAP
	mg/g dw	mg GAEs/g dw	mg QEs/g dw	mg TEs/g dw	μmol FeSO$_4$·7H$_2$O/g dw
SFE_10	28.6 ± 0.36 [a]	0.77 ± 0.02 [a]	0.47 ± 0.02 [a]	2.48 ± 0.13 [a]	8.31 ± 0.24 [a]
SFE_20	30.7 ± 0.96 [a]	1.24 ± 0.07 [b]	1.05 ± 0.15 [b]	3.08 ± 0.16 [b]	10.01 ± 0.81 [b]
SFE_40	31.2 ± 0.21 [a]	2.50 ± 0.03 [c]	1.75 ± 0.10 [c]	5.29 ± 0.04 [c]	25.49 ± 0.66 [c]

Values are means of three replications ± standard deviation. Values in the same column followed by different superscript letters differ significantly ($p < 0.05$). SFE: supercritical fluid extraction. GAEs: gallic acid equivalents; QEs: quercetin equivalent; TEs: Trolox equivalent; FeSO$_4$·7H$_2$O: ferrous sulfate heptahydrate; ABTS: 2,2′-azinobis (3-ethylbenzothiazoline-6-sulfonic acid). FRAP: ferric reducing antioxidant power.

Concerning the TEY, the amount of ethanol did not influence the extraction yield. Comparable results were recorded, even though some differences in the chemical composition of the extract were found. As can be seen in the Table 1, the gradual increase of ethanol in the SFE statistically improves the extraction capacity of phenolic compounds, flavonoids, and antioxidant activity of spruce extracts [28]. In particular, TPC steadily increased as the ethanol concentration increased (from 0.77 ± 0.02 to 2.50 ± 0.03 mg GAEs/g dw), and a similar behavior was also observed for TFC, with the highest value 1.75 ± 0.10 mg QEs/g dw in the assay SFE_40. Similarly, SFE_40 shows the radical scavenging ABTS and FRAP capacity respectively 2 and 3 times higher than that obtained with SFE_10. It is well known that ethanol promotes the recovery of polar compounds from the samples due to changes in the extractive properties (diffusivity, density, and viscosity); in fact, the purpose of ethanol is to swell plant cells, thus allowing both solvent penetration and diffusion of the solute in the solid matrix. In this way, both the increase in polarity of supercritical CO_2 and the rapid formation of interactions with the analyte of interest are promoted [29]. A similar behavior was also reported in several previous studies that demonstrated how ethanol enhanced the extraction of bioactive compounds and consequently improved the antioxidant activity. Conde et al. [30] used supercritical CO_2 to extract phenolic compounds from *Pinus pinaster* wood and noted that the extraction yield and the phenolic concentration increased when ethanol was used as co-solvent. Fabrowska et al. [31] also developed a supercritical fluid extraction process in order to revalorize different freshwater green macro-algae species, demonstrating that the increase in the concentration of ethanol from 0% to 15% resulted in an increase in the extraction yield and in bioactive compound concentrations. Our assay with 70% of ethanol solution allowed recording extract statistically poorer in phenol and flavonoid content compared with the assay with lower ethanol concentrations (data not shown). As also reported in the literature, the use of high concentration of co-solvent can sometimes provoke reduction of target bioactive compounds, due to interactions between CO_2 and co-solvent [32].

3.2. Pressurized Liquid and Ultrasound-Assisted Extraction

In Table 2 are reported the TEY, TPC, TFC, and the antioxidant activity for Norway spruce bark extracts obtained by PLE and UAE techniques.

Table 2. Total extraction yield (TEY), total phenolic content (TPC), total flavonoid content (TFC), and antioxidant activity (ABTS and FRAP) of PLE and UAE spruce extracts with water (PLE_H_2O) and absolute ethanol (PLE_EtOH and UAE_EtOH).

	TEY	TPC	TFC	ABTS	FRAP
	mg/g dw	mg GAEs/g dw	mg QEs/g dw	mg TEs/g dw	μmol $FeSO_4 \cdot 7H_2O$/g dw
PLE_H_2O	130.7 ± 8.62 [a]	33.45 ± 1.44 [a]	19.03 ± 0.98 [a]	69.87 ± 1.46 [c]	389.10 ± 16.87 [b]
PLE_EtOH	127.9 ± 2.52 [a]	46.32 ± 2.17 [b]	21.14 ± 1.42 [a]	257.11 ± 13.31 [a]	506.10 ± 31.37 [a]
UAE_EtOH	123.3 ± 5.77 [a]	54.97 ± 2.00 [c]	14.44 ± 1.31 [b]	128.47 ± 8.61 [b]	580.25 ± 25.18 [a]

[a–c] Values are means of three replications ± standard deviation. Values in the same column followed by different superscript letters differ significantly ($p < 0.05$). PLE: pressurized liquid extraction. UAE: ultrasound-assisted extraction.

As regards PLE extract, from a general point of view, the technique is more efficient than SFE, because the working conditions of PLE generally allow protecting of bioactive compounds [33]. In particular, Rostagno et al. [34] described how isoflavones can be extracted by PLE from soybeans without degradation. As can be observed in Table 2, the two different tested solvents did not influence significantly the yield. On the contrary, higher content in TPC, TFC, and radical scavenging ABTS and FRAP were obtained for PLE extraction with absolute ethanol compared with water. Howard and Pandjaitan [35] reported that the flavonoids extracted from spinach by PLE with ethanol were more effective when compared with the same compounds extracted by different conditions. The significant increase in terms of antioxidant capacity, measured by ABTS or FRAP is also in accordance with literature data. Zhao et al. [36] also reported high DPPH (2,2-diphenyl-1-picryl-hydrazyl-hydrate) radical scavenging capacity and total phenolic content in barley extract with ethanol, compared with water extract. In the PLE technique, the application of high temperature during the extraction significantly decreases the dielectric constant of water and cuts down the surface tension, thus promoting the extraction of bioactive compounds, but PLE water extraction can damage some thermolabile compounds, such as polyphenols and flavonoids, thus justifying the preference for ethanol as co-solvent [23].

Results obtained from UAE extraction highlighted that UAE yield was higher than SFE_40 yield and comparable to that of the PLE_EtOH extract. As can be seen in the Table 2, UAE extract exerted the maximum phenolic concentration among the various green techniques adopted (54.97 ± 2.00 mg GAEs/g dw). The literature also confirms that UAE applied to various natural matrices significantly increases the phenolic compounds extracted, compared with alternative extraction methods [19,23]. As a fact, the production of cavitation bubbles promotes better extraction yield and increases the antioxidant activity of these extracts [37]. A similar trend was also observed when the comparison was made in terms of FRAP-antioxidant activity.

In order to highlight some possible relationships among the values reported in Table 2, a remarkable correlation between TPC and FRAP assay can be found with both the extraction methods adopted [22]. Thaipong et al. [38], studying the comparison among ABTS, DPPH, FRAP, and ORAC assays for estimating the antioxidant activity of guava fruit extracts, also demonstrated that FRAP test showed high correlation with total phenolic content. Youn et al. [39] also investigated the relationship between antioxidant activity and polyphenol or flavonoid contents in leaf extracts obtained from *Dendropanax morbifera* LEV. and showed that FRAP value was strongly correlated with polyphenols. A similar correlation can be also observed between total flavonoid compounds and ABTS. Similar trends were also found in other literature data carried out on various medicinal plants [40,41].

The different mechanism of action between FRAP and ABTS assay justifies the different values recorded between them. As a fact, the FRAP assay is based on the singlet electron transfer, while ABTS is based on the mixed mode with singlet electron transfer and hydrogen atom transfer [42].

3.3. Chromatographic Identification of Trans-RSV

The spruce bark extracts obtained by the two best extraction techniques in terms of total phenol content (PLE_EtOH and UAE) were also analyzed by chromatographic identification in order to achieve a quantitative and complete characterization of *trans*-RSV. In Figure 1 is reported the *trans*-RSV content in PLE_EtOH and UAE extracts. As can be observed, higher levels of *trans*-RSV were found in UAE extract (0.29 mg/g dw). Extraction conditions and isomerization to *cis* isomer can be the explanation to justify the low content of *trans*-RSV in PLE extract [43]. Zupancic et al. [44] also highlight that pH, temperature, and different extraction methods influenced *trans*-RSV stability. Garcìa-Pèrez et al. [45] found resveratrol as the only stilbenes in *Picea marina* bark extract with ethyl acetate. Differently, Co et al. [3] identified resveratrol in spruce extract with PLE by nuclear magnetic resonance and mass spectrometry detection.

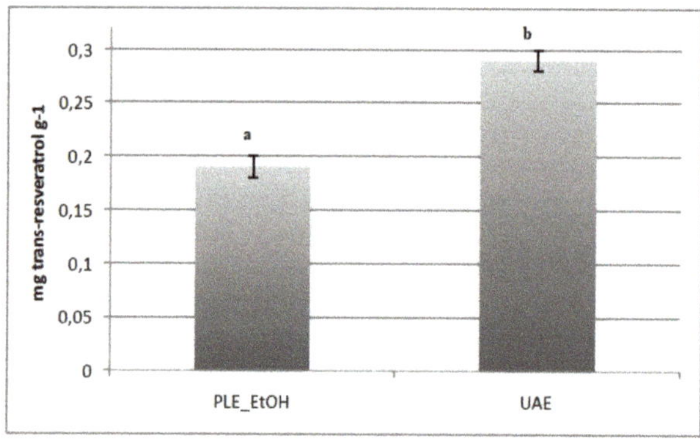

Figure 1. *Trans*-resveratrol content in PLE and UAE extracts. Samples with different superscript letters differ significantly ($p < 0.05$).

4. Conclusions

The results presented in this research show valid means to obtain antioxidant compounds, with a better focus on resveratrol, from Norway spruce bark using different environmental-friendly extraction techniques. In particular, the best results in terms of total phenolic compounds and antioxidant capacity measured by FRAP assays were obtained for UAE extract. PLE extract obtained with absolute ethanol shows the highest total flavonoid content and the best antioxidant capacity measured by ABTS assays. The highest *trans*-RSV content identified by chromatography was recorded for the UAE extract (0.29 mg/g dw), thus suggesting the potential of ultrasound-assisted extraction with ethanol (70%

v/v) for the valorization of waste, to record antioxidant compounds that can be applied to food and pharmaceutical sectors.

Author Contributions: S.S., C.C., and L.P. carried out experimental data. A.C. and M.A.D.N. elaborated data and contributed to the manuscript writing. N.L.P. contributed to the to the study proposal, to the results evaluation and to the manuscript writing.

Funding: This research was partially co-financed by the Grant P1611034I provided by the Fondazione Edmund Mach.

Conflicts of Interest: The authors declare no conflict of interest.

References

1. Jablonsky, M.M.; Vernarecovà, A.; Haz, L.; Dubinyovà, S.; Andrea, A.; Slàdkovà, I.; Surina, J. Extraction of phenolic and lipophilic compounds from spruce (*Picea abies*) bark using accelerated solvent extraction by ethanol. *Wood Res.* **2015**, *6*, 583–590.
2. Ekman, A.; Campos, M.S.; Lindhal, M.; Co, P.; Borjesson, E.N.; Karlsson, E.N.; Turner, C. Bioresource utilisation by sustainable technologies in new value-added biorefinery concepts -two case studies from food and forest industry. *J. Clean. Prod.* **2013**, *57*, 46–58. [CrossRef]
3. Co, M.; Fagerlund, A.; Engman, L.; Sunnerheim, K.; Sjoberg, P.J.; Turner, C. Extraction of antioxidants from spruce (*Picea abies*) bark using eco-friendly solvents. *Phytochem. Anal.* **2012**, *23*, 1–11. [CrossRef] [PubMed]
4. Kahkonen, M.P.; Hopia, A.I.; Vuorela, H.J.; Rauha, J.P.; Pihlaja, K.; Kujala, T.S.; Heinonen, M. Antioxidant activity of plant extracts containing phenolic compounds. *J. Agric. Food Chem.* **1999**, *47*, 3954–3962. [CrossRef]
5. Pan, H.; Lundgren, L.N. Phenolic extractives from root bark of *Picea abies*. *Phytochemical* **1995**, *39*, 1423–1428. [CrossRef]
6. Gabaston, J.; Richard, T.; Biais, B.; Waffo-Teguo, P.; Pedrot, E.; Jourdes, M.; Corio-Costet, M.F.; Mèrillon, M.J. Stilbenes from common spruce (*Picea abies*) bark as natural antifungal agent against downy mildew (*Plasmopara viticola*). *Ind. Crops Prod.* **2017**, *103*, 267–273. [CrossRef]
7. Hasan, M.; Bae, H. An overview of stress-induced resveratrol synthesis in grapes: Perspectives for resveratrol-enriched grape products. *Molecules* **2017**, *22*, 294. [CrossRef]
8. Smoliga, J.M.; Baur, J.A.; Hausenblas, H. Resveratrol and health: A comprehensive review of human clinical trials. *Mol. Nutr. Food Res.* **2011**, *55*, 1129–1141. [CrossRef]
9. Kursvietiene, L.; Staneviciene, I.; Morgidiene, A.; Bernatoniene, J. Multiplicity of effects and health benefits of resveratrol. *Medicina* **2016**, *52*, 148–155. [CrossRef]
10. Szkudelska, K.; Szkudelski, T. Resveratrol, obesity and diabetes. *Eur. J. Pharm.* **2010**, *635*, 1–8. [CrossRef]
11. Fogacci, F.; Tocci, G.; Presta, V.; Fratter, A.; Borghi, C.; Cicero, A.F. Effect of resveratrol on blood pressure: A systematic review and meta-analysis of randomized, controlled, clinical trials. *Crit. Rev. Food Sci. Nutr.* **2018**, *59*, 1–14. [CrossRef]
12. Rauf, A.; Imran, M.; Butt, M.S.; Nadeem, M.; Peters, D.G.; Mubarak, M.S. Resveratrol as an anti-cancer agent: A review. *Crit. Rev. Food Sci. Nutr.* **2018**, *58*, 1428–1447. [CrossRef] [PubMed]
13. Ratz-Łyko, A.; Arct, J. Resveratrol as an active ingredient for cosmetic and dermatological applications: A review. *J. Cosmet. Laser Ther.* **2019**, *21*, 84–90. [CrossRef] [PubMed]
14. Talmaciu, A.I.; Ravber, M.; Volf, I.; Knez, Z.; Popa, V. Isolation of bioactive compounds from spruce bark waste using sub- and supercritical fluids. *J. Supercrit. Fluids* **2016**, *117*, 243–251. [CrossRef]
15. Ghitescu, R.E.; Volf, I.; Carausu, C.; Buhlmann, A.M.; Glica, I.A.; Popa, V. Optimization of ultrasound-assisted extraction of polyphenols from spruce wood bark. *Ultrason. Sonochem.* **2015**, *22*, 535–541. [CrossRef]
16. Hatti-Kaul, R.; Tornvall, U.; Gustafsson, L.; Borjesson, P. Industrial biotechnology for the production of bio-based chemicals—A cradle-to-grave perspective. *Trends Biotechnol.* **2007**, *25*, 119–124. [CrossRef]
17. Herrero, M.; Cifuentes, A.; Ibanez, E. Sub- and supercritical fluid extraction of functional ingredients from different natural sources: Plants, food-by-products, algae and microalgae: A review. *Food Chem.* **2006**, *98*, 136–148. [CrossRef]
18. Chemat, F.; Rombaut, N.; Sicaire, A.G.; Meullemiestre, A.; Fabiano-Tixier, A.S.; Abert-Vian, M. Ultrasound assisted extraction of food and natural products. Mechanisms, techniques, combinations, protocols and applications. A review. *Ultrason. Sonochem.* **2017**, *34*, 540–560. [CrossRef]

19. Garcìa-Salas, P.; Morales-Soto, A.; Segura-Carretero, A.; Fernández-Gutiérrez, A. Phenolic compound extraction systems for fruit and vegetable samples. *Molecules* **2010**, *15*, 8813–8826. [CrossRef]

20. Pereira, C.G.; Meireles, M.A.A. Supercritical fluid extraction of bioactive compounds: Fundamentals, applications and economic perspectives. *Food Bioprocess Technol.* **2010**, *3*, 340–372. [CrossRef]

21. Nautyal, O.H. Food processing by supercritical carbon dioxide-review. *EC Chem.* **2016**, *21*, 111–135.

22. Spinelli, S.; Conte, A.; Lecce, L.; Padalino, L.; Del Nobile, M.A. Supercritical carbon dioxide extraction of brewer's spent grain. *J. Supercrit. Fluids* **2016**, *107*, 69–74. [CrossRef]

23. Mustafa, A.; Turner, C. Pressurized liquid extraction as a green approach in food and herbal plants extraction: A review. *Anal. Chim. Acta* **2011**, *703*, 8–18. [CrossRef] [PubMed]

24. Santos, D.T.; Veggi, P.C.; Meireles, M.A.A. Optimization and economic evaluation of pressurized liquid extraction of phenolic compounds from jabuticaba skins. *J. Food Eng.* **2012**, *108*, 444–452. [CrossRef]

25. Marinelli, V.; Padalino, L.; Nardiello, D.; Del Nobile, M.A.; Conte, A. New approach to enrich pasta with polyphenols from grape marc. *J. Chem.* **2015**, *8*, 1–8. [CrossRef]

26. Lucera, A.; Costa, C.; Marinelli, V.; Saccotelli, M.A.; Del Nobile, M.A.; Conte, A. Fruit and Vegetable By-Products to Fortify Spreadable Cheese. *Antioxidants* **2018**, *7*, 61. [CrossRef]

27. Sun, Y.; Qi, Y.; Mu, Z.; Wang, K. Quantitative determination of resveratrol in *Polygonum cuspidatum* and its anti-proliferative effect on melanoma A375 cells. *Biomed. Res.* **2015**, *26*, 750–754.

28. Veggi, P.C.; Prado, J.M.; Bataglioni, G.A.; Eberlin, M.N.; Meireles, M.A.A. Obtaining phenolic compounds from jatoba (*Hymenaea courbaril* L.) bark by supercritical fluid extraction. *J. Supercrit. Fluids* **2014**, *89*, 68–77. [CrossRef]

29. Salleh, L.M.; Rahman, R.A.; Selamat, J.; Hamid, A.; Sarker, M.Z.I. Optimization of extraction condition for supercritical carbon dioxide (SC-CO_2) extraction of *Strobhilantes Crispus* (Pecah Kaca) leaves by response surface methodology. *J. Food Process Technol.* **2013**, *4*, 197–201.

30. Conde, E.; Hemming, J.; Smeds, A.; Reinoso, B.D.; Moure, A.; Willfor, S.; Domìnguez, H.; Parajò, J.C. Extraction of low-molar-mass phenolics and lipophilic compounds from Pinus pinaster wood with compressed CO_2. *J. Supercrit. Fluids* **2013**, *81*, 193–199. [CrossRef]

31. Fabrowska, J.; Ibanez, E.; Leska, B.; Herrero, M. Supercritical fluid extraction as a tool to valorize underexploited freshwater green algae. *Algal Res.* **2016**, *19*, 237–245. [CrossRef]

32. Khaw, K.Y.; Parat, M.O.; Shaw, P.N.; Falconer, J.R. Solvent Supercritical Fluid Technologies to Extract Bioactive Compounds from Natural Sources: A Review. *Molecules* **2017**, *22*, 1186. [CrossRef] [PubMed]

33. Zaidulism, J.A.; Rahman, M.M.; Sharis, K.M.; Mohamed, A.; Sahena, F.; Jahurul, M.H.A.; Ghafoor, K.; Noriliani, N.A.N.; Omar, A.K.M. Techniques for extraction of bioactive compounds from plant materials: A review. *J. Food Eng.* **2013**, *117*, 426–436.

34. Rostagno, M.A.; Palma, M.; Barroso, C.G. Pressurized liquid extraction of isoflavones from soybeans. *Anal. Chim. Acta* **2004**, *522*, 169–177. [CrossRef]

35. Howard, L.; Pandjaitan, N. Pressurized liquid extraction of flavonoids from spinach. *J. Food Sci.* **2008**, *73*, 151–157. [CrossRef]

36. Zhao, H.; Dong, J.; Lu, J.; Chen, J.; Li, Y.; Shan, L.; Lin, Y.; Fan, W.; Gu, G. Effects of extraction solvent mixtures on antioxidant activity evaluation and their extraction capacity and selectivity for free phenolic compounds in barley (*Hordeum vulgare* L.). *J. Agric. Food Chem.* **2006**, *54*, 7277–7286. [CrossRef]

37. Hemwimol, S.; Pavasant, P.; Shotipruk, A. Ultrasound-assisted extraction of anthraquinones from roots of *Morinda citrifolia*. *Ultrason. Sonochem.* **2006**, *13*, 543–548. [CrossRef]

38. Thaiponga, K.; Boonprakoba, U.; Crosbyb, K.; Cisneros-Zevallosc, L.; Byrne, D.H. Comparison of ABTS, DPPH, FRAP, and ORAC assays for estimating antioxidant activity from guava fruit extracts. *J. Food Compos. Anal.* **2006**, *19*, 669–675. [CrossRef]

39. Youn, J.S.; Kim, Y.J.; Na, H.J.; Jung, H.R.; Song, C.K.; Kang, S.Y.; Kim, J.Y. Antioxidant activity and contents of leaf extracts obtained from *Dendropanax morbifera* LEV are dependent on the collecting season and extraction conditions. *Food Sci. Biotechnol.* **2019**, *28*, 201–207. [CrossRef]

40. Teixeira, T.S.; Vale, R.C.; Almeida, R.R.; Ferreira, T.P.S.; Guimarães, L.G.L. Antioxidant Potential and its Correlation with the Contents of Phenolic Compounds and Flavonoids of Methanolic Extracts from Different Medicinal Plants. *Rev. Virtual Quim.* **2017**, *9*, 1546–1559. [CrossRef]

41. Chaudhari, G.M.; Mahajan, R.T. Comparative Antioxidant Activity of Twenty Traditional Indian Medicinal Plants and its Correlation with Total Flavonoid and Phenolic Content. *Int. J. Pharm. Sci. Rev. Res.* **2015**, *30*, 105.

42. Kareem, H.S.; Ariffin, A.; Nordin, N.; Heidelberg, T.; Abdul-Aziz, A.; Kong, K.W.; Yehye, W.A. Correlation of antioxidant activities with theoretical studies for new hydrazone compounds bearing a 3,4,5-trimethoxy benzyl moiety. *Eur. J. Med. Chem.* **2015**, *103*, 497–505. [CrossRef] [PubMed]

43. Du, F.Y.; Xiao, X.H.; Li, J.K. Application of ionic liquids in the microwave-assisted extraction of *trans*-resveratrol from *Rhizma Polygoni Cuspidati*. *J. Chromatogr. A* **2007**, *1140*, 56–62. [CrossRef] [PubMed]

44. Zupancic, S.; Lavric, Z.; Kristl, J. Stability and solubility of *trans*-resveratrol are strongly influenced by pH and temperature. *Eur. J. Pharm. Biopharm.* **2015**, *93*, 196–204. [CrossRef] [PubMed]

45. Garcìa-Pèrez, M.E.; Royer, M.; Herbette, G.; Desjardins, Y.; Pouliot, R.; Stevanovic, T. *Picea mariana* bark: A new source of *trans*-resveratrol and other bioactive polyphenols. *Food Chem.* **2012**, *135*, 1173–1182. [CrossRef] [PubMed]

Review

A New Insight on Cardoon: Exploring New Uses besides Cheese Making with a View to Zero Waste

Cássia H. Barbosa [1], Mariana A. Andrade [1,2], Fernanda Vilarinho [1,*], Isabel Castanheira [1], Ana Luísa Fernando [3], Monica Rosa Loizzo [4] and Ana Sanches Silva [5,6]

[1] Department of Food and Nutrition, National Institute of Health Dr. Ricardo Jorge, Av. Padre Cruz, 1649-016 Lisbon, Portugal; cassia.barbosa@insa.min-saude.pt (C.H.B.); mariana.andrade@insa.min-saude.pt (M.A.A.); isabel.castanheira@insa.min-saude.pt (I.C.)

[2] REQUIMTE/LAQV, Faculty of Pharmacy, University of Coimbra, Coimbra, Azinhaga de Santa Comba, 3000-548 Coimbra, Portugal

[3] MEtRICs, Departamento de Ciências e Tecnologia da Biomassa, Faculdade de Ciências e Tecnologia, FCT, Universidade Nova de Lisboa, Campus de Caparica, 2829-516 Caparica, Portugal; ala@fct.unl.pt

[4] Department of Pharmacy, Health and Nutritional Sciences, University of Calabria, 87036 Arcavacata di Rende, Italy; monica_rosa.loizzo@unical.it

[5] National Institute for Agricultural and Veterinary Research (INIAV), I.P., Rua dos Lagidos, Lugar da Madalena, Vairão, 4485-655 Vila do Conde, Portugal; anateress@gmail.com

[6] Center for Study in Animal Science (CECA), ICETA, University of Oporto, 4051-401 Oporto, Portugal

* Correspondence: fernanda.vilarinho@insa.min-saude.pt

Received: 25 March 2020; Accepted: 22 April 2020; Published: 2 May 2020

Abstract: Cardoon, *Cynara cardunculus* L., is a perennial plant whose flowers are used as vegetal rennet in cheese making. Cardoon is native from the Mediterranean area and is commonly used in the preparation of salads and soup dishes. Nowadays, cardoon is also being exploited for the production of energy, generating large amount of wastes, mainly leaves. These wastes are rich in bioactive compounds with important health benefits. The aim of this review is to highlight the main properties of cardoon leaves according to the current research and to explore its potential uses in different sectors, namely the food industry. Cardoon leaves are recognized to have potential health benefits. In fact, some studies indicated that cardoon leaves could have diuretic, hepato-protective, choleretic, hypocholesterolemic, anti-carcinogenic, and antibacterial properties. Most of these properties are due to excellent polyphenol profiles, with interesting antioxidant and antimicrobial activities. These findings indicate that cardoon leaves can have new potential uses in different sectors, such as cosmetics and the food industry; in particular, they can be used for the preparation of extracts to incorporate into active food packaging. In the future, these new uses of cardoon leaves will allow for zero waste of this crop.

Keywords: *Cynara cardunculus* L.; cardoon leaves; by-products; antioxidant activity; antimicrobial activity

---

1. Introduction

Cynara cardunculus L. is a perennial plant belonging to the family Asteraceae, which is native to the Mediterranean area [1–5]. Commonly known as cardoon or artichoke thistle, *Cynara cardunculus* L. is a complex species comprising three botanical varieties: the globe artichoke (var. *scolymus* (L.) Fiori), the cultivated cardoon (var. *altilis* DC.), and the wild cardoon (var. *sylvestris* (Lamk) Fiori) [5,6]. This review is focused on the cultivated and wild cardoons that herein are referred to as cardoon.

Cardoon is considered to be a valuable crop as it shows high yields, drought tolerance, and low input needs, and it provides benefits regarding soil properties, erodibility, and biological and landscape diversity [7]. The edible parts of cardoon arc the fleshy stems and the immature heads, which, traditionally, are used in Mediterranean cuisine, mainly in salads and soup dishes [1,8]. In Italy, for example, the edible parts are used in many dishes as a vegetable and are sold canned in olive oil. Besides, they are also used in other Mediterranean traditional dishes such as Tunisian or Algerian "couscous", and the flower (the pistils) is used in cheese making as a vegetable rennet substitute [4,9–13]. Cardoon by-products are mainly composed of leaves, stems, and seeds. They have been used to produce biomass for energy; and oil for human consumption, biodiesel, and animal feed [14,15]. The leaves are used in traditional medicine due to their high content in bioactive compounds such as cynarin and silymarin [2,4,14–19]. Moreover, the cultivation of cardoon in large areas is gaining interest as feedstock for novel industrial bio-based products (e.g., conversion into biopolymers or as a source of cellulose for nanometric technological applications) [20,21].

The world population is growing every day, and consequently, increasing the demand for food. In consequence, food production has been intensified, generating tons of by-products, which, if not discarded in a sustainable and responsible way, can represent a serious environmental problem, depending on the composition of the by-product itself. Despite its known powerful biological properties, unfortunately, the majority of these by-products are not being applied for other purposes, even though studies have announced their potential in different areas. For instance, several research studies in progress are incorporating fruit by-products in active food packages to delay the natural lipid oxidation phenomenon and microbial deterioration of foods [22–24]. Additionally, the increased demand for bioproducts, biomaterials, and bioenergy may result in a higher accumulation of by-products from the value chains. Therefore, this review aims to systematize the current knowledge of cardoon and emphasize the main composition of the by-products of the plant. Specifically, this review highlights the main bioactive compounds of cardoon leaves and their functional properties. Finally, their potential uses are also addressed, with a special focus on active food packaging to extend food shelf life.

2. Cardoon Botanical Description, Distribution, and Cultivation

Although native from the Mediterranean area (southern Europe and North Africa), cardoon has been spread to several other countries like the United States of America, Mexico, Australia, and New Zealand [1]. Due to its natural habitat, cardoon can grow in adverse climate conditions, with high temperatures, severe drought, and in thin unproductive and stony soils [1,14,25,26]. Moreover, cardoon is also a pollinator-supporting industrial crop, with all the associated benefits in terms of biodiversity [7,26–28].

Cardoon is a perennial plant that can grow up to two meters high with thick and rigid stems. It has an annual development cycle, and the reproductive cycle is completed by summer. With adequate soil moisture and temperature, cardoon development can start during autumn or spring with seed germination [5,14,29]. The seeds are light grey, brown, or black and can be up to 8 mm long [1]. Cardoon starts by developing a root that can grow one meter down and regenerates each year [1,5,14,29]. Simultaneously, the leaves grow to originate a leaf rosette. The leaf rosette is large and strong, with over 40 leaves that can be up to 120 by 30 cm, while the leaf from the upper stem can be up to 50 cm in length [1,5,14]. By late spring, the plant develops its inflorescence at the top of a branch on the stem, which can be up to 3 m high and about 2–4 cm of diameter. The inflorescence can be called capitula or heads and have an almost round shape. The cardoon inflorescence consists of several hermaphrodites and tubular flowers (florets) fitted in a well-developed receptacle. The florets are usually blue–violet colored [1,5,14,29]. By summertime, the aboveground plant parts dry, but the underground parts (the roots and perennating buds) remain alive until weather conditions become milder, and the perennating buds sprout, and a new development cycle starts [14,29].

3. Cardoon Flower—Cheese Making

Cardoon flowers (Figure 1) are used as milk clotting in cheese making, producing a cheese with a creamy soft texture and a genuine and slightly piquant aroma [12,30–32]. The clotting activity is due to the stigma and style of the inflorescence. Two proteases have been identified as responsible for this activity: cardosins A and B [12,31–33]. Both enzymes are responsible for the milk clotting and have proteolytic activities. Specifically, cardosin A is responsible for the clotting activities by the hydrolysis of the k-casein, and cardosin B is responsible for proteolysis, similar to pepsin activity [12,31].

Figure 1. Cardoon flower [34].

Cardoon flower, as a coagulant agent for cheese making, is used mostly in Portugal, Spain, and Italy [13,30–33]. In these countries, some cheeses, classified as protected designation of origin (PDO), are produced using cardoon flower as a coagulant [11,30,32]. The PDO denomination is legally defined as a product that has a specific origin and/or production that takes place in a specific geographic area and endorses the quality and authenticity of the product [11,35]. For instance, for the "Serra da Estrela" cheese, a well-known Portuguese semi-soft cheese made with raw ewe's milk of a native breed (Bordaleira da Serra da Estrela), the usage of cardoon flower is crucial for its authenticity and its high quality [30,36,37].

Gomes et al. [32] described a process to produce the cardoon flower extract to use as a natural rennet for cheese making. For this purpose, the flower is collected during the flower season. Then, the flower pistils are cut and separated from the rest of the plant and then dried under room temperature (23 ± 1 °C), protected from sunlight. After that, and according to the traditional method, the dried pistils are macerated by hand in a mortar with acetate buffer and then filtrated to a final volume of 50 mL at room temperature [30,32,33].

Moreover, since cardoon flower is used directly in dairy products, it is important to know its chemical composition, in particular, its content in phenolic compounds and its bioactive properties. Therefore, several authors have studied the cardoon flower composition. Dias et al. [36] characterized the phenolic composition of the inflorescence of cardoon flower and evaluated the antioxidant and antibacterial properties. The authors concluded that flavonoids are the major family of compounds present in cardoon flower; these include apigenin (9.9 ± 0.5 mg/g of extract) and caffeoylquinic acid (7.6 ± 0.2 mg/g of extract), which were the major compounds identified. Regarding the antibacterial activity, the cardoon flower presented a good outcome against Gram-positive bacteria such as *Listeria monocytogenes*, methicillin-sensitive *Staphylococcus aureus* (MSSA), and methicillin-resistant *Staphylococcus aureus* (MRSA); and Gram-negative bacteria such as *Morganella morganii* and *Pseudomonas aeruginosa*. Overall, the authors concluded that cardoon flowers, for use as rennet in cheese making, is a highly valuable ingredient as a source of bioactive compounds [36].

4. Cardoon By-Products and their Current Applications

4.1. Biomass

Cardoon crops are capable of growing in adverse environments, and they have been identified as potential crops for energy production [27]. Cardoon crop by-products are mainly used to produce biomass for different applications. At an industrial level, cardoon crops represent a great interest in the production of solid biofuel, seed oil, biodiesel, paper pulp, green forage, and pharmacologically active compounds [14,38–43]. Mauromicale et al. [27] studied the potential ability of cultivated and wild cardoons to produce energy in terms of biomass, achenes, and energy yield. The authors concluded that both cultivated and wild cardoon are potential energy crops and improved the soil fertility characteristics by increasing organic matter, total nitrogen, available phosphorus, and exchangeable potassium content. The annual average outcome of cultivated cardoon presented was 14.6 t/ha of dry biomass, 550 kg/ha of achenes, and 275 Gj/ha yields, while for wild cardoon, the outcome was 7.4 t/ha of dry biomass, 240 kg/ha of achenes, and 138 Gj/ha of energy yield. Several authors [43–45] studied cardoon for biomass production and demonstrated that this crop can be grown as an energy crop. These authors concluded that the cardoon aboveground dry biomass yield continuously increased during the seasons and that the low yield in the first season may be due to establishment difficulties, especially under the low input management in very dry conditions [27,43–45].

4.1.1. Solid Biofuel

Cardoon production is mainly based on the production of biomass for the production of biofuel. The solid biofuel is generally used for heating applications and power generation [14,46]. This is mainly due to the biomass characteristics that have high productivity in severe climate conditions; it has low moisture content, and it has high lignocellulose content [14]. Several studies have been made to assess the characteristics of cardoon as a solid biofuel [14,29,47].

4.1.2. Seed Oil and Biodiesel

Cardoon biomass can also be used to produce biodiesel and seed oil, which in turn can be used for human nutrition or biodiesel production [14,48]. The cardoon flower produces a fruit that acts as a dispersal unit, generally called seeds. This is a common trait for cardoon and sunflower since they belong to the same botanical family (Asteraceae) [14,43,44,48,49]. Although constituting a small percentage of the biomass, the seeds have been widely studied due to their important bioactive properties and potential use for energy [14,18,39]. The seeds also have an important nutritional value, having high fat (23.7 g/100 g dry weight) and protein (30.4 g/100 g dry weight) contents. The cardoon seed oil has an interesting lipid profile, composed, on average, by 11% palmitic, 4% stearic, 25% oleic, and 60% linoleic fatty acids. The oil from cardoon can be used for food applications or the production of biodiesel [14,18,39].

4.1.3. Paper Pulp

As cardoon can grow and flourish in hot and dry climate conditions and has high biomass productivity, its utilization in the production of paper pulp has been suggested [25,40]. Additionally, due to its similarities to eucalypt pulp, cardoon can be a good alternative in the hardwood pulp sector [14,25,45]. In a study by Gominho et al. [25], cardoon pulp was produced with good yields, with very little residual lignin, high bulk, and remarkable tensile strength properties. The authors concluded that the variables in production could be improved for even better results [25]. These results confirmed the potential of cardoon in the hardwood pulp sector, as it originated a good pulp with high yields, low rejection rate, and very good strength properties [25].

4.1.4. Green Forage

As the cardoon crop may generate large amounts of biomass by-products when cultivated in large areas, it is important to find alternative ways to use them. Therefore, authors in this field have suggested the use of cardoon by-products as green forage for animal feed. Some authors have studied the nutrition value of green forage and have considered it safe and acceptable for animal feed [14,41,42,45,50]. Cajarville et al. [42,50] evaluated the nutrition value of cardoon for animal feed and concluded that it is a very good forage with good content in protein (156 g/kg dry matter), high digestibility coefficients (86.1 ± 1.3% for organic matter), high energy value (82.7 ± 1.6 MJ/kg dry matter), and suitability for ruminant feed. Cabiddu et al. [41] also evaluated cardoon seed pressed cakes to be used for the feeding of small ruminants. The authors concluded that, due to a high source of protein (18.52 ± 0.08% dry matter), fiber (1.22 ± 0.30%), phenolic compounds (32.7 ± 0.02 mg/tannic acid equivalent), and polyunsaturated fatty acid (PUFA), namely linoleic acid (1.7% dry matter) and oleic acid (0.7% dry matter), it is suitable for animal feed. The authors also suggested that the high content of phenolic compounds and PUFA of the green forage might be transferred to dairy products, increasing their nutritional value. It is important to develop more studies regarding this field because the utilization of crop biomass for animal feed can reduce the environmental problems associated with livestock production [51].

4.2. Source of Bioactive Compounds

4.2.1. Stems

Stems and leaves, the most abundant waste regarding cardoon crops [16], may represent also a source of bioactive compounds. Studies concerning cardoon stems are scarce, but it has been identified as a source of caffeoylquinic acids [16,52,53]. Caffeoylquinic acids are natural antioxidants associated with the structural support of the plant since they establish bridges with the polymeric compounds of the cell wall [16,52]. Caffeoylquinic acids have been suggested to decrease the risk of chronic diseases including cancer and cardiovascular disease [6]. Pandino et al. [52] studied the cardoon stems and corroborated the richness in caffeoylquinic acids. Besides caffeoylquinic acids (17.7 g/kg dry matter), luteolin (3.0 g/kg dry matter) and apigenin (4.7 g/kg dry matter) were also identified in the cardoon stems. Several authors [53–55] also studied the globe artichoke stems and also concluded that they may represent a good source of caffeoylquinic acids. Globe artichoke stem presented, on average, a lower content of caffeoylquinic acids (13 g/kg dry matter) and lutein (0.4 g/kg dry matter), and apigenin was not detected [55]. Yet, the variability of the results may be attributed to genetic variation, crop management, post-harvest processing options, and environmental factors [6,53–55].

The stem presents good antioxidant activity, mainly due to its richness in bioactive compounds. More studies should be performed for a better knowledge of the total composition of phenolic compounds in cardoon stems.

4.2.2. Leaves

Cardoon leaves represent, on average, about 60% of total cardoon waste [56]. Since the initial stage of the plants, the leaves showed beneficial properties, such as diuretic, hepato-protective, choleretic, hypocholesterolemic, anti-carcinogenic, and antibacterial effects [8,14,57,58]. Such properties are due to the high content in bioactive compounds presented by the leaves, such as chlorogenic acid, cynarine, and luteolin [8,16,59,60]. Cardoon leaves present also a high content of sesquiterpene lactones. The sesquiterpene lactones are responsible for the phytotoxic, cytotoxic, fungicidal, antiviral, and antimicrobial activity of cardoon. Thus, cardoon leaf extracts can be used in the development of herbicides of natural origin [61]. Cardoon leaves can as well be used in the preparation of alcoholic beverages as a flavoring agent [47].

5. Bioactive Properties of Cardoon Leaves

5.1. Nutritional Value

As far as the authors know, there is just one study regarding the nutritional value characterization of the leaves. The study indicates that cardoon leaves are a good source of carbohydrates and fat [59]. Cardoon leaves have an energetic contribution of 360.03 ± 1.09 kcal/100 g dry weight [59]. In terms of soluble sugars, cardoon leaves presented fructose, glucose, sucrose, and trehalose in its constitution [59]. For the organic acid profile, Chihoub et al. [59] identified oxalic, quinic, shikimic, citric, and fumaric acid in cardoon leaves. Oxalic acid was found to be the major organic acid, and fumaric acid was only found in trace amounts. The authors also studied the lipid profile of cardoon leaves and identified several fatty acids. The major fatty acid identified was α-linolenic acid (C18:3 n-3, PUFA) followed by linoleic acid (C18:2 n-6c, PUFA). According to the same study, cardoon leaves can be considered of high nutritional value as they have a high ratio of PUFA and saturated fatty acids (SFA) (PUFA/SFA = 3.80). Regarding the tocopherols, cardoon leaves present all four isoforms (α, β, γ, and δ), δ-tocopherol being the most abundant isoform. Tocopherols are known for their biological properties, for example, anti-inflammatory properties, and for their high antioxidant potential, when present in different isoforms [59].

5.2. Phenolic Composition and Antioxidant Properties

Several authors studied the antioxidant capacity of the cardoon leaves and stated the presence of polyphenolic compounds in high content [3,8,59,62]. Furthermore, traditionally leaves have been used in European medicine because of their known pharmacological properties, mainly due to the presence of cynarin and silymarin [14]. These compounds enhance liver and gallbladder function by stimulating the secretion of digestive juices [19].

So far, several polyphenol antioxidant compounds have been identified in cardoon leaves, chlorogenic acid and flavonoids being the most common ones [60]. Flavonoid presence in the leaves is suggested to be strategic due to its properties, since flavonoids are compounds that can absorb ultraviolet (UV) rays, specifically UV-B, so they can protect the plant from the harm that the radiation can cause [52,60]. Additionally, Pinelli et al. [60] compared the flavonoid content of cardoon that was and was not submitted to UV-light and concluded that the samples that were kept away from UV-light presented a lower number of flavonoids, confirming that flavonoids are essential to protect the plant from damage caused by UV rays. Besides, flavonoids have important antioxidant properties with important health benefits. Table 1 presents the main bioactive compounds that have been identified in cardoon leaves and their respective contents. Caffeoylquinic acid is the most common bioactive compound quantified in cardoon leaves. Pandino et al. [16] indicated that this compound possesses a potential inhibitory capacity of the development of cancers, exacerbated by the presence of reactive oxygen species. The bioactive content of cardoon leaves depends on several intrinsic and extrinsic factors such as genetic, environmental, handling, and storage, as well as the maturity of the plant [8,63,64]. Pandino et al. [65] studied the environmental effect on polyphenol content of the F1 population bred from a cross between a globe artichoke and a cultivated cardoon. The authors selected eight segregants that accumulated more caffeoylquinic acid in their leaves than did those of either of their parental genotypes. The selections were grown over two seasons to assess their polyphenol profile. The leaves of the first season had a higher content of polyphenols than those of the second season, considering that it rained less and had overall a higher average relative humidity. The authors concluded that the growing environment exerted a strong effect on polyphenol content.

Table 1. Main bioactive compounds of cardoon leaves.

Species/Variety	Presentation form and Extraction Procedure	Main Bioactive Compounds and Levels Found	Main Conclusion	Reference
C. cardunculus L. var. *sylvestris*	Hydroethanolic extract: 1 g of dried samples was added to 30 mL of solvent (ethanol:water, 80:20, *v/v*) for 1 h. It was filtered and reextracted in 30 mL of solvent for 1 h. The combined extracts were evaporated until dryness of the ethanol at 35 °C and the aqueous phase was frozen and lyophilized.	3-O-Caffeoylquinic acid (0.48 ± 0.01 mg/g of extract) 5-Hydroxyferuloylglycoside (1.3 ± 0.1 mg/g of extract) 4-O-Caffeoylquinic acid (13.6 ± 0.1 mg/g of extract) 5-O-Caffeoylquinic acid (0.179 ± 0.001 mg/g of extract) 3-O-Feruloylquinic acid (0.32 ± 0.01 mg/g of extract) 5-O-Feruloylquinic acid (0.53 ± 0.01 mg/g of extract) Luteolin-O-hexoside (5.5 ± 0.1 mg/g of extract) Pinoresinol-O-hexoside (0.37 ± 0.01 mg/g of extract) 3,4-O-Dicaffeoylquinic acid (6.492 ± 0.002 mg/g of extract) 3,5-O-Dicaffeoylquinic acid (0.331 ± 0.001 mg/g of extract) Luteolin-O-malonylhexoside (7.39 ± 0.02 mg/g of extract) Acetylapigenin-O-hexoside (1.85 ± 0.01 mg/g of extract) Total phenolic acids (23.6 ± 0.2 mg/g of extract) Total flavonoids (14.7 ± 0.1 mg/g of extract) Total phenolic compounds (38.3 ± 0.3 mg/g of extract)	The authors concluded that the hydroethanolic was more effective than the infusion preparation in the extraction of active compounds.	[59]
	Infusion preparation: 1 g of dried samples was added to 100 mL of boiling distilled water, left to stand for 5 min, filtered, and then frozen and lyophilized.	3-O-Caffeoylquinic acid (0.66 ± 0.01 mg/g of extract) 5-Hydroxyferuloylglycoside (0.95 ± 0.01 mg/g of extract) 4-O-Caffeoylquinic acid (10.2 ± 0.1 mg/g of extract) 5-O-Caffeoylquinic acid (0.185 ± 0.003 mg/g of extract) 3-O-Feruloylquinic acid (0.25 ± 0.1 mg/g of extract) 5-O-Feruloylquinic acid (0.39 ± 0.01 mg/g of extract) Luteolin-O-hexoside (1.6 ± 0.1 mg/g of extract) 3,4-O-Dicaffeoylquinic acid (3.5 ± 0.2 mg/g of extract) 3,5-O-Dicaffeoylquinic acid (0.319 ± 0.001 mg/g of extract) Luteolin-O-malonylhexoside (5.8 ± 0.2 mg/g of extract) Acetylapigenin-O-hexoside (4.73 ± 0.04 mg/g of extract) Total phenolic acids (16.0 ± 0.3 mg/g of extract) Total flavonoids (12.1 ± 0.2 mg/g of extract) Total phenolic compounds (29 ± 1 mg/g of extract)		

Table 1. *Cont.*

Species/Variety	Presentation form and Extraction Procedure	Main Bioactive Compounds and Levels Found	Main Conclusion	Reference
C. cardunculus L. var. *sylvestris*	Dried samples were extracted in 1 mL of 70% methanol, containing butylated hydroxytoluene and hesperetin, for 1 h at room temperature with shaking. After centrifugation, the supernatant was transferred to a microfuge tube, and the sample was centrifuged once more with 0.25 mL of 70% methanol. The supernatants were combined and kept at −20 °C until analysis.	Luteolin glucoside (0.8 g/kg DM) Luteolin glucuronide (1.9 ± 0.1 g/kg DM) Luteolin (0.3 g/kg DM) Total luteolin (3.2 g/kg DM) Apigenin rutinoside (0.2 g/kg DM) Apigenin glucuronide (3.3 ± 0.1 g/kg DM) Apigenin (1.3 g/kg DM) Total apigenin (4.8 g/kg DM) Total measured polyphenols (8.0 g/kg DM)	Both wild and cultivated cardoon presented a good profile of polyphenols, but the cultivated cardoon profile was richer and more variable than the wild cardoon. The apigenin derivates were the most abundant compounds in both cases.	[52]
C. cardunculus L. var. *altilis*		5-Caffeoylquinic acid (0.3 g/kg DM) Total caffeoylquinic acid (0.3 g/kg DM) Luteolin glucoside (0.1 g/kg DM) Luteolin glucuronide (2.4 ± 0.3 g/kg DM) Luteolin (0.9 ± 0.1 g/kg DM) Total luteolin (3.4 g/kg DM) Apigenin rutinoside (0.5 g/kg DM) Apigenin glucuronide (3.3 ± 0.3 g/kg DM) Apigenin malonylglucoside (0.4 g/kg DM) Apigenin (1.3 ± 0.2 g/kg DM) Total apigenin (5.6 g/kg DM) Total measured polyphenols (9.3 g/kg DM)		
C. cardunculus	Leaves were dried at room temperature for two weeks. The extract was made with 2.5 g of dry powder with 25 mL of solvent (methanol) under stirring for 30 min. Then the extract was filtered, evaporated to dryness under vacuum, and stored at 4 °C until analysis.	Polyphenol content (14.79 (mg/GAE g DW) Flavonoid content (9.08 mg/CE g DW) Tannin content (1.96 mg/CE g DW)	The sample exhibit a very good bioactive compounds profile, greater than several other species, mainly due to the hard climate conditions of development.	[8]

Table 1. *Cont.*

Species/Variety	Presentation form and Extraction Procedure	Main Bioactive Compounds and Levels Found	Main Conclusion	Reference
C. cardunculus L. var. *sylvestris*	The samples were lyophilized, after comminuted to a powder. The powder was extracted in quadruplicate with 3 × 50 mL of ethanol (70% *v/v*) at room temperature, under stirring. The extracts were completely defatted with n-hexane (4 × 20 mL), then concentrated under vacuum and rinsed with the extraction solvent to a final volume of 25 mL. Hydro-alcoholic extracts were stored at −20 °C until use.	1-O-Caffeoylquinic acid (8.23 ± 0.68 µmol/g d wt) Chlorogenic acid (61.84 ± 2.09 µmol/g d wt) Luteolin 7-O-rutinoside (1.10 ± 0.21 µmol/g d wt) Luteolin 7-O-glucoside (27.29 ± 1.87 µmol/g d wt) Dicaffeoylquinic acids (114.57 ± 4.25 µmol/g d wt) Luteolin 7-O-malonylglucoside (14.62 ± 0.41 µmol/g d wt) Apigenin 7-O-rutinoside (1.11 ± 0.11 µmol/g d wt) Luteolin (0.80 ± 0.01 µmol/g d wt) Luteolin 7-O-glucuronide (34.61 ± 1.48 µmol/g d wt) Apigenin 7-O-glucuronide (23.72 ± 0.06 µmol/g d wt) Apigenin (4.74 ± 0.14 µmol/g d wt) Total polyphenols (292.63 µmol/g d wt)	Both wild and cultivated cardoon presented a good profile of polyphenols, but the wild cardoon profile was richer and more variable than the cultivated cardoon. The samples were both rich in different flavonoids compounds, and the authors emphasized their role in the protection of the plant against UV-radiation.	[60]
C. cardunculus L. var. *altilis*		1-O-Caffeoylquinic acid (9.53 ± 2.63 µmol/g d wt) Chlorogenic acid (73.68 ± 4.83 µmol/g d wt) Luteolin 7-O-glucoside (33.55 ± 8.21 µmol/g d wt) Dicaffeoylquinic acids (29.17 ± 9.26 µmol/g d wt) Succinyldicaffeoylquinic acid (8.67 ± 0.41 µmol/g d wt) Luteolin 7-O-malonylglucoside (43.00 ± 0.50 µmol/g d wt) Succinyldicaffeoylquinic acid (2.20 ± 0.16 µmol/g d wt) Luteolin (1.68 ± 0.08 µmol/g d wt) Luteolin 7-O-glucuronide (13.70 ± 2.40 µmol/g d wt) Total polyphenols (215.18 µmol/g d wt)		

Table 1. *Cont.*

Species/Variety	Presentation form and Extraction Procedure	Main Bioactive Compounds and Levels Found	Main Conclusion	Reference
C. cardunculus L. var. *scolymus*	Infusion: 20 g of dried chopped leaves was added to 1000 mL of ultra-pure water at 95 °C, and the mixture was left to stand for 10 min and then filtered through cotton. Extract were frozen and freeze-dried.	Chlorogenic acid (64 ± 2 mg/g extract) *p*-Coumaroylquinic acid (1.1 ± 0.1 mg/g extract) 5-Feruloylquinic acid (1.6 ± 0.3 mg/g extract) Luteolin-7-rutinoside (7.6 ± 0.1 mg/g extract) Luteolin-7-glucoside (cynaroside) (3.0 ± 0.1 mg/g extract) 3,4-Dicaffeoylquinic acid (2.1 ± 0.1 mg/g extract) 1,3-Dicaffeoylquinic acid (cynarin) (22.4 ± 0.1 mg/g extract) Luteolin-7-malonyl-hexoside (1.7 ± 0.1 mg/g extract) 4,5-Dicaffeoylquinic acid (5.1 ± 0.1 mg/g extract) Phenolic contents (108 ± 2 mg/g extract)	All extracts presented good phenolic content. The infusion extract presented higher phenolic content. Chlorogenic acid was the major phenolic compound identified in all extracts.	[66]
	Decoction: The dried chopped leaves (20 g) were added to 1000 mL of ultrapure water, heated, and boiled for 10 min, and then the mixture was removed from the heat and left to stand for 5 min to be filtered through cotton. The extract was frozen and freeze-dried.	Chlorogenic acid (40 ± 3 mg/g extract) *p*-Coumaroylquinic acid (1.1 ± 0.1 mg/g extract) 5-Feruloylquinic acid (0.9 ± 0.1 mg/g extract) Luteolin-7-rutinoside (7.4 ± 0.8 mg/g extract) Luteolin-7-glucoside (cynaroside) (2.9 ± 0.4 mg/g extract) 3,4-Dicaffeoylquinic acid (0.9 ± 0.3 mg/g extract) 1,3-Dicaffeoylquinic acid (cynarin) (6.5 ± 0.5 mg/g extract) Luteolin-7-malonyl-hexoside (1.3 ± 0.1 mg/g extract) 4,5-Dicaffeoylquinic acid (1.9 ± 0.4 mg/g extract) Phenolic content (63 ± 5 mg/g extract)		
	Hydroalcoholic extract: 20 g of dried chopped leaves was added to 1000 mL of a mixture of ethanol/water (70:30, *v/v*) and stirred on an orbital shaker (70 rpm) for 12 h at 25 °C. The hydroalcoholic mixture was filtered through cotton, concentrated under reduced pressure in a rotary evaporator (40 °C), and then freeze-dried.	Chlorogenic acid (43 ± 2 mg/g extract) 5-Feruloylquinic acid (0.6 ± 0.1 mg/g extract) Luteolin-7-rutinoside (9.3 ± 0.4 mg/g extract) Luteolin-7-glucoside (cynaroside) (3.8 ± 0.3 mg/g extract) 3,4-Dicaffeoylquinic acid (0.03 ± 0.01 mg/g extract) 1,3-Dicaffeoylquinic acid (cynarin) (14 ± 1 mg/g extract) Luteolin-7-malonyl-hexoside (1.0 ± 0.1 mg/g extract) 4,5-Dicaffeoylquinic acid (1.1 ± 0.1 mg/g extract) Phenolic content (73 ± 4 mg/g extract)		

Table 1. *Cont.*

Species/Variety	Presentation form and Extraction Procedure	Main Bioactive Compounds and Levels Found	Main Conclusion	Reference
C. cardunculus L. var. altilis	Bidistilled water extract: Dried leaves were soaked in bidistilled water in the ratio 1:10 *w/v*. Then, the mixture was kept under dark conditions for 72 h at room temperature (20 °C = 1) and filtered to eliminate the solid fraction.	5-O-caffeoylquinic acid (51.3 ± 0.2 mg/L) 1,5-O-dicaffeoylquinic acid (119.3 ± 33.3 mg/L) Monosuccinildicaffeoylquinic acid (37.6 ± 1.0 mg/L) Total caffeoylquinic acid (208 mg/L) Luteolin 7-O-glucoronide (10.9 ± 0.4 mg/L) Luteolin (53.2 ± 0.4 mg/L) Total luteolin (64 mg/L) Cynaropicrin (5.4 ± 0.2 mg/L) Total measured polyphenols (272 mg/L)	Caffeoylquinic acid represents more than 50% of the total phenolic compounds present in the extracts. The methanolic extract was more efficient in extracting the compounds, followed by the ethanolic and water extracts.	[63]
	Ethanolic extract: Dried leaves were soaked in 80% ethanol in the ratio 1:10 *w/v*. Then, the mixture was kept under dark conditions for 72 h at room temperature (20 °C ±) and filtered to eliminate the solid fraction. The ethanolic solution was evaporated at 35 °C with a rotary evaporator, and the residue was dissolved in bidistilled water to maintain the same ratio.	5-O-caffeoylquinic acid (340.0 ± 0.9 mg/L) 1,5-O-dicaffeoylquinic acid (230.5 ± 1.5 mg/L) Monosuccinildicaffeoylquinic acid (36.4 ± 0.3 mg/L) Total caffeoylquinic acid (607 mg/L) Luteolin 7-O-glucoronide (189.4 ± 0.06 mg/L) Luteolin 7-O-malonylglucoside (50.5 ± 0.01 mg/L) Total luteolin (240 mg/L) Apigenin 7-O-glucoside (45.4 ± 0.4 mg/L) Apigenin 7-O-glucoronide (87.7 ± 0.8 mg/L) Apigenin malonylglucoside (62.0 ± 1.6 mg/L) Apigenin (3.8 ± 0.1 mg/L) Total apigenin (199 mg/L) Cynaropicrin (10.7 ± 0.9 mg/L) Total measured polyphenols (1046 mg/L)		
C. cardunculus L. var. altilis	Methanolic extract: Dried leaves were soaked in 70% methanol in the ratio 1:10 *w/v*. Then, the mixture was kept under dark conditions for 72 h at room temperature (20 °C ± 1) and filtered to eliminate the solid fraction. The methanolic solution was evaporated at 35 °C with a rotary evaporator, and the residue was dissolved in bidistilled water to maintain the same ratio.	5-O-caffeoylquinic acid (632.0 ± 0.1 mg/L) 1,5-O-dicaffeoylquinic acid (206.4 ± 0.3 mg/L) Monosuccinildicaffeoylquinic acid (59.4 ± 0.01 mg/L) Total caffeoylquinic acid (898 mg/L) Luteolin 7-O-glucoronide (22.7 mg/L) Luteolin 7-O-malonylglucoside (83.3 ± 0.01 mg/L) Total luteolin (106 mg/L) Apigenin 7-O-glucoside (61.7 ± 0.04 mg/L) Apigenin 7-O-glucoronide (115.0 ± 0.2 mg/L) Apigenin malonylglucoside (72.1 ± 0.04 mg/L) Total apigenin (249 mg/L) Cynaropicrin (15.8 ± 0.1 mg/L) Total measured polyphenols (1253 mg/L)	Caffeoylquinic acid represents more than 50% of the total phenolic compounds present in the extracts. The methanolic extract was more efficient in extracting the compounds, followed by the ethanolic and water extract.	[63]

CE—catechin equivalent; DM—dry matter; DW—dry weight; GAE—gallic acid equivalent.

These compounds have important properties in different fields of human wellbeing, namely antioxidant, antimicrobial, anti-inflammatory, anticancer, and anti-anxiety [36,52,58,59].

Several assays for measuring the antioxidant activity of the leaves or their potential to inhibit lipid oxidation can be performed. The ferric reducing-antioxidant power (FRAP) assay, DPPH free radical inhibition assay, β-carotene bleaching assay, thiobarbituric acid reactive substances (TBARS) inhibition, and the reducing power assay are some of these most commonly used techniques to evaluate the antioxidant activity either directly or indirectly. Numerous authors [8,52,59] executed these assays and confirmed the antioxidant activity of the cardoon leaf extracts [62] due to the presence of many bioactive compounds previously described. For example, Chihoub et al. [59] determined the antioxidant activity of cardoon leaf hydroethanolic extract (5 mg/mL) through the DPPH free radical inhibition assay (0.07 ± 0.00 mg/mL), reducing power (0.27 ± 0.09 mg/mL), β-carotene bleaching assay (0.19 ± 0.05 mg/mL), and TBARS inhibition (0.05 ± 0.01 mg/mL). The authors compared these results with cardoon leaf infusion preparations (DPPH free radical inhibition assay (0.22 ± 0.03 mg/mL), reducing power (0.22 ± 0.00 mg/mL), β-carotene bleaching assay (0.30 ± 0.04 mg/mL), and TBARS inhibition (0.40 ± 0.04 mg/mL)) and hydroethanolic extracts of turnip (DPPH free radical inhibition assay (1.57 ± 0.06 mg/mL), reducing power (1.07 ± 0.29 mg/mL), β-carotene bleaching assay (0.67 ± 0.15 mg/mL), and TBARS inhibition (0.60 ± 0.12 mg/mL)) and radish (DPPH free radical inhibition assay (0.14 ± 0.01 mg/mL), reducing power (0.58 ± 0.07 mg/mL), β-carotene bleaching assay (0.48 ± 0.02 mg/mL), and TBARS inhibition (0.43 ± 0.07 mg/mL)) and, assuming that lower results indicated higher antioxidant activity, concluded that cardoon leaf hydroethanolic extracts presented higher antioxidant activity.

5.3. Antimicrobial and Antifungal Properties

Cardoon leaves are known for having antimicrobial activity [8,63]. Falleh et al. [8] pointed out the antimicrobial activities of cardoon leaves due to their high content in phenolic compounds. The antimicrobial activities of cardoon leaves have been tested against several microorganisms of high importance for human beings due to their pathogenicity [8,63]. Cardoon leaves exhibit good antimicrobial activity against Gram-positive and Gram-negative bacteria, such as *Escherichia coli* and *Staphylococcus aureus*. However, they did not show antimicrobial activity against *Salmonella Typhimurium*. Table 2 features the studied species and their respective minimum inhibitory concentrations (MIC, which corresponds to the minimal extract concentration that inhibited the bacterial growth) or diameter of growth inhibition obtained by different authors. Scavo et al. [63] pointed out that Gram-positive bacteria are more susceptible than Gram-negative bacteria to plant extracts due to the permeability of the bacteria cell barrier, as well as the presence in the periplasmic space of enzymes that can break down molecules introduced from outside.

Authors have linked the antimicrobial activity of cardoon leaves to its richness in phenolic compounds, mainly due to the presence of lutein [8]. Nonetheless, more studies are needed for a better evaluation of the potential antimicrobial activities of cardoon leaves. Kukíc et al. [17] studied the antifungal activities of *C. cardunculus* involucral bract extracts, which showed potential antifungal activities. This may suggest that cardoon leaves may have antifungal activities too. Thus, more studies should be made to assess this issue.

Table 2. Main antimicrobial activity of cardoon leaves.

Species/Variety	Presentation form and Extraction Procedure	Main Bioactive Compounds and Levels Found	Main Conclusion	Reference
C. cardunculus L. var. *sylvestris*	Hydroethanolic extract: 1 g of dried samples in 30 mL of solvent (ethanol:water, 80:20, *v/v*) for 1 h. Filtered and re-extracted in 30 mL of solvent for 1 h. The combined extracts were evaporated to dryness of the ethanol at 35 °C, and the aqueous phase was frozen and lyophilized.	*Escherichia coli* (MIC = 2.5 mg/mL) *Escherichia coli* ESBL (MIC = 10 mg/mL) *Klebsiella pneumoniae* (MIC = 20 mg/mL) *Klebsiella pneumoniae* ESBL (MIC = 20 mg/mL) *Morganella morganii* (MIC = 10 mg/mL) *Pseudomonas aeruginosa* (MIC = 10 mg/mL) *Enterococcus faecalis* (MIC = 5 mg/mL) *Listeria monocytogenes* (MIC = 10 mg/mL) MRSA (MIC = 5 mg/mL) MSSA (MIC = 5 mg/mL)	Both extracts were demonstrated to have good antimicrobial activities, with low MIC values. The authors' concluded that the ethanolic extract was more effective when compared with other results obtained in this study.	[59]
	Infusion preparation: 1 g of dried samples was added to 100 mL of boiling distilled water, left to stand for 5 min, filtered and then frozen and lyophilized.	*Escherichia coli* (MIC = 2.5 mg/mL) *Escherichia coli* ESBL (MIC = 5 mg/mL) *Klebsiella pneumoniae* (MIC = 20 mg/mL) *Klebsiella pneumoniae* ESBL (MIC = 20 mg/mL) *Morganella morganii* (MIC = 2.5 mg/mL) *Pseudomonas aeruginosa* (MIC = 20 mg/mL) *Enterococcus faecalis* (MIC = 10 mg/mL) *Listeria monocytogenes* (MIC = 10 mg/mL) MRSA (MIC = 5 mg/mL) MSSA (MIC = 5 mg/mL)		
C. cardunculus	Extract: Leaves were dried at room temperature for two weeks. The extract was made with 2.5 g of dry powder with 25 mL of solvent (methanol), under stirring for 30 min. Then the extract was filtered and evaporated to dryness under vacuum and stored at 4 °C until analysis.	*Staphylococcus aureus* ATCC25923 (DGI = 25.7 ± 0.6 mm) *Staphylococcus epidermidis* CIP106510 (DGI = 20.3 mm) *Micrococcus luteus* NCIMB 8166 (DGI = 21.7 ± 0.6 mm) *Escherichia coli* ATCC 35,218 (DGI = 22.3 mm) *Enterococcus faecalis* ATCC29212 (DGI = 16.3 ± 0.6 mm) *Listeria monocytogenes* ATCC19115 (DGI = 9.3 ± 0.6 mm) *Pseudomonas aeruginosa* ATCC 27,853 (DGI = 13.7 ± 0.6 mm) *Salmonella typhimurium* LT2 (DGI = 0 mm)	The extract was effective against several human pathogenic bacteria but unfortunately had no activity against *Salmonella typhimurium* LT2. Antimicrobial activities could be related to the presence of phenolic compounds.	[8]

Table 2. *Cont.*

Species/Variety	Presentation form and Extraction Procedure	Main Bioactive Compounds and Levels Found	Main Conclusion	Reference
C. cardunculus L. var. *altilis*	Bidistilled water extract: Dried leaves were soaked in bidistilled water in the ratio 1:10 *w/v*. Then, the mixture was kept under dark conditions for 72 h at room temperature (20 °C ± 1) and filtered to eliminate the solid fraction.	*Bacillus cereus* (DGI = 0.7 cm) *Bacillus megaterium* (DGI = 0.8 ± 0.1 cm) *Listeria innocua* (DGI = 0.8 cm) *Pseudomonas syringae pv.* Tomato (DGI = 1.2 cm) *Rhodococcus fascians* (DGI = 0.6 cm) *Staphylococcus aureus* (DGI = 0.7 cm) *Xanthomonas perforans* (DGI = 1.5 ± 0.1 cm)	Water extract was effective against Gram-positive bacteria, although methanolic and ethanolic extracts controlled the growth more effectively. Regarding Gram-negative bacteria, the methanolic extract was not effective, and the ethanolic extract showed detectable antibacterial activity. Overall, the ethanolic extract was more efficient against the studied bacteria when compared to the other two extracts.	[63]
	Ethanolic extract: Dried leaves were soaked in ethanol 80% in the ratio 1:10 *w/v*. Then, the mixture was kept under dark conditions for 72 h at room temperature (20 °C ± 1) and filtered to eliminate the solid fraction. The ethanolic solution was evaporated at 35 °C with a rotary evaporator, and the residue was dissolved in bidistilled water to maintain the same ratio.	*Bacillus cereus* (DGI = 0.9 ± 0.1 cm) *Bacillus megaterium* (DGI = 2.3 ± 0.1 cm) *Bacillus subtilis* (DGI = 0.8 cm) *Listeria innocua* (DGI = 0.8 ± 0.1 cm) *Pseudomonas fluorescens* (DGI = 0.7 ± 0.1 cm) *Pseudomonas syringae pv.* Tomato (DGI = 0.6 cm) *Rhodococcus fascians* (DGI = 1.2 ± 0.1 cm) *Staphylococcus aureus* (DGI = 1.1 ± 0.1 cm) *Xanthomonas perforans* (DGI = 0.4 cm)		
	Methanolic extract: Dried leaves were soaked in methanol 70% in the ratio 1:10 *w/v*. Then, the mixture was kept under dark conditions for 72 h at room temperature (20 °C ± 1) and filtered to eliminate the solid fraction. The methanolic solution was evaporated at 35 °C with a rotary evaporator and the residue was dissolved in bidistilled water to maintain the same ratio.	*Bacillus cereus* (DGI = 1.3 ± 0.1 cm) *Bacillus megaterium* (DGI = 1.3 ± 0.1 cm) *Bacillus subtilis* (DGI = 0.8 cm) *Listeria innocua* (DGI = 1.2 cm) *Rhodococcus fascians* (DGI = 1.2 cm) *Staphylococcus aureus* (DGI = 1 ± 0.1 cm)		

DGI—diameter of growth inhibition; ESBL—extended spectrum β-lactamases; MIC values correspond to the minimal extract concentration that inhibited the bacterial growth; MRSA—methicillin-resistant *Staphylococcus aureus*; MSSA—methicillin-susceptible *Staphylococcus aureus*.

5.4. Phytotoxic and Allelopathic Properties

Cardoon leaf extracts are also known for their allelopathic activities. Allelopathy is the direct or indirect effect of a plant over a target species (plants, algae, bacteria, or fungus) through the release of chemical compounds, allelochemicals, into the environment [61,67,68]. The allelochemicals can be integrated with weed management and can be introduced into the environment through volatilization from the aboveground parts of the plant, root exudation, foliar leaching, or plant residue decomposition [61,67–70]. Phenolic compounds and sesquiterpene lactones are the most common examples of allelochemicals and can be found in different parts of plants, such as leaves, roots, stems, rhizomes, seeds, flowers, and pollen [69,71]. Cardoon has been found to manifest phytotoxic activity on weeds and standard target species, and sesquiterpene lactones were identified as the most relevant allelochemicals [61,67–72]. Rial et al. [61] studied the potential phytotoxic activity of cardoon allelochemicals against the development of standard target species (lettuce, watercress, tomato, and onion) and weeds (barnyard grass and brachiaria). The authors used different solvents to perform cardoon extracts that were tested in phytotoxic bioassays, and the ethyl acetate extract had the highest inhibitory activity. The extract was very active on root growth in both standard target species and weeds, with values close to 80% inhibition in most species. The extract was also active on the germination and the shoot length of the cress. Moreover, the authors isolated six sesquiterpene lactones (aguerin B, grosheimin, 8α-acetoxyzaluzanin C, dehydromelitensin, cynaropicrin, and 11,13-dihydroxy-8-deoxy-grosheimin). Aguerin B, grosheimin, and cynaropicrin showed strong phytotoxicity against standard target species and weeds [61]. Other compounds such as caffeoylquinic acid, luteolin, and apigenin derivatives have been reported to show allelopathic activity against several crops [67–69,71]. More recently, studies by Kaab et al. [70] and Scavo et al. [69] confirmed that cardoon leaf extracts could present a suitable source of natural compounds for a good potential bioherbicide.

6. Cardoon Leaves and Potential Applications

Cardoon leaves are a source of several bioactive compounds with important health benefits. Therefore, it is important to exploit the potential uses of cardoon leaves for better use. Cardoon leaves, constituted by several bioactive compounds with antioxidant and antimicrobial activity, can be considered a potential ingredient in the food industry. The leaves could be used as a food additive or as an ingredient in the development of a novel food with functional properties and health benefits [9,73]. Plant extracts have been incorporated as an antioxidant in meat and meat products. The extracts inhibit lipid oxidation and meat degradation, which improved the nutritional quality and extended meat' shelf life [74,75]. Thus, it can be foreseen that cardoon leaf extracts could also be incorporated into meat products to enhance their quality.

Another potential application of cardoon leaf is in the cosmetic industry. Several members of the Asteraceae family are used in the cosmetic industry for their bioactive compounds [76]. *Cynara scolymus* L., also known as globe artichoke, belongs to the Asteraceae family and is used in the cosmetic area for its richness in polyphenols content with antioxidant activity that can prevent aging and oxidative stress-related diseases and protect against UV-rays [77,78]. Due to its similarity, cardoon leaves may also be interesting to be used in this field.

Nonetheless, more studies should be made to assess the potential use of cardoon leaves in the food and cosmetics industries.

Food Packaging

The primary function of food packaging is to protect foods during the transportation process until it reaches the final consumer, increasing the shelf life of food [79]. Mainly manufactured from non-biodegradable matrices, such as polyethylene, these packages act as a defense barrier against insects and microorganisms, some of which are pathogenic. They can also protect foods against possible impacts that may occur during the transportation process, as well as temperature variations and radiation [80,81].

Active and intelligent food packaging are relatively new packaging technologies that may increase even further food shelf life or help increase food quality. Intelligent food packaging is designed to monitor at least one condition of the packaged food, such as temperature (e.g., breaks in the cold chain) and leaks (e.g., leaks in vacuum packaging and modified atmosphere packaging). These packages normally have an indicator, visible to the consumer (e.g., color) that changes when the monitored condition reaches non-tolerable values [82].

Regarding active food packaging, although it is a relatively new concept, it was inspired by the "packages" used by human civilization in ancient times, where the primitive man kept his food in leaves, which were composed of active compounds and that could help to increase the shelf life of food. Nowadays, there are two types of active food packaging. The absorption/adsorption active packaging absorb/adsorb possible deterioration gases and liquids that are generated through the natural food's degradation process [83,84]. The emission active packaging directly interacts with foods, emitting substances or compounds to the packaged foods. Normally, these substances or compounds have powerful antioxidant and antimicrobial activities that will help retard the food's degradation process, thus increasing the shelf life of food [84–86].

Food additives also help to delay natural food degradation. Normally, these substances are applied directly into foods, and their application is regulated by the Food and Drug Administration in the United States of America and through the Regulation no. 1333/2008 and its amendments in the European Union [87,88]. A food additive is a substance, not usually consumed as a food itself, added to foods for technological purposes [87]. Although their applications and limits are highly controlled by the respective institutes and entities, their long-term effects on human health are still unknown. Actually, in recent years, some synthetic food antioxidants, such as butylated hydroxytoluene (BHT) and butylated hydroxyanisole (BHA) have been associated with the promotion of carcinogenesis and the emergence of neurodegenerative diseases, such as Parkinson's and Alzheimer's diseases [89–91]. That being said, active food packaging can help decrease the applied quantities of these compounds and, consequently, the consumer's ingestion of these compounds. Additionally, the investigation and application of new natural compounds from food by-products can be an answer to this problem, as these by-products are rich in active compounds, with powerful antimicrobial, antifungal, and antioxidant activities.

So far, plant extracts and plant essential oils with known antioxidant and antimicrobial activity have been used in food packaging to control lipid oxidation and microbial deterioration [79,83,92–95]. For instance, Rizzo et al. [96] conducted a study on globe artichoke slices to evaluate dipping in locust bean gum edible coating with *Foeniculum vulgare* essential oil to extend its shelf life. The locust bean gum edible coating with added *Foeniculum vulgare* essential oil effectively decreased all microbiological groups tested and improved physical, chemical, and sensory qualities during 11 days of storage. Mazzaglia et al. [97] studied the effect of cardoon extract in reducing microbiological contamination and increasing the shelf life of aubergine-based burgers. Two concentrations of extract were used (1% and 3%) for the preparation of burgers, and the microbial load and sensory changes of vacuum-packed burgers were analyzed at processing day and after 30 and 105 days of storage. The burger with 3% of cardoon extract presented the best results, as it significantly reduced the growth of bacteria up to 30 and 105 days of refrigerated storage. Taking this into account, and knowing the bioactive characterization of cardoon leaf extract, it can present itself as a potential candidate in the production of active packaging. Therefore, studies should be done following this perspective.

7. Conclusion and Future Perspectives

Recent studies have indicated that cardoon leaves are rich in several polyphenol compounds, with several health benefits. Additionally, cardoon leaves are pointed out to have potential antimicrobial activity. Caffeoylquinic acids, which are the major bioactive compounds identified in cardoon leaves, are naturally-occurring antioxidant compounds that have been suggested for use as natural additives for extending the shelf life of food products.

Moreover, as leaves are considered cardoon by-products, they can have economic benefits if their natural antioxidants, with benefits to human health, are extracted and applied in food packaging to increase shelf life.

Nevertheless, cardoon by-products and their potential for application in several industrial fields such as cosmetics, food, and food packaging is still not entirely known and should be investigated further for a better comprehension of the potential uses of this valuable Mediterranean crop.

Author Contributions: The review paper was planned and written with contributions of all the authors. Conceptualization, A.S.S. and A.L.F., M.A.A., F.V.; methodology, C.H.B., M.A.A., F.V.; resources, C.H.B., M.A.A., F.V.; writing—original draft preparation, C.H.B., M.A.A., F.V.; writing—review and editing, I.C., A.L.F., M.R.L., A.S.S.; supervision, A.S.S. and A.L.F.; funding acquisition, A.S.S. and A.L.F. All authors have read and agreed to the published version of the manuscript.

Funding: The work was supported by UIDB/04077/2020 and UIDB/00211/2020 with funding from FCT/MCTES through national funds.

References

1. Centre of Agriculture and Biosciences International (CABI) Cynara cardunculus (Cardoon). Available online: https://www.cabi.org/isc/datasheet/17584 (accessed on 4 February 2020).
2. Kelly, M.; Pepper, A. Controlling Cynara Cardunculus (Artichoke Thistle, Cardoon, etc.). *California Exotic Pest Plant Council* **1996**, Symposium Proceedings. 1–5.
3. Avio, L.; Maggini, R.; Ujvári, G.; Incrocci, L.; Giovannetti, M.; Turrini, A. Scientia Horticulturae Phenolics content and antioxidant activity in the leaves of two artichoke cultivars are differentially affected by six mycorrhizal symbionts. *Sci. Hortic. (Amst.)* **2020**, *264*, 109153. [CrossRef]
4. Gostin, A.I.; Waisundara, V.Y. Edible flowers as functional food: A review on artichoke (*Cynara cardunculus* L.). *Trends Food Sci. Technol.* **2019**, *86*, 381–391. [CrossRef]
5. Pesce, G.; Mauromicale, G. Cynara cardunculus L.: Historical and Economic Importance, Botanical Descriptions, Genetic Resources and Traditional Uses. In *The Globe Artichoke Genome, Compendium of Plant Genomes*; Springer: Berlin/Heidelberg, Germany, 2019; pp. 1–19.
6. Pandino, G.; Lombardo, S.; Williamson, G.; Mauromicale, G. Polyphenol profile and content in wild and cultivated *Cynara cardunculus* L. *Ital. J. Agron.* **2012**, *7*, 254–261. [CrossRef]
7. Fernando, A.L.; Costa, J.; Barbosa, B.; Monti, A.; Rettenmaier, N. Environmental impact assessment of perennial crops cultivation on marginal soils in the Mediterranean Region. *Biomass Bioenergy* **2018**, *111*, 174–186. [CrossRef]
8. Falleh, H.; Ksouri, R.; Chaieb, K.; Karray-Bouraoui, N.; Trabelsi, N.; Boulaaba, M.; Abdelly, C. Phenolic composition of Cynara cardunculus L. organs, and their biological activities. *Comptes Rendus Biol.* **2008**, *331*, 372–379. [CrossRef]
9. Petropoulos, S.A.; Karkanis, A.; Martins, N.; Ferreira, I.C.F.R. Edible halophytes of the Mediterranean basin: Potential candidates for novel food products. *Trends Food Sci. Technol.* **2018**, *74*, 69–84. [CrossRef]
10. Khaldi, S.; Sonnante, G.; El Gazzah, M. Analysis of molecular genetic diversity of cardoon (*Cynara cardunculus* L.) in Tunisia. *Comptes Rendus Biol.* **2012**, *335*, 389–397. [CrossRef]
11. Almeida, C.M.; Simões, I. Cardoon-based rennets for cheese production. *Appl. Microbiol. Biotechnol.* **2018**, *102*, 4675–4686. [CrossRef]
12. Liburdi, K.; Emiliani Spinelli, S.; Benucci, I.; Lombardelli, C.; Esti, M. A preliminary study of continuous milk coagulation using Cynara cardunculus flower extract and calf rennet immobilized on magnetic particles. *Food Chem.* **2018**, *239*, 157–164. [CrossRef]
13. Ben Amira, A.; Makhlouf, I.; Flaviu Petrut, R.; Francis, F.; Bauwens, J.; Attia, H.; Besbes, S.; Blecker, C. Effect of extraction pH on techno-functional properties of crude extracts from wild cardoon (*Cynara cardunculus* L.) flowers. *Food Chem.* **2017**, *225*, 258–266. [CrossRef] [PubMed]
14. Fernández, J.; Curt, M.D.; Aguado, P.L. Industrial applications of Cynara cardunculus L. for energy and other uses. *Ind. Crops Prod.* **2006**, *24*, 222–229. [CrossRef]

15. Ierna, A.; Mauromicale, G. Cynara cardunculus L. genotypes as a crop for energy purposes in a Mediterranean environment. *Biomass Bioenergy* **2010**, *34*, 754–760. [CrossRef]
16. Pandino, G.; Lombardo, S.; Mauromicale, G. Globe artichoke leaves and floral stems as a source of bioactive compounds. *Ind. Crops Prod.* **2013**, *44*, 44–49. [CrossRef]
17. Kukić, J.; Popović, V.; Petrović, S.; Mucaji, P.; Ćirić, A.; Stojković, D.; Soković, M. Antioxidant and antimicrobial activity of Cynara cardunculus extracts. *Food Chem.* **2008**, *107*, 861–868. [CrossRef]
18. Petropoulos, S.; Fernandes, Â.; Pereira, C.; Tzortzakis, N.; Vaz, J.; Soković, M.; Barros, L.; Ferreira, I.C.F.R. Bioactivities, chemical composition and nutritional value of *Cynara cardunculus* L. seeds. *Food Chem.* **2019**, *289*, 404–412. [CrossRef]
19. Wahba, H.E.; Sarhan, A.Z.; Salama, A.B.; Sharaf-Eldin, M.A.; Gad, H.M. Growth response and active constituents of Cynara cardunculus plants to the number of leaves harvests. *Eur. J. Agron.* **2016**, *73*, 118–123. [CrossRef]
20. De Corato, U.; De Bari, I.; Viola, E.; Pugliese, M. Assessing the main opportunities of integrated biorefining from agro-bioenergy co/by-products and agroindustrial residues into high-value added products associated to some emerging markets: A review. *Renew. Sustain. Energy Rev.* **2018**, *88*, 326–346. [CrossRef]
21. Pires, J.R.A.; Souza, V.G.L.; Fernando, A.L. Valorization of energy crops as a source for nanocellulose production—Current knowledge and future prospects. *Ind. Crops Prod.* **2019**, *140*, 111642. [CrossRef]
22. Sganzerla, W.G.; Rosa, G.B.; Ferreira, A.L.A.; da Rosa, C.G.; Beling, P.C.; Xavier, L.O.; Hansen, C.M.; Ferrareze, J.P.; Nunes, M.R.; Barreto, P.L.M.; et al. Bioactive food packaging based on starch, citric pectin and functionalized with Acca sellowiana waste by-product: Characterization and application in the postharvest conservation of apple. *Int. J. Biol. Macromol.* **2020**, *147*, 295–303. [CrossRef]
23. Nur Hanani, Z.A.; Aelma Husna, A.B.; Nurul Syahida, S.; Nor Khaizura, M.A.B.; Jamilah, B. Effect of different fruit peels on the functional properties of gelatin/polyethylene bilayer films for active packaging. *Food Packag. Shelf Life* **2018**, *18*, 201–211. [CrossRef]
24. Genskowsky, E.; Puente, L.A.; Pérez-Álvarez, J.A.; Fernandez-Lopez, J.; Muñoz, L.A.; Viuda-Martos, M. Assessment of antibacterial and antioxidant properties of chitosan edible films incorporated with maqui berry (*Aristotelia chilensis*). *LWT-Food Sci. Technol.* **2015**, *64*, 1057–1062. [CrossRef]
25. Gominho, J.; Fernandez, J.; Pereira, H. *Cynara cardunculus* L.—A new fibre crop for pulp and paper production. *Ind. Crops Prod.* **2001**, *13*, 1–10. [CrossRef]
26. Von Cossel, M.; Lewandowski, I.; Elbersen, B.; Staritsky, I.; Van Eupen, M.; Iqbal, Y.; Mantel, S.; Scordia, D.; Testa, G.; Cosentino, S.L.; et al. Marginal agricultural land low-input systems for biomass production. *Energies* **2019**, *12*, 3123. [CrossRef]
27. Mauromicale, G.; Sortino, O.; Pesce, G.R.; Agnello, M.; Mauro, R.P. Suitability of cultivated and wild cardoon as a sustainable bioenergy crop for low input cultivation in low quality Mediterranean soils. *Ind. Crops Prod.* **2014**, *57*, 82–89. [CrossRef]
28. Francaviglia, R.; Bruno, A.; Falcucci, M.; Farina, R.; Renzi, G.; Russo, D.E.; Sepe, L.; Neri, U. Yields and quality of Cynara cardunculus L. wild and cultivated cardoon genotypes. A case study from a marginal land in Central Italy. *Eur. J. Agron.* **2016**, *72*, 10–19. [CrossRef]
29. Gominho, J.; Curt, M.D.; Lourenço, A.; Fernández, J.; Pereira, H. Cynara cardunculus L. as a biomass and multi-purpose crop: A review of 30 years of research. *Biomass Bioenergy* **2018**, *109*, 257–275. [CrossRef]
30. Louro-Martins, A.P.; Pestana De Vasconcelos, M.M.; De Sousa, R.B. Thistle (*Cynara cardunculus* L.) flower as a coagulant agent for cheesemaking. Short characterization. *Lait* **1996**, *76*, 473–477. [CrossRef]
31. Aquilanti, L.; Babini, V.; Santarelli, S.; Osimani, A.; Petruzzelli, A.; Clementi, F. Bacterial dynamics in a raw cow's milk Caciotta cheese manufactured with aqueous extract of Cynara cardunculus dried flowers. *Lett. Appl. Microbiol.* **2011**, *52*, 651–659. [CrossRef]
32. Gomes, S.; Belo, A.T.; Alvarenga, N.; Dias, J.; Lage, P.; Pinheiro, C.; Pinto-Cruz, C.; Brás, T.; Duarte, M.F.; Martins, A.P.L. Characterization of Cynara cardunculus L. flower from Alentejo as a coagulant agent for cheesemaking. *Int. Dairy J.* **2019**, *91*, 178–184. [CrossRef]
33. Sales-Gomes, M.; Lima-Costa, M.E. Immobilization of endoproteases from crude extract of Cynara cardunculus L. flowers. *Food Sci. Technol. Int.* **2008**, *14*, 271–276. [CrossRef]
34. Spadefoot Nursery Artichoke and Cardoon. Available online: https://www.spadefootnursery.com/blog/2019/12/29/artichoke-and-cardoon (accessed on 25 February 2020).

35. European Union. European Parliament and the Council of the European Union Regulation (EU) No. 1151/2012 of 21 November 2012 on quality schemes for agricultural products and foodstuffs. *Off. J. Eur. Union* **2012**, *343*, 1–29.

36. Dias, M.I.; Barros, L.; Barreira, J.C.M.; Alves, M.J.; Barracosa, P.; Ferreira, I.C.F.R. Phenolic profile and bioactivity of cardoon (*Cynara cardunculus* L.) inflorescence parts: Selecting the best genotype for food applications. *Food Chem.* **2018**, *268*, 196–202. [CrossRef] [PubMed]

37. Macedo, A.C.; Tavares, T.G.; Malcata, F.X. Influence of native lactic acid bacteria on the microbiological, biochemical and sensory profiles of Serra da Estrela cheese. *Food Microbiol.* **2004**, *21*, 233–240. [CrossRef]

38. Vergara, P.; Ladero, M.; García-Ochoa, F.; Villar, J.C. Valorization of Cynara Cardunculus crops by ethanol-water treatment: Optimization of operating conditions. *Ind. Crops Prod.* **2018**, *124*, 856–862. [CrossRef]

39. Petropoulos, S.; Fernandes, Â.; Calhelha, R.C.; Danalatos, N.; Barros, L.; Ferreira, I.C.F.R. How extraction method affects yield, fatty acids composition and bioactive properties of cardoon seed oil? *Ind. Crops Prod.* **2018**, *124*, 459–465. [CrossRef]

40. Gominho, J.; Lourenço, A.; Palma, P.; Lourenço, M.E.; Curt, M.D.; Fernández, J.; Pereira, H. Large scale cultivation of Cynara cardunculus L. for biomass production-A case study. *Ind. Crops Prod.* **2011**, *33*, 1–6. [CrossRef]

41. Cabiddu, A.; Contini, S.; Gallo, A.; Lucini, L.; Bani, P.; Decandia, M.; Molle, G.; Piluzza, G.; Sulas, L. In vitro fermentation of cardoon seed press cake - A valuable byproduct from biorefinery as a novel supplement for small ruminants. *Ind. Crops Prod.* **2019**, *130*, 420–427. [CrossRef]

42. Cajarville, C.; González, J.; Repetto, J.L.; Alvir, M.R.; Rodríguez, C.A. Nutritional evaluation of cardoon (*Cynara cardunculus*) seed for ruminants. *Anim. Feed Sci. Technol.* **2000**, *87*, 203–213. [CrossRef]

43. Raccuia, S.A.; Melilli, M.G. Biomass and grain oil yields in *Cynara cardunculus* L. genotypes grown in a Mediterranean environment. *F. Crop. Res.* **2007**, *101*, 187–197. [CrossRef]

44. Pesce, G.R.; Negri, M.; Bacenetti, J.; Mauromicale, G. The biomethane, silage and biomass yield obtainable from three accessions of Cynara cardunculus. *Ind. Crops Prod.* **2017**, *103*, 233–239. [CrossRef]

45. Cravero, V.; Martin, E.; Crippa, I.; Anido, F.L.; García, S.M.; Cointry, E. Fresh biomass production and partitioning of aboveground growth in the three botanical varieties of *Cynara cardunculus* L. *Ind. Crops Prod.* **2012**, *37*, 253–258. [CrossRef]

46. Oliveira, I.; Gominho, J.; Diberardino, S.; Duarte, E. Characterization of *Cynara cardunculus* L. stalks and their suitability for biogas production. *Ind. Crops Prod.* **2012**, *40*, 318–323. [CrossRef]

47. Foti, S.; Mauromicale, G.; Raccuia, S.A.; Fallico, B.; Fanella, F.; Maccarone, E. Possible alternative utilization of Cynara spp. I. Biomass, grain yield and chemical composition of grain. *Ind. Crops Prod.* **1999**, *10*, 219–228. [CrossRef]

48. Fernández, J.; Curt, M. Low-cost biodiesel from Cynara oil. In Proceedings of the 2nd World Conference and Exhibition on Biomass for Energy, Industry Climate Protection, Rome, Italy, 10–14 May 2004; pp. 12–15.

49. Curt, M.D.; Sánchez, G.; Fernández, J. The potential of Cynara cardunculus L. for seed oil production in a perennial cultivation system. *Biomass Bioenergy* **2002**, *23*, 33–46. [CrossRef]

50. Cajarville, C.; González, J.; Repetto, J.L.; Rodríguez, C.A.; Martínez, A. Nutritive value of green forage and crop by-products of Cynara cardunculus. *Ann. Zootech.* **1999**, *48*, 353–365. [CrossRef]

51. Salami, S.A.; Luciano, G.; O'Grady, M.N.; Biondi, L.; Newbold, C.J.; Kerry, J.P.; Priolo, A. Sustainability of feeding plant by-products: A review of the implications for ruminant meat production. *Anim. Feed Sci. Technol.* **2019**, *251*, 37–55. [CrossRef]

52. Pandino, G.; Lombardo, S.; Mauromicale, G.; Williamson, G. Phenolic acids and flavonoids in leaf and floral stem of cultivated and wild *Cynara cardunculus* L. genotypes. *Food Chem.* **2011**, *126*, 417–422. [CrossRef]

53. Romani, A.; Pinelli, P.; Cantini, C.; Cimato, A.; Heimler, D. Characterization of Violetto di Toscana, a typical Italian variety of artichoke (*Cynara scolymus* L.). *Food Chem.* **2006**, *95*, 221–225. [CrossRef]

54. Dabbou, S.; Dabbou, S.; Flamini, G.; Pandino, G.; Gasco, L.; Helal, A.N. Phytochemical Compounds from the Crop Byproducts of Tunisian Globe Artichoke Cultivars. *Chem. Biodivers.* **2016**, *13*, 1475–1483. [CrossRef]

55. Pandino, G.; Lombardo, S.; Antonino, L.M.; Ruta, C.; Mauromicale, G. In vitro micropropagation and mycorrhizal treatment influences the polyphenols content profile of globe artichoke under field conditions. *Food Res. Int.* **2017**, *99*, 385–392. [CrossRef]

56. Esposito, M.; Di Pierro, P.; Dejonghe, W.; Mariniello, L.; Porta, R. Enzymatic milk clotting activity in artichoke (*Cynara scolymus*) leaves and alpine thistle (*Carduus defloratus*) flowers. Immobilization of alpine thistle aspartic protease. *Food Chem.* **2016**, *204*, 115–121. [CrossRef]

57. Grammelis, P.; Malliopoulou, A.; Basinas, P.; Danalatos, N.G. Cultivation and characterization of Cynara cardunculus for solid biofuels production in the mediterranean region. *Int. J. Mol. Sci.* **2008**, *9*, 1241–1258. [CrossRef]

58. Stumpf, B.; Künne, M.; Ma, L.; Xu, M.; Yan, F.; Piepho, H.P.; Honermeier, B. Optimization of the extraction procedure for the determination of phenolic acids and flavonoids in the leaves of globe artichoke (*Cynara cardunculus* var. scolymus L.). *J. Pharm. Biomed. Anal.* **2020**, *177*, 112879. [CrossRef]

59. Chihoub, W.; Dias, M.I.; Barros, L.; Calhelha, R.C.; Alves, M.J.; Harzallah-Skhiri, F.; Ferreira, I.C.F.R. Valorisation of the green waste parts from turnip, radish and wild cardoon: Nutritional value, phenolic profile and bioactivity evaluation. *Food Res. Int.* **2019**, *126*, 108651. [CrossRef]

60. Pinelli, P.; Agostini, F.; Comino, C.; Lanteri, S.; Portis, E.; Romani, A. Simultaneous quantification of caffeoyl esters and flavonoids in wild and cultivated cardoon leaves. *Food Chem.* **2007**, *105*, 1695–1701. [CrossRef]

61. Rial, C.; Novaes, P.; Varela, R.M.; José, J.M.; Macias, F.A. Phytotoxicity of cardoon (*Cynara cardunculus*) allelochemicals on standard target species and weeds. *J. Agric. Food Chem.* **2014**, *62*, 6699–6706. [CrossRef]

62. Brás, T.; Guerreiro, O.; Duarte, M.F.; Neves, L.A. Impact of extraction parameters and concentration by nanofiltration on the recovery of phenolic compounds from Cynara cardunculus var. altilis: Assessment of antioxidant activity. *Ind. Crops Prod.* **2015**, *67*, 137–142. [CrossRef]

63. Scavo, A.; Pandino, G.; Restuccia, C.; Parafati, L.; Cirvilleri, G.; Mauromicale, G. Antimicrobial activity of cultivated cardoon (*Cynara cardunculus* L. var. altilis DC.) leaf extracts against bacterial species of agricultural and food interest. *Ind. Crops Prod.* **2019**, *129*, 206–211. [CrossRef]

64. Lombardo, S.; Pandino, G.; Mauromicale, G. The influence of pre-harvest factors on the quality of globe artichoke. *Sci. Hortic. (Amst.)* **2018**, *233*, 479–490. [CrossRef]

65. Pandino, G.; Lombardo, S.; Moglia, A.; Portis, E.; Lanteri, S.; Mauromicale, G. Leaf polyphenol profile and SSR-based fingerprinting of new segregant Cynara cardunculus genotypes. *Front. Plant Sci.* **2015**, *5*, 1–7. [CrossRef]

66. Pistón, M.; Machado, I.; Branco, C.S.; Cesio, V.; Heinzen, H.; Ribeiro, D.; Fernandes, E.; Chisté, R.C.; Freitas, M. Infusion, decoction and hydroalcoholic extracts of leaves from artichoke (*Cynara cardunculus* L. subsp. cardunculus) are effective scavengers of physiologically relevant ROS and RNS. *Food Res. Int.* **2014**, *64*, 150–156.

67. Scavo, A.; Rial, C.; Varela, R.M.; Molinillo, J.M.G.; Mauromicale, G.; Macias, F.A. Influence of Genotype and Harvest Time on the Cynara cardunculus L. Sesquiterpene Lactone Profile. *J. Agric. Food Chem.* **2019**, *67*, 6487–6496. [CrossRef]

68. Scavo, A.; Rial, C.; Molinillo, J.M.G.; Varela, R.M.; Mauromicale, G.; Macias, F.A. The extraction procedure improves the allelopathic activity of cardoon (*Cynara cardunculus* var. altilis) leaf allelochemicals. *Ind. Crops Prod.* **2019**, *128*, 479–487. [CrossRef]

69. Scavo, A.; Pandino, G.; Restuccia, A.; Mauromicale, G. Leaf extracts of cultivated cardoon as potential bioherbicide. *Sci. Hortic. (Amst.)* **2020**, *261*, 109024. [CrossRef]

70. Kaab, S.B.; Rebey, I.B.; Hanafi, M.; Hammi, K.M.; Smaoui, A.; Fauconnier, M.L.; De Clerck, C.; Jijakli, M.H.; Ksouri, R. Screening of Tunisian plant extracts for herbicidal activity and formulation of a bioherbicide based on Cynara cardunculus. *S. Afr. J. Bot.* **2020**, *128*, 67–76. [CrossRef]

71. Scavo, A.; Restuccia, A.; Pandino, G.; Onofri, A.; Mauromicale, G. Allelopathic effects of Cynara cardunculus L. Leaf aqueous extracts on seed germination of some mediterranean weed species. *Ital. J. Agron.* **2018**, *13*, 119–125. [CrossRef]

72. Scavo, A.; Pandino, G.; Restuccia, A.; Lombardo, S.; Roberto Pesce, G.; Mauromicale, G. Allelopathic potential of leaf aqueous extracts from Cynara cardunculus L. On the seedling growth of two cosmopolitan weed species. *Ital. J. Agron.* **2019**, *14*, 78–83. [CrossRef]

73. Ruiz-Cano, D.; Pérez-Llamas, F.; Frutos, M.J.; Arnao, M.B.; Espinosa, C.; López-Jiménez, J.Á.; Castillo, J.; Zamora, S. Chemical and functional properties of the different by-products of artichoke (*Cynara scolymus* L.) from industrial canning processing. *Food Chem.* **2014**, *160*, 134–140. [CrossRef]

74. Shah, M.A.; Bosco, S.J.D.; Mir, S.A. Plant extracts as natural antioxidants in meat and meat products. *Meat Sci.* **2014**, *98*, 21–33. [CrossRef]

75. Nikmaram, N.; Budaraju, S.; Barba, F.J.; Lorenzo, J.M.; Cox, R.B.; Mallikarjunan, K.; Roohinejad, S. Application of plant extracts to improve the shelf-life, nutritional and health-related properties of ready-to-eat meat products. *Meat Sci.* **2018**, *145*, 245–255. [CrossRef] [PubMed]

76. Charles Dorni, A.I.; Amalraj, A.; Gopi, S.; Varma, K.; Anjana, S.N. Novel cosmeceuticals from plants—An industry guided review. *J. Appl. Res. Med. Aromat. Plants* **2017**, *7*, 1–26. [CrossRef]

77. Faria-Silva, C.; Ascenso, A.; Costa, A.M.; Marto, J.; Carvalheiro, M.; Ribeiro, H.M.; Simões, S. Feeding the skin: A new trend in food and cosmetics convergence. *Trends Food Sci. Technol.* **2020**, *95*, 21–32. [CrossRef]

78. Marques, P.; Marto, J.; Gonçalves, L.M.; Pacheco, R.; Fitas, M.; Pinto, P.; Serralheiro, M.L.M.; Ribeiro, H. Cynara scolymus L.: A promising Mediterranean extract for topical anti-aging prevention. *Ind. Crops Prod.* **2017**, *109*, 699–706. [CrossRef]

79. Andrade, M.A.; Ribeiro-Santos, R.; Guerra, M.; Sanches-Silva, A. Evaluation of the oxidative status of salami packaged with an active whey protein film. *Foods* **2019**, *8*, 387. [CrossRef]

80. Risch, S.J. Food Packaging History and Innovations. *J. Agric. Food Chem.* **2009**, *57*, 8089–8092. [CrossRef]

81. Trinetta, V. Definition and Function of Food Packaging. In *Reference Module in Food Science*; Elsevier: Amsterdam, The Netherlands, 2016; ISBN 978-0-08-100596-5.

82. Ribeiro-Santos, R.; Andrade, M.; Sanches-Silva, A. Application of encapsulated essential oils as antimicrobial agents in food packaging. *Curr. Opin. Food Sci.* **2017**, *14*, 78–84. [CrossRef]

83. Andrade, M.A.; Ribeiro-Santos, R.; Costa Bonito, M.C.; Saraiva, M.; Sanches-Silva, A. Characterization of rosemary and thyme extracts for incorporation into a whey protein based film. *LWT Food Sci. Technol.* **2018**, *92*, 497–508. [CrossRef]

84. Dainelli, D.; Gontard, N.; Spyropoulos, D.; Zondervan-van den Beuken, E.; Tobback, P. Active and intelligent food packaging: Legal aspects and safety concerns. *Trends Food Sci. Technol.* **2008**, *19*, S103–S112. [CrossRef]

85. Ribeiro-Santos, R.; Andrade, M.; de Melo, N.R.; Sanches-Silva, A. Use of essential oils in active food packaging: Recent advances and future trends. *Trends Food Sci. Technol.* **2017**, *61*, 132–140. [CrossRef]

86. Souza, V.G.L.; Pires, J.R.A.; Vieira, É.T.; Coelhoso, I.M.; Duarte, M.P.; Fernando, A.L. Activity of chitosan-montmorillonite bionanocomposites incorporated with rosemary essential oil: From in vitro assays to application in fresh poultry meat. *Food Hydrocoll.* **2019**, *89*, 241–252. [CrossRef]

87. European Union European Parliament and the Council of the European Union Regulation (EC) No 1333/2008. *Off. J. Eur. Union* **2008**, *354*, 16–33.

88. Food and Drug Administration (FDA) 21CFR181.27. *Title 21 Food Drugs*; FDA: Silver Spring, MD, USA, 2017.

89. Pereira de Abreu, D.A.; Losada, P.P.; Maroto, J.; Cruz, J.M. Evaluation of the effectiveness of a new active packaging film containing natural antioxidants (from barley husks) that retard lipid damage in frozen Atlantic salmon (*Salmo salar* L.). *Food Res. Int.* **2010**, *43*, 1277–1282. [CrossRef]

90. Carocho, M.; Morales, P.; Ferreira, I.C.F.R. Natural food additives: Quo vadis? *Trends Food Sci. Technol.* **2015**, *45*, 284–295. [CrossRef]

91. Sanches-Silva, A.; Costa, D.; Albuquerque, T.G.; Buonocore, G.G.; Ramos, F.; Castilho, M.C.; Machado, A.V.; Costa, H.S. Trends in the use of natural antioxidants in active food packaging: A review. *Food Addit. Contam. Part A* **2014**, *31*, 374–395. [CrossRef] [PubMed]

92. Martins, C.; Vilarinho, F.; Sanches Silva, A.; Andrade, M.; Machado, A.V.; Castilho, M.C.; Sá, A.; Cunha, A.; Vaz, M.F.; Ramos, F. Active polylactic acid film incorporated with green tea extract: Development, characterization and effectiveness. *Ind. Crops Prod.* **2018**, *123*, 100–110. [CrossRef]

93. Castro, F.V.R.; Andrade, M.A.; Sanches Silva, A.; Vaz, M.F.; Vilarinho, F. The Contribution of a Whey Protein Film Incorporated with Green Tea Extract to Minimize the Lipid Oxidation of Salmon (*Salmo salar* L.). *Foods* **2019**, *8*, 327. [CrossRef]

94. Pires, J.R.A.; de Souza, V.G.L.; Fernando, A.L. Chitosan/montmorillonite bionanocomposites incorporated with rosemary and ginger essential oil as packaging for fresh poultry meat. *Food Packag. Shelf Life* **2018**, *17*, 142–149. [CrossRef]

95. Pascoal, A.; Quirantes-Piné, R.; Fernando, A.L.; Alexopoulou, E.; Segura-Carretero, A. Phenolic composition and antioxidant activity of kenaf leaves. *Ind. Crops Prod.* **2015**, *78*, 116–123. [CrossRef]

Foods **2020**, *9*, 564

96. Rizzo, V.; Lombardo, S.; Pandino, G.; Barbagallo, R.N.; Mazzaglia, A.; Restuccia, C.; Mauromicale, G.; Muratore, G. Shelf-life study of ready-to-cook slices of globe artichoke 'Spinoso sardo': Effects of anti-browning solutions and edible coating enriched with Foeniculum vulgare essential oil. *J. Sci. Food Agric.* **2019**, *99*, 5219–5228. [CrossRef]

97. Mazzaglia, A.; Licciardello, F.; Aurelio, S.; Muratore, G.; Giovanni, M.; Restuccia, C. Effect of Cynara Cardunculus Extract on the shelf life of aubergine burgers. *Ital. J. Food Sci.* **2018**, *30*, 19–24.

MDPI

St. Alban-Anlage 66

4052 Basel

Switzerland

Tel. +41 61 683 77 34

Fax +41 61 302 89 18

www.mdpi.com

Foods Editorial Office

E-mail: foods@mdpi.com

www.mdpi.com/journal/foods